ADVANCES IN
GEOPHYSICS

VOLUME 14

Contributors to This Volume

A. K. Ångström
E. E. Barr
R. E. Bedford
A. J. Drummond
D. E. Erminy
E. J. Gillham
R. A. Hanel
A. T. Hattenburg

J. R. Hickey
A. R. Karoli
H. J. Kostkowski
P. M. Kuhn
E. H. Putley
G. D. Robinson
R. Stair
J. Strong

Advances in
GEOPHYSICS

Edited by

H. E. LANDSBERG

*Institute for Fluid Dynamics and Applied Mathematics
University of Maryland, College Park, Maryland*

J. VAN MIEGHEM

*Royal Belgian Meteorological Institute
Uccle, Belgium*

Editorial Advisory Committee

BERNARD HAURWITZ   R. STONELEY
ROGER REVELLE   URHO A. UOTILA

VOLUME 14

Precision Radiometry

Edited by

A. J. DRUMMOND

*Eppley Laboratory, Inc.
Newport, Rhode Island*

1970

Academic Press • New York and London

Copyright © 1970, by Academic Press, Inc.

ALL RIGHTS RESERVED

NO PART OF THIS BOOK MAY BE REPRODUCED IN ANY FORM,
BY PHOTOSTAT, MICROFILM, RETRIEVAL SYSTEM, OR ANY
OTHER MEANS, WITHOUT WRITTEN PERMISSION
FROM THE PUBLISHERS.

ACADEMIC PRESS, INC.
111 Fifth Avenue
New York, New York 10003

United Kingdom Edition
Published by
ACADEMIC PRESS, INC. (London) Ltd.
Berkeley Square House, London W1X 6BA

Library of Congress Catalog Card Number: 52–12266

PRINTED IN THE UNITED STATES OF AMERICA

LIST OF CONTRIBUTORS

Numbers in parentheses indicate the pages on which the authors' contributions begin.

A. K. ÅNGSTRÖM, *Bromma, Stockholm, Sweden* (269)

E. E. BARR, *IRI Thin Films Products, Waltham, Massachusetts* (391)

R. E. BEDFORD, *National Research Council, Ottawa, Canada* (165)

A. J. DRUMMOND, *Eppley Laboratory, Inc., Newport, Rhode Island* (1)

D. E. ERMINY, *National Bureau of Standards, Washington, D.C.* (111)

E. J. GILLHAM, *National Physical Laboratory, Teddington, Middlesex, England* (53)

R. A. HANEL, *NASA Goddard Space Flight Center, Greenbelt, Maryland* (359)

A. T. HATTENBURG, *National Bureau of Standards, Washington, D.C.* (111)

J. R. HICKEY, *Eppley Laboratory, Inc., Newport, Rhode Island* (227)

A. R. KAROLI, *Eppley Laboratory, Inc., Newport, Rhode Island* (203)

H. J. KOSTKOWSKI, *National Bureau of Standards, Washington, D.C.* (111)

P. M. KUHN, *ESSA, Boulder, Colorado* (331)

E. H. PUTLEY, *Royal Radar Establishment, Great Malvern, Worcestershire, England* (129)

G. D. ROBINSON, *Travelers Research Center, Hartford, Connecticut* (285)

R. STAIR, *Hunter Associates Laboratory, Fairfax, Virginia* (83)

J. STRONG, *University of Massachusetts, Amherst, Massachusetts* (307)

FOREWORD

This volume can be credited principally to the enterprise and initiative of the Eppley research staff. It was conceived as a possible contribution to this series in 1967, when the Eppley Laboratory presented its third group of lectures on precision radiometry, at Newport, Rhode Island. These contributions by a number of authorities were critical reviews of the state-of-the-art in sectors to which each of the contributors had made original contributions. It was decided to revise these and place them into a somewhat more formal frame but not lose the original flavor of spontaneity and opinion.

There is little need to justify the importance of the theme treated here for many of the geophysical disciplines. Meteorologists, in spite of a few devoted individuals, are only beginning to awaken to the importance of radiative factors in the atmosphere. The problem of the "solar constant" deserves much study yet. Radiative sensing from satellites for meteorological, oceanographic, and resource studies, is still in its infancy. Radiative exchanges in the biosphere have yet to be fully exposed.

The treatment of a single subject in depth is not entirely new. Volume 6 was devoted entirely to problems of turbulence. We hope this volume will stimulate as much interest as that volume did.

The next volume will return to our usual practice of having a series of authoritative reviews covering a broad spectrum of the geophysical sciences. Some of the articles now envisaged will deal with tsunamis and sea ice.

H. E. LANDSBERG
J. VAN MIEGHEM

PREFACE

For a long time, workers in the field of experimental precision radiometry have been acutely conscious of the need for adequate text material. Too little literature is presently available on most practical problems in radiometry. The rapid increase in interest in this subject, largely brought about through the impressive and expanding programs in the atmospheric and space sciences, was the reason for the introduction, in 1965, of the Eppley Lecture Series entitled "Fundamental Radiometry for Experimental Scientists." Its main objective has been to bring people with differing backgrounds together and to acquaint them with some of the limitations and ground rules of precise radiometric practices.

This book is an outgrowth of the lectures given during the period 1965–1967. Wherever necessary, the chapters are updated and where delivered more than once, the most recent form is included. The book provides a comprehensive presentation of the basic concepts and the practical implications inherent in the accurate measurement of electromagnetic radiation, and defines clearly what it is that needs to be measured, what are the fundamental limitations on the measurements, and of what use the measurements are when they are made.

We wish to express our appreciation for making possible this volume of "Advances in Geophysics" to all of the authors, to Dr. Helmut Landsberg, and to Academic Press. Thanks are due to Mr. Roy Anderson, President of the Eppley Laboratory and of the Eppley Foundation for Research, for generously providing Eppley facilities to supplement our private efforts; and also to Mrs. Angie Medeiros who typed our texts and assisted in additional work on the other manuscripts.

A. J. DRUMMOND, Volume Editor
J. R. HICKEY, Assistant Editor
A. R. KAROLI, Assistant Editor

The Eppley Laboratory
Newport, Rhode Island

CONTENTS

LIST OF CONTRIBUTORS v
FOREWORD vii
PREFACE ix

1. Precision Radiometry and Its Significance in Atmospheric and Space Physics
A. J. DRUMMOND

1.1.	Introduction	1
	1.1.1. Evolution of (Thermal) Radiometry	3
1.2.	Historical Review of Precision Radiometry	6
	1.2.1. Ångström Pyrheliometric Scale	7
	1.2.2. Smithsonian Pyrheliometric Scale	9
	1.2.3. International Pyrheliometric Scale	12
	1.2.4. National Physical Laboratory (U.K.) Radiometric Reference	16
	1.2.5. National Bureau of Standards (U.S.A.) Radiometric Reference	18
	1.2.6. National Research Council (Canada) Radiometric Reference	20
	1.2.7. Investigations Concerning the Spectrum of the "Blackbody"	21
1.3.	Precision and Accuracy of Radiometric Devices	22
1.4.	The Extraterrestrial Solar Fluxes	24
	1.4.1. Problems in Solar-Constant Determination (Extrapolation Methods)	28
	1.4.2. Direct Measurement—Eppley and Jet Propulsion Laboratories Research Project	33
	1.4.3. Field Calibration of Radiometers for Extraterrestrial Solar Measurements	42
1.5.	Instrumental Problems in Precision Measurements of Solar and Terrestrial Radiative Fluxes above the Earth's Surface.......	44
	1.5.1. Radiation Sensing	44
	1.5.2. Readout Instrumentation	48
References		49

2. Radiometry from the Viewpoint of the Detector

E. J. Gillham

2.1.	Introduction	53
2.2.	Spectral Distribution	59
2.3.	The Measurement of Radiation—Basic Principles	60
2.4.	The Thermopile	62
2.5.	Use of Radiation Thermopiles—Measurement of Total Radiation	65
2.6.	Establishment of the Radiation Scale	70
2.7.	Measurement of Monochromatic Radiation	72
2.8.	The Absolute Measurement of Radiant Power	75
2.9.	The Absolute Measurement of Radiation	76
References		81

3. Sources as Radiometric Standards

R. Stair

3.1.	Introduction	83
3.2.	Standards of Total Irradiance	84
3.3.	Standards of Spectral Radiance	90
	3.3.1. Apparatus and Method	91
	3.3.2. Use of the Standards of Spectral Radiance	95
3.4.	Standards of Spectral Irradiance	97
	3.4.1. Use of the Standards of Spectral Irradiance	101
3.5.	Low-Intensity Standards of Spectral Irradiance	101
3.6.	Concluding Remarks	103
References		108

4. High-Accuracy Spectral Radiance Calibration of Tungsten-Strip Lamps

H. J. Kostkowski, D. E. Erminy, and A. T. Hattenburg

4.1.	Introduction	111
4.2.	Accuracy Requirements	112
	4.2.1. Accuracy of C_1	112
	4.2.2. Accuracy of C_2	112
	4.2.3. Wavelength Accuracy	112
	4.2.4. Accuracy of Blackbody Temperature	113
	4.2.5. Blackbody Quality	113
	4.2.6. Accuracy of Comparison	114

4.3.	The Blackbody	114
	4.3.1. Description	114
	4.3.2. Temperature Uniformity	114
	4.3.3. Emissivity	114
	4.3.4. Absorption	116
4.4.	The Spectroradiometer	118
	4.4.1. Design Criteria	118
	4.4.2. Description	118
4.5.	Temperature Calibration	120
	4.5.1. Low-Range Calibration	120
	4.5.2. High-Range Calibration	123
	4.5.3. Correction to the Thermodynamic Scale	124
4.6.	Comparison of Blackbody and Tungsten-Strip Lamp	125
4.7.	Summary of Uncertainties	126
References		127

5. New Infrared Detectors

E. H. PUTLEY

5.1.	Introduction	129
5.2.	The Performance of an Ideal Detector	131
5.3.	Intrinsic Photoconductive Detectors	138
5.4.	Extrinsic Photoconductive Detectors	147
5.5.	New Types of Thermal Detectors	154
	5.5.1. Low-Temperature Bolometers	154
	5.5.2. The Pyroelectric Detector	158
5.6.	Laser Detectors	160
References		162

6. Blackbodies as Absolute Radiation Standards

R. E. BEDFORD

6.1.	Introduction	165
6.2.	Principles of Absolute Total Blackbody Radiometry	169
6.3.	Temperature Measurements	171
6.4.	Stefan–Boltzmann Constant	172
6.5.	Emissivities of Blackbody Cavities	177
6.6.	Configuration Factors	182
6.7.	Low-Temperature Blackbody as an Absolute Standard of Total Radiation	184
6.8.	Experimental Difficulties in the Transfer of a Total Radiation Scale to Secondary Devices	188

6.9.	Accuracy of Total Irradiance Standards	196
6.10.	Applications	198
References		199

7. Experimental Blackbody (Absolute) Radiometry

A. R. Karoli

7.1.	Historical		203
	7.1.1.	Investigation of the Stefan–Boltzmann Constant	203
	7.1.2.	Verification of Planck's Law	204
7.2.	Blackbody Design		206
	7.2.1.	Features of Early Blackbodies	206
	7.2.2.	Development of High-Temperature Sources	209
7.3.	Sources of Error		211
	7.3.1.	Associated with Blackbody Design	212
	7.3.2.	Originating in the Source-Detector Arrangement	213
7.4.	Current Status of Cavity Sources		215
	7.4.1.	Types Commercially Available	216
	7.4.2.	Development of Large-Area, Low-Temperature Blackbodies	218
	7.4.3.	Future Requirements	220
References			223

8. Laboratory Methods of Experimental Radiometry Including Data Analysis

J. R. Hickey

8.1.	Introduction		227
8.2.	Total Radiation Scales in Practice		228
	8.2.1.	NBS Scale	228
	8.2.2.	IPS Scale	231
	8.2.3.	Cross Referencing the Scales	232
8.3.	Calibration of a Radiometer		237
	8.3.1.	Use of the Calibrated Radiometer	239
8.4.	Sensitivity Definitions for Thermopiles		240
8.5.	Filter Radiometry with Thermopile-Type Detectors		242
8.6.	Correlation of Broadband and Narrowband Filter Radiometer Measurements		248
8.7.	Monochromator Measurements		250
	8.7.1.	Status of Spectral Standards	250
	8.7.2.	Basic Compatibility Considerations in Using Spectral Standards with Monochromators	255

8.7.3. Measurement of Source Emission by Monochromator Methods 257
8.8. Use of Monochromator Data in Calculating Filter Factors 258
8.9. Correlation of Monochromator and Filter Radiometer Measurement Results 263
8.10. Other Uses of Spectral Standards for Calibration Purposes ... 265
    8.10.1. Use of Spectral Radiance Standards as Spectral Irradiance Standards 265
    8.10.2. Use of Irradiance Standards as Radiance Standards .. 265
References 267

9. On Determinations of the Atmospheric Turbidity and Their Relation to Pyrheliometric Measurements

A. K. Ångström

9.1. Introduction 269
9.2. Scattering of Solar Radiation within the Atmosphere 270
9.3. Determination of the Ångström Turbidity Coefficient 272
9.4. Influence of Circumsolar Sky Radiation on Pyrheliometric Measurements 275
9.5. Absorption of Solar Radiation by Atmospheric Water Vapor .. 277
9.6. Influence of Atmospheric Turbidity on the Planetary Albedo of the Earth 280
Appendix 282
References 283

10. Some Meteorological Aspects of Radiation and Radiation Measurement

G. D. Robinson

10.1. General Introduction: Magnitudes 285
10.2. Computation of Radiative Transfer in the Atmosphere 287
    10.2.1. Transmission of Direct Solar Radiation in a Clean Cloudless Atmosphere 289
    10.2.2. Diffuse Solar Radiation in a Clean Cloudless Atmosphere 289
    10.2.3. Transfer of Terrestrial Radiation in a Clean Cloudless Atmosphere 289
    10.2.4. Solar Radiation in a Cloudy Atmosphere 290
    10.2.5. Terrestrial Radiation in a Cloudy Atmosphere 290
    10.2.6. Radiation Computations as an Adjunct to Dynamical Meteorology 291

10.3.	Measurements of Solar Radiation		292
	10.3.1.	Available Measurements	292
	10.3.2.	Atmospheric Haze and Solar Radiation	296
	10.3.3.	Volcanic Dust and Solar Radiation	298
	10.3.4.	Cloud and Solar Radiation	299
	10.3.5.	Solar Radiation and the Earth's Energy Balance	300
10.4.	Measurements of Terrestrial Radiation		300
	10.4.1.	Measurements of Terrestrial Radiation at the Earth's Surface	301
	10.4.2.	Radiative Flux Divergence in the Free Atmosphere	302
10.5.	Indirect Sensing of Atmospheric Properties by Observations of Radiation		303
References			305

11. Measurement of Radiation Flux and Equivalent Radiation Temperature in the Atmosphere

J. STRONG

11.1.	Introduction	307
11.2.	A New MRT Meter	312
11.3.	Four-Sphere MRT Meter	318
11.4.	Observations with the Four-Sphere MRT Meter	320
11.5.	Experiments with the Four-Sphere Model	320
	11.5.1. Case Where $a/e \neq 1.0$	321
	11.5.2. Thermistor MRT Meter	321
11.6.	Equivalent Radiation Temperature in the Upper Air	322
11.7.	Future Problems	324
11.8.	Absolute Emissivity Measurement of Flat Surfaces	324
11.9.	Pyrheliometer Development	328
References		328

12. Applications of Thermal Radiation Measurements in Atmospheric Science

P. M. KUHN

12.1.	Role of Radiative Energy Exchange in Weather Processes	331
	12.1.1. Weather Processes	333
12.2.	Instrumentation for Radiation Measurements in the Free Atmosphere	333
	12.2.1. Balloon Radiometersonde Observations of Atmospheric Infrared Irradiance	333

	12.2.2. Rocket IR Instrumentation	343
	12.2.3. Satellite IR Instrumentation Capability	343
12.3.	Effects on Radiative Atmospheric Heat Budget during Artificially Induced Cirrus Metamorphoses	345
12.4.	Observed and Calculated Irradiances for the Real Atmosphere	352
12.5.	Infrared Remote Sensing of the Atmosphere	353
	12.5.1. Background	353
	12.5.2. Radiant Power Computations	354
References		356

13. Recent Advances in Satellite Radiation Measurements

R. A. HANEL

13.1.	Brief Summary of Satellite Radiometry	359
13.2.	Physical Principles of Vertical Sounding	368
	13.2.1. Temperature Profile	368
	13.2.2. Water Vapor and Ozone Distribution	371
	13.2.3. Minor Atmospheric Constituents	372
13.3.	Vertical Sounding Experiments	372
13.4.	Intercomparison of the Instruments	377
13.5.	Calibration of the Interferometer	382
13.6.	Future Experiments	385
Appendix		386
References		388

14. The Design and Construction of Evaporated Multilayer Filters for Use in Solar Radiation Technology

E. E. BARR

14.1.	Introduction	391
14.2.	Theory of the Fabry–Perot Interferometer	392
14.3.	Filter Contrast and Bandwidth	401
14.4.	Temperature Dependence of Band Position	404
14.5.	Wide Bandpass Filters	405
14.6.	Physical Properties of Narrow Bandpass Filters	410
References		412

Appendix: Radiation Terminology, Symbols, Units, and Conversion Factors 413

ADVANCES IN
GEOPHYSICS

VOLUME 14

—1—
PRECISION RADIOMETRY AND ITS SIGNIFICANCE IN ATMOSPHERIC AND SPACE PHYSICS

A. J. Drummond

1.1. Introduction

Radiation is the transfer of energy by either emitted waves or particles. Here the discussion will be limited to the type of energy which travels in the form of electromagnetic waves from some material substance excited by heat, electrical discharge, or other means, and is finally absorbed by another material substance. This electromagnetic spectrum is a continuum consisting of the ordered arrangement of energy according to wavelength or frequency. It has been established, experimentally, that the electromagnetic spectrum extends from waves as short as a few angstroms (i.e., of the order of a millionth of a millimeter) to those of many kilometers in length at the other end of the scale. However, the region of interest in this chapter is that extending from 2000 Å (200 nanometers—nm), at the ultraviolet end, to about 100 micrometers (μm) in the infrared. For convenience, we may here subdivide this spectral domain into several ranges, viz., the ultraviolet $\lambda < 380$ nm, the visible $\lambda = 380$–750 nm, the near infrared $\lambda = 750$ nm (i.e., 0.75 μm)–3 μm, the infrared $\lambda = 3$–15 μm, and the far infrared $\lambda > 15$ μm. In natural sunlight, there is measurable energy, above the terrestrial atmosphere, at a wavelength as short as 200 nm, but at the earth's surface it is rare that the radiation penetrates in wavelengths below about 300 nm; for practical purposes, we may adopt a value of 4 μm as the longwave limit of the sun's energetic emission. The infrared radiation of the terrestrial system (i.e., earth plus atmosphere) is largely contained within the limits of 5 and 50 μm. There are, of course, many studies of the far ultraviolet radiation and the X rays beyond and also of the solar energy emitted in the form of radio waves, but these topics fall outside the scope of this presentation.

The importance of measurements of radiant flux in the ultraviolet, visible, and infrared regions needs little emphasis, especially when regard is paid to the complexity of new requirements imposed by the geo-astrophysical programs of the evolving space age in which we live. Energy in the two shorter wavelength intervals is capable of inducing photochemical reactions, such as those responsible for plant growth and animal vision, and also of accelerating the degradation of materials. The infrared radiation generally comprises the greatest portion of the thermal emission by bodies at normally encountered temperatures and thus (with the shorter wavelength radiation) plays a considerable role in heat transfer processes. As a practical example, the thermal equilibrium of a space vehicle is mainly dependent upon its rate of absorption of incident solar radiation and the consequent emission, to space, of such energy with change of wavelength. This brings up the particular form of heat radiation which is in equilibrium thermally with matter at a given temperature—the so-called "temperature or blackbody radiation." Here, the intensity and spectral composition of the energy depend solely on temperature. The measurement of this radiation is intimately related to the establishment of the temperature scale and thus to practical methods of temperature determination.

With regard to the measurement of thermal radiation, this involves the choice of detectors (as will become apparent in the associated chapters which follow). Such detectors have to be designed not only for exposure to the natural sources (i.e., sun, sky, atmosphere, and ground) but also to artificial sources like blackbody radiators, tungsten-filament lamps, and arc systems. Further, there is generally another source present in determinations of radiation intensity, which is frequently ignored, and this is the background; for example, it is not possible, in practice, to isolate the sun's direct radiation from the immediate circumsolar radiation scattered downwards by the sky—and, indeed, in many applications it is not desirable to separate the aureole from the parent sun. The contribution from background radiations usually increases in significance as this temperature approaches that of the principal source, which may considerably complicate the measurement of the desired flux. In the philosophy of thermal radiometry, it is often not appreciated that we are not dealing with a single stream of radiation but with two streams, the second emanating from the detector itself. Thus, all radiometric measurements entail essentially the derivation of a net result of the energy exchange between source and detector. The meteorologist who is accustomed to record the solar radiation does not concern himself with the outgoing radiation from his sensor, and rightly so, on account of the great difference between its temperature and that of the sun. But this he cannot do when he has to evaluate incoming or outgoing longwave radiation of terrestrial origin. He then is faced with problems similar to those posed by the physicist in the laboratory. In this connection, it is clear that the circumstances of the experiment determine the appropriate procedures and this matter is treated in detail elsewhere in this book.

Another aspect involved in radiation measurements is the treatment of the medium intervening between source and detector. In the case of studies conducted out-of-doors, the atmosphere either wholly or a part thereof is the prime absorbing and scattering agent in radiant energy transfer. However, filters or window materials may be introduced into the system, to modify the source emission or the detector response. In the laboratory, recourse may be necessary to flush the system with an inert gas, especially where it is desired to eliminate effects of absorption by water vapor and carbon dioxide in the path of the energy beam.

1.1.1. Evolution of (Thermal) Radiometry

There are fundamental physical laws governing the radiation from blackbodies, i.e., perfect absorbers and radiators at all temperatures and for all wavelengths. These laws are discussed in the appropriate texts on the subject. Here, it will suffice to outline their introduction, in time, in association with the principal discoveries made and the advancements which have occurred in this field. Table I is therefore a historical summary of the development of radiometry since the time of Newton, who first resolved the visible spectrum, up to present.

Two comments on the table are necessary. In the first place, this series of lectures is largely concerned with the precise or absolute use of detectors which are fully black, i.e., with a response which is independent of the wavelength of the incident energy. Hence, the listing, in the table, of information relating to (selective) photodetectors is confined to the dates of the more important introductions. In this connection, too, it should be pointed out that the developments in photocells, phototubes, and photomultipliers which have occurred since the Second World War are too numerous to deal with here, as is also the case with the many new filters and transparent windows (especially for the infrared regions) which have similarly emerged.

In the second place, it will be noticed that no mention is made in the table of Beer's well-known law of absorption. This is on purpose: In 1905, it was found by K. Ångström that the absorption of radiation within a given amount of carbon dioxide, enclosed in a tube of given length, is changed (and generally increased) if the pressure is raised through the addition of another gas which does not absorb in the wavelength interval under investigation. The result demonstrated that Beer's law does not hold for mixtures of gases. This finding subsequently gave rise to many associated investigations. In general, it may be stated that the absorption of a gas is not only dependent upon the partial pressure of its constituents, but also upon the total pressure.

TABLE I. A survey of the important developments in thermal radiometry.

1666	Newton	Visible spectrum first investigated
1722 to 1800	Petit Scheele Priestley Rumford	Demonstrations of chemical effects attributed to light
1752	Melville	Spectrum of sodium flame observed
1760	Bougher Lambert	Law of extinction of monochromatic radiation in a homogeneous medium (independently, but usually associated with Bougher)
1760	Lambert	Cosine law
1800	Herschel	Infrared radiation discovered
1802	Young	Wave nature of ultraviolet radiation proved
1814	Fraunhofer	Dark (absorption) lines in solar spectrum observed. Application of the diffraction grating to spectroscopic studies
1826	Seebeck	Thermoelectricity discovered (followed in 1834 by Peltier's complimentary heating or cooling effect)
1830	Nobile	Radiation thermocouple invented
1833	Melloni	Thermopile (extension of thermocouple) introduced
1837	Pouillet	Construction of pyrheliometer on calorimeter principle
1839	Becquerel, E.	Photovoltaic effect observed (sunlight on a voltaic cell)
1843	Becquerel, E.	Photographic effects of thermal radiation demonstrated
1859	Kirchhoff	Law for absorption–emission radiation relationship
1860	Kirchhoff Bunsen	Development of the first practical spectroscope
1862	Maxwell	Classical electromagnetic theory of radiation predicting the existence of electrical waves (associated is such work as Fresnel's wave theory of light, Clausius' development of thermodynamics, and Faraday's investigations into the electrical nature of matter)
1873	Smith, W.	Photoconductive effect observed (resistance of selenium decreased on exposure to light)
1879 to 1884	Stefan Boltzmann	Law governing temperature dependence of integral radiation from a blackbody source
1881	Langley	Introduction of the bolometer (sensitive electrical thermometer)
1877	Hertz	Photoemissive effect observed (during investigation of electric waves)
1888	Hertz	Verification of Maxwell's theory, through demonstration experimentally that electrically produced very long (radio) wavelength radiation is similar in its properties to light waves
1893	Ångström, K.	First construction of electrical compensation pyrheliometer (final form in 1896). Introduction of Ångström Pyrheliometric Scale
1894	Wien	Law relating wavelength of maximum emission with blackbody temperature

TABLE I (cont.)

1895	Röntgen	Discovery of X rays, followed (in 1900) by that of gamma radiation by Villard; shown subsequently to be short electromagnetic waves
1900	Rubens Kurlbaum	Study of "rest-radiation" over wide ranges of wavelength and temperature (establishing experimentally the fundaments for the spectral energy distribution of blackbody radiation in the far infrared region)
1900	Planck	Law for wavelength distribution of blackbody emission as a function of temperature (based upon his quantum theory)
1902	Langley	Commencement of Smithsonian Institution's "solar constant" observing program (subsequently continued by Abbot and his collaborators)
1905	International Meteorological organization	Recommended universal adoption of Ångström pyrheliometer as a standard in solar radiation networks and associated Ångström scale
1905	Coblentz	Introduction of improved spectroscopic techniques (and resulting establishment of spectra of many solid, liquid, and gaseous materials)
1908	Abbot Aldrich Fowle	First construction of Smithsonian water-flow pyrheliometer (improved form in 1932)
1908	Abbot Fowle	Commencement of comparison of Ångström and Smithsonian pyrheliometric scales
1909	Abbot	First construction of Smithsonian silver-disk pyrheliometer (modified to long tube in 1927)
1910	Callendar	Construction of radio-balance radiometer. Introduction of British National Physical Laboratory (NPL) radiometric reference
1913	Abbot	First revision of Smithsonian pyrheliometric scale, based upon the water-flow pyrheliometer
1914	Coblentz	Introduction of U.S. National Bureau of Standards (NBS) reference of total radiation (blackbody). Commencement of regular issue of calibrated secondary standards of total irradiance (carbon-filament lamp)
1932	Abbot Aldrich	Smithsonian pyrheliometric scale further revised, according to improved water-flow pyrheliometer (and confirmed in 1934, 1947, and 1952), but such revisions never applied in meteorological practice
1937	Guild	Comparison, indirectly, of NPL, NBS, Ångström, and Smithsonian scales of radiation. Introduction of absolute drift radiometers to maintain the NPL radiometric reference
1940	(approx.) to date	Intense development in many countries of photocell, phototube, photomultiplier, and thermistor bolometer detectors. Introduction of great variety of infrared windows and lens materials
1947	Golay	Construction of improved type of infrared pneumatic detector (sensitive from uv to microwave region)

TABLE I (cont.)

1956	International Radiation Commission	Introduction of International Pyrheliometric Scale (IPS), based upon Ångström and Smithsonian Rreferences
1960	Bedford	Construction of low-temperature standard of total radiation (blackbody)
		Introduction and maintenance, independently, of Canadian National Research Council (NRC) radiometric reference
1960	Stair Johnson Halbach	Establishment of NBS reference of spectral radiance Commencement of regular issue of such secondary standards (tungsten-quartz lamp)
1962	Gillham	Comparison of Guild (drift) and Gillham (cavity: electrical compensation) absolute radiometers. Verification of NPL radiometric reference
1963	Stair Schneider Jackson	Establishment of NBS reference of spectral irradiance Commencement of regular issue of such secondary standards (tungsten-quartz-iodine lamp)
1965	Stair Schneider Fussell	Revision of NBS reference of total radiation. Replacement of former secondary standard by new-type standards of total irradiance (tungsten lamps)
1968	Kostkowski Erming Hattenburg	Introduction of (NBS) high-accuracy radiance standards
1968	Schneider Waters Jackson	Development of a NBS high-intensity standard of total and spectral irradiance

1.2. Historical Review of Precision Radiometry

The principal efforts in the 20th century to establish references of thermal radiation intensity were due to K. Ångström (1893, 1899, 1900, 1903), in Sweden; Callendar (1910), followed by Guild (1937) and Gillham (1959, 1961, 1962, 1968), in England; Abbot *et al.*, (1900–1954) and Coblentz (*et al.*, 1915, 1916, 1933), followed by Stair *et al.* (1954, 1960, 1963, 1967) and Kostkowski *et al.* (1970), in the United States; and Bedford (1960, 1968), in Canada.

Two fundamentally different approaches have been made and, in each instance, distinctly different types of instrumentation introduced. Until recently, the viewpoint, generally, in Europe has been to attempt realization of the true reference of radiometry, and its reproduction, through the utilization of so-called absolute standard detectors. In North America, this conception has also found favor in meteorological application, with the sun as source: However, in laboratory practice, a standard blackbody source has been preferred. It is perhaps also pertinent to mention that Mulders (1935), in Germany, employed a blackbody as his reference in a determination of the extraterrestrial solar flux (i.e., solar constant of radiation—see later section). However, as will be apparent subsequently here and elsewhere in this book, any improvement in the fixation of the absolute reference of radiometry must result from the dual use of detector and source as standards.

In meteorological services and geophysical institutes, radiation measurements have normally been standardized by one of two types of instrument, viz. the Ångström electrical-compensation pyrheliometer or the Abbot silver-disk pyrheliometer. Since the methods used in both cases to arrive at absolute radiometric determination are not independent of the specimen instruments, it has been customary to designate these references the Ångström and Smithsonian Scales of Pyrheliometry.

1.2.1. Ångström Pyrheliometric Scale

The electrical-compensation pyrheliometer of K. Ångström is one of the best-known instruments for measuring solar intensity and is perhaps the most reliable. Although it is generally employed as a secondary relative instrument, it is capable of being used absolutely. In operation, the absorption of heat by a thin blackened manganin strip, exposed to the sun's rays, can be determined by measuring the electrical energy necessary to warm a similar but shielded strip to the same temperature. The rate of generation of heat in the electrical circuit is then equal to the rate of absorption of heat by the exposed strip. The equality of the temperatures of the strips is verified by means of thermocouples attached to the backs of the strips and connected in opposition through a sensitive null detector. Figure 1.1 shows the modern Eppley version of the instrument with its auxiliary apparatus for precise regulation of the compensating electrical current. Figure 1.2 is an electrical circuit diagram of the pyrheliometer.

The intensity I of direct solar radiation is calculated by means of the formula $I = ki^2$, where i is the heating current in amperes and k is a constant typical of each Ångström pyrheliometer. When the pyrheliometer is used absolutely, the constant k is derived as follows:

$$k = r/ba \qquad (I \ in \ W \ cm^{-2})$$

or

(1.1) $$k = 60r/Jba \qquad (I \ in \ cal \ cm^{-2} \ min^{-1})$$

where r is the resistance of unit length of strip ($\Omega \ cm^{-1}$), b the mean width of strip (cm), a the absorption coefficient of the blackened surface of the strip (e.g., 0.985 for Eppley–Parsons black lacquer), and J the electrical heat equivalent (1 cal sec^{-1} = 4.185 W).

In most of the earlier types of this instrument, including the standard models maintained, originally, at Uppsala and, latterly, at Stockholm, the

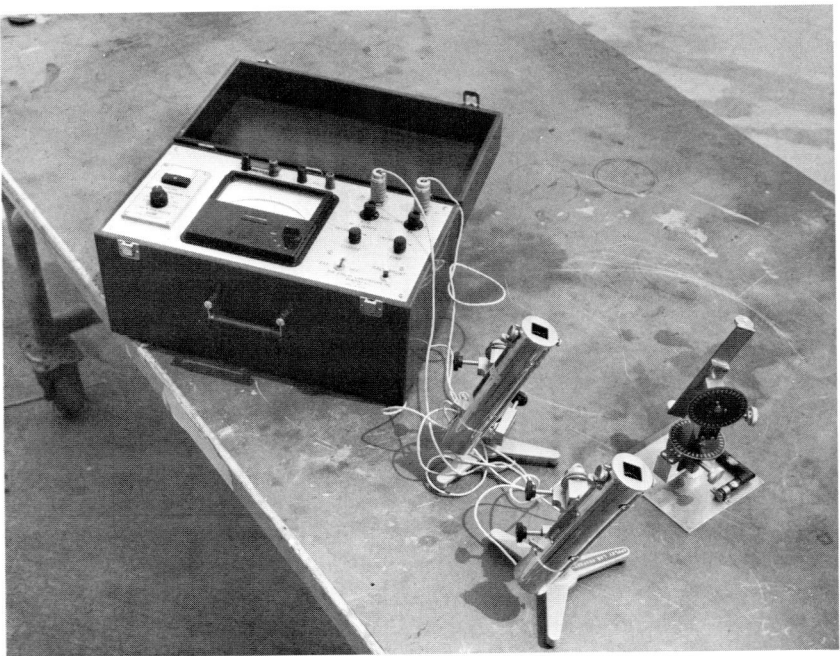

Fig. 1.1. Ångström electrical-compensation pyrheliometers (Eppley model) with auxiliary apparatus for precise regulation of the compensating electrical current.

exposed strip is shaded at the ends by the innermost diaphragm while the electrically heated strip has the same temperature over its whole length. This introduced the so-called "edge effect" which, it is estimated, has caused measurements made with these pyrheliometers to be too low by approximately 2 %. The correction involved has never been applied in meteorological

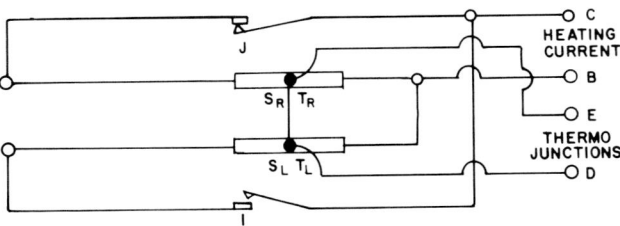

Fig. 1.2. The electrical circuit diagram of the Ångström pyrheliometer (when the S_r strip is shaded).

practice on account of construction differences in the individual instruments. Prior to the introduction of the International Pyrheliometric Scale of 1956 (where an attempt is made to correct for this and other possible errors—see below), all such measurements were referred to the original (uncorrected) Ångström scale. In the recent designs of this pyrheliometer, originated at Stockholm and at Newport, an effort is made to eliminate such edge effects through carefully locating the relevant diaphragm, so that the screening occurs exactly at the terminations of the strip. Of the other error sources, the most important may arise through the inability, still inherent, to heat the shielded strip in the same manner as the exposed one. Also, small optical effects may be introduced within the diaphragm system of the pyrheliometer tube and behind the strips.

The aperture (rectangular) angles of the older models are about $6 \times 24°$; the corresponding values for the newer designs are $4.5 \times 10.5°$. When account is taken of the fact that the strip screened from the direct solar component also views a portion of the circumsolar sky (but not the same portion as that viewed by the exposed strip), there is a measure of compensation introduced for the circumsolar contribution to the exposed strip. This has the effect of modifying the geometrical apertures of the tube system to effective pyrheliometric apertures equivalent to a circle of about $8°$ whole angle, in the case of the earlier instruments, and $5°$ in the case of the later specimens.

1.2.2. Smithsonian Pyrheliometric Scale

The history of the Smithsonian scale of radiation may be studied in the Annals of the Astrophysical Observatory of the Smithsonian Institution (1900–1954). It is important to note that, in 1932, following the introduction of an improved (water-flow) calorimeter, the Institution announced that its basic scale of 1913 was in error by nearly 2.5 %. This was confirmed in 1934, 1947, and 1952. Although this correction has been applied to the evaluations of the solar constant (the prime object then of the observing program of the Astrophysical Observatory), it has never been introduced anywhere in the routine meteorological measurements based upon the Smithsonian reference. This may have been for the sake of preserving continuity in the many existing series of records. These revisions of this scale, therefore, have had no official acceptance in the presentation of pyrheliometric measurements.

The silver-disk pyrheliometer, developed by Abbot, was for many years the principal instrument employed to reproduce the Smithsonian 1913 scale. However, within the solar-constant observing network (disbanded about 1960) of the Astrophysical Observatory, a modified version of the Ångström pyrheliometer was introduced, about 1935, mainly because of greater ease of operation than is the case with the silver-disk pyrheliometer. Due to its construction principles, this type of electrical-compensation pyrheliometer is incapable of use in an absolute manner. Likewise, the silver-disk pyrheliometer is strictly a secondary instrument; but on account of its high reliability over long periods of years, it has functioned as a working standard pyrheliometer and has been extensively used for the calibration of secondary routinely operated pyrheliometers.

The silver-disk pyrheliometer (see Fig. 1.3) consists, essentially, of a blackened, relatively massive disk of silver positioned at the lower end of a diaphragmed tube, the aperture (circular) angle of which is approximately 6° (i.e., in the present working standard specimens). A bent mercurial thermometer, graduated in tenths °C and read to hundredths, is immersed from the side of the tube in a mercury-filled hole bored in the silver disk. A thin steel jacket separates the mercury from the silver, preventing amalgamation but permitting rapid transfer of heat from the receiving surface of the disk to the thermometer. The unit is well insulated thermally. In normal practice, a triple shutter is alternately opened and closed every 2 min, systematically exposing the disk to, and isolating it from, the solar beam. It is important that the procedures described by Abbot and his co-workers be closely followed, as the results obtained from the instrument may change somewhat according

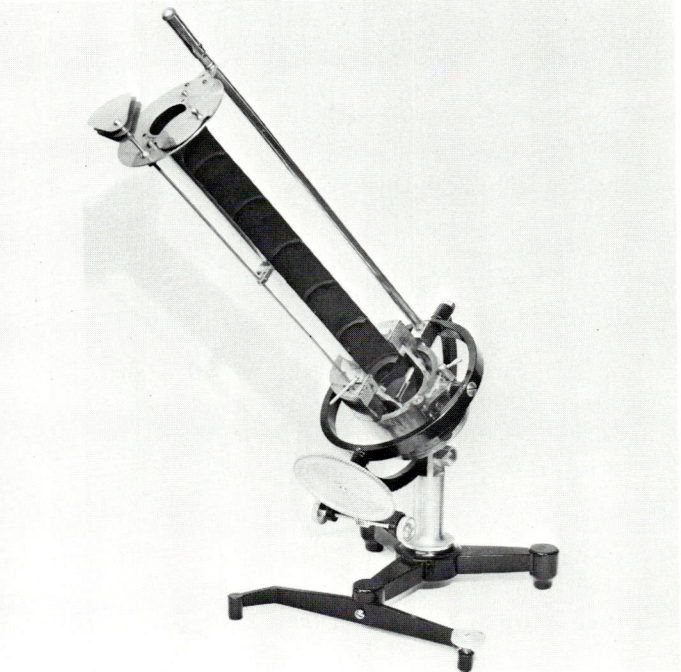

Fig. 1.3. The Abbot silver-disk pyrheliometer (courtesy of the Smithsonian Institution).

to the number of cycles used in each observation series (i.e., the routine measurements should follow the method prescribed for standardization). A major difficulty is the required time accuracy, and a number of automatic and other means have been devised to minimize such an error which can be of the order of 1 % for an interval timing uncertainty of 1 sec. The readings obtained by exposure and shading of the pyrheliometer require corrections for air, stem, and bulb temperature, prior to multiplication by the appropriate instrument constant to yield values in energy units.

Fundamently, the series of working standard Abbot silver-disk pyrheliometers were calibrated by reference to a calorimeter maintained by the Astrophysical Observatory, at its Table Mountain station, the sun being used as source. Hence, the fixation of the Smithsonian Scale(s) depended on the adequacy of this calorimeter (water-flow pyrheliometer) as a primary standard radiometer. Originally, the instrument comprised a single deep test-tube-like blackened chamber of metal with hollow walls. In the side walls, in the extreme rear wall, and in the walls of a hollow cone not quite at the rear (on which all the sun's rays fell directly), a current of water constantly flowed to carry off the solar heat as fast as it was absorbed. A carefully designed diaphragm system was incorporated to admit the direct sunlight and, at the same time, to minimize the effects of air currents and to obstruct stray light. The rise of temperature between the entrance and the exit of the stream of water and its flow rate contributed the basic measurements. The aperture angle was similar to that adopted for the silver-disk instruments.

The improved form of this basic radiometric reference is shown in some detail, in Fig. 1.4. "The two chambers abc and a'b'c' are almost exactly the same. Distilled water enters at d and divides into two streams at e and e'. The water flows around the receiver and out of the instrument at j and j'. i and i' are the thermoelectric junctions used to determine the equality of temperature of the water streams outflowing from the two chambers. m is the water bath for the two receivers, water entering at p and being discharged at o. n is a wooden case surrounding the instrument. The heating coils are indicated by k and k'. Not shown is a shutter for alternating the chambers exposed to solar and electric heating. A detailed description of one of the receivers is given in Vol. 3 of the Annals of the Astrophysical Observatory. In order to keep the water bath surrounding the two receivers and the distilled water entering the instrument at the same temperature, a 50-gal drum of water was used as a source of water for the water bath. A circulating pump continually stirred the water in the drum and a bypass on the pump circulated some of the water through the water bath. The distilled water flowed through a coil in the drum before entering the instrument. Thus, the bath water and the distilled water were always at the same temperature when leaving the drum" (Hoover and Froiland, 1953).

and a circumsolar sky radiance rarely uniform within the instrument aperture. Bossy and Pastiels (1948, 1959) have treated this problem in considerable detail, mathematically, and have shown how the theory, first originated by Lambert, in 1760, may be applied to pyrheliometers in actual use.

As a means of standardizing the aperture conditions of a radiometer, the following parameters were recommended, by the Radiation Commission of the International Association of Meteorology and Atmospheric Physics (IAMAP), in 1956:

$$\alpha = \tan^{-1}(a/b)$$
$$\alpha_1 = \tan^{-1}[(a-1)/b]$$
$$\alpha_2 = \tan^{-1}[(a+1)/b]$$

where $a = R/r$ and $b = l/r$.

It was realized that it would be impractical to propose an aperture angle so small that the circumsolar radiation falling upon the receiver would be negligible. It was, however, recognized that measurements made with different working standard instruments ought to be capable of precise intercomparison, as far as possible, and that a certain standardization of the pyrheliometer aperture conditions is highly advisable. For this reason, the following instrument design parameters were recommended:

(1) slope angle $2° > \alpha > 1°$;

(2) ratio $l/r \geqslant 15$.

These conditions imply that the opening (i.e., $\frac{1}{2}$ aperture) angle should not be greater than 4°.

The intensity of the solar aureole, as measured by instruments of varying aperture, exposed under different conditions of atmospheric turbidity, has been the subject of a number of investigations. Considering measurements by Abbot, Dorno, Voltz, and Schüepp, data analysis by van de Hulst, and theoretical treatment by Linke and Ulmitz (1940) regarding the angular distribution of circumsolar sky radiation, curves showing the estimated average proportion of sky radiation entering specific pyrheliometric apertures, as compared with the direct solar radiation for typical values of the Ångström turbidity coefficient β (Ångström, 1929, 1930) as a function of the optical air mass m (IGY Instruction Manual, 1957), have been presented by Ångström and Rodhe (1966)—see this volume, Chapter 9.

For observing locations near sea level, values of the $m\beta$ product generally are of the order $0.15 > m\beta > 0.08$. At high altitude stations $m\beta$ may be 0.05. In July 1966 two checks were made using a radiometer equipped with a variable aperture. At Table Mountain (2250 m, California), with very clear skies and a $m\beta$ value of 0.07, the observed increase in sky radiation for an aperture difference of 5–15° was 1.2%, as compared with the computed figure of 1.4 %. Filter measurements were included; the results indicated a slightly higher value for the shorter wavelengths. At Newport, with hazy conditions ($m\beta = 0.14$), the measured and computed values for the same 10° aperture difference were, respectively, 2.5 and 2.4 %.

That it is the turbidity conditions, and not the altitude, which influence the aperture correction is clear from Drummond's South African study of data assembled on the high southern plateau at heights of about 1500 m (Drummond, 1956a). Here, the turbidity level is practically the same as that at sea level and the Smithsonian/Ångström relationship is in perfect agreement with the generally established value (see below).

Many comparisons have been made between representative instruments of the two traditional pyrheliometric scales, with the sun as source. The most general value of the Smithsonian/Ångström ratio is 1.035; but on account of the factors discussed above, differences of up to 6 % have been observed on occasion. In an attempt to eliminate the geometry inherent in the design of the standard pyrheliometers, Guild (1937) investigated the relationship between both pyrheliometer types, in the laboratory, with the aid of an artificial source of radiation. This work was undertaken in connection with the verification of the reference of thermal radiation maintained at the National Physical Laboratory, London (see below). Guild's results were subsequently closely confirmed by those of Eldridge (1952) and Drummond (1956a). These laboratory examinations of the pyrheliometers, under conditions when they are unaffected by the inclusion of different amounts of skylight as is the case with solar comparisons, indicated a ratio value of 1.028.

Before the advent of the Third International Year (IGY), in 1957, much confusion existed in the publications of solar radiation data, especially when serial values were presented side by side, often without reference to the relevant scale. As a result of a proposal by Drummond (1956b), a universal reference was recommended, in 1956, by the Radiation Commission of IAMAP and accepted by the World Meteorological Organization (WMO), to coincide with the commencement of the IGY. For more information regarding the background to this proposal, Courvoisier's (1963) study should be consulted.

The various pyrheliometric comparisons are summarized in Fig. 1.6.

It is considered that measurements on the uncorrected Ångström scale increased by 1.5 % will almost certainly be within ± 1 % and may be within ± 0.5 % of the best approximation which can at present be made of the true absolute scale of radiation.

In 1959 and also in 1964, international comparisons between primary working standard pyrheliometers were carried out at Davos, Switzerland (World Meteorological Organization, 1960, 1964). Some typical results for the two series of Ångström pyrheliometer measurements are given in Table II. The values tabulated are the ratios with respect to the Stockholm reference.

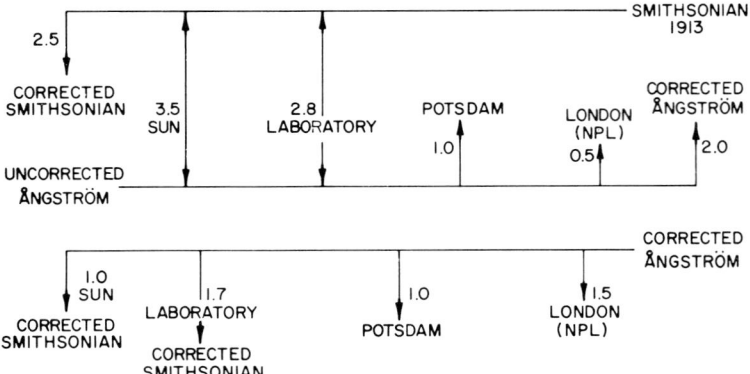

Fig. 1.6. Derivation of the International Pyrheliometric Scale (IPS) from the relationships of the basic radiometric references (differences as percentage).

Table II. Comparison, in Switzerland (Davos Observatory), of national standard pyrheliometers (WMO, 1960 and 1964).

1959	Sweden 1.00	Switzerland 1.006	USA 1.002
	Belgium 0.999	Congo 0.999	S. Africa 0.994
1964	Sweden 1.000	Switzerland 1.002	USA 1.000
	India 1.017	Norway 0.998	S. Africa 0.984
	Congo 1.001	Australia 0.987	Tunisia 1.005
	USSR 0.992		

1.2.4. National Physical Laboratory (U.K.) Radiometric Reference

The most important British work undertaken with the object of establishing a reference of radiation intensity is that of Callendar (1910) during the early years of this century. His radio-balance instrument was used into the 1930's to calibrate secondary detectors. As has been pointed out, the design and operation of certain types of absolute radiometer entail a knowledge of the reflectance of the radiation sensing surface over the wavelength interval of interest. However, the accomplishment of such examinations, especially in the infrared region, can be extremely difficult where the object of the study is the actual surface of the sensor to be employed. At the National Physical Laboratory (NPL) a different approach is used, namely, that of effectively controlling this reflection factor of the reference instrument rather than measuring it. This is realized through the use of a thermopile which has a receiver of the blackbody (cavity) type, i.e., with the receiver in the form of a hollow receptacle containing a small hole through which the radiation can be admitted.

The first blackbody thermopile used for absolute radiometric measurements was Callendar's detector. It is of the compensated type with two similar receivers side by side, each consisting of a spun copper tube of about 3 mm bore and about 9 mm depth. A diaphragm of 2 mm diameter mounted in front of the open end of the receiver limits the incident radiation, so that it clears the side walls of the receiver and falls only on the closed end. The black with which the interior is coated is diffusely reflecting, and since the solid angle subtended by the open end of the receiver at the closed end is approximately 0.1 then only a fraction $0.1/\pi \sim 1/30$ of the radiation reflected from the black escapes from the receiver. The reflectance of the black probably does not exceed 0.05 over the whole spectral range of interest, and therefore the effective absorptance of the receiver is thus always within about 0.002 of unity. Since part of the radiation entering the receiver is absorbed at the closed end, and part is absorbed at the side walls after reflection from the closed end, it is essential for correct operation that the receiver should have uniform sensitivity, i.e., a given quantity of radiant power should produce the same response from the temperature-sensitive element no matter where in the receiver it is absorbed. The walls are thus made of sufficiently thick material to give reasonable temperature uniformity throughout the receiver, and the series-connected wire thermocouples (which constitute the temperature-sensitive element) are attached to the receiver in a disposition which gives the best indication of the average temperature rise.

On account of its low sensitivity, this instrument is only suitable for the measurement of high radiant flux density. It is also somewhat awkward to use because of the relatively large thermal capacity of the receiver. With the purpose of checking the radio-balance, after Callendar's death, and also to investigate the relationship of the Ångström and Smithsonian scales in the laboratory, Guild (1937) constructed two different models of temperature-drift radiometers (1934–1936). The aim of these designs was to operate temperature-drift instruments by alternate exposure to radiant and electrically generated heat, so approaching as nearly as possible the ideal of pure substitution. In principle, the measurements, in each case, are made by connecting the thermocouple leads to the two similar receivers (front one only exposed) to a null detector, and then adjusting the electrical power in the resistance element until its heating effect is equal to that of the radiation being sampled. A very full account of the construction of these calorimeters, the underlying theory, and the operating procedures has been provided by Guild. The main results of his laboratory comparisons, using a 1500-W tungsten-filament lamp, were as follows:

(a) The agreement between the Callendar and the Guild references was of the order of 0.5 %.

(b) The agreement between this latter NPL reference and that employed by NBS (Washington), as transferred through carbon-filament lamps, was within a few parts per thousand.

(c) The 1913 Smithsonian pyrheliometric scale overrates radiation intensity by 2.3 % and the Ångström scale underrates radiation by approximately 0.5 %. (These results, however, are based on a comparison with only one specimen instrument of each type on which the two traditional pyrheliometric scales were based.)

Until a few years ago, the NPL scale of radiation intensity was founded entirely on measurements made with Guild's radiometer. Gillham (1959, 1961, 1962) has recently described refinements in the method of using these instruments. He has also described the development of three new absolute radiometers intended to verify the accuracy of the British radiometric reference and to facilitate its maintenance. Each works on the electrical substitution principle and consists, in effect, of a radiation thermopile with an electrical heating element attached to the receiver. The first instrument is similar to the Guild drift radiometer; the second is a small fast-response detector which has a plane receiver coated with a glossy, i.e. specularly reflecting black, for which the absorbance can be readily determined; the third instrument is again small, but has a cavity (blackbody) receiver which is almost completely absorbing (see this volume, Chapter 2 for further information). With regard to the determined error sources in absolute radiometric measurements, Gillham concluded that his new detectors were mutually consistent in their performance to within 2 to 3 parts per thousand, but divergence between these results and those using the Guild instrument could be as high as 0.6 %.

1.2.5. National Bureau of Standards (U.S.A.) Radiometric Reference

A convenient standard of thermal radiation was early recognized in the U.S.A., as an important requirement in scientific laboratories engaged in research involving the use of radiant energy. Towards this end, such a reference of total radiation intensity was set up, at the National Bureau of Standards (NBS), by Coblentz (1915, 1916) many years ago. This standard has continued to be in operation (but see below and this volume, Chapters 3 and 4).

The fundamental reference is a blackbody operated at temperatures between about 1000 and 1150°C. This blackbody was earlier of a ceramic type heated through a wire-wound element. In its current form, however, it consists of an oxidized nickel-chromium alloy cavity. The low reflectance of this cavity, considered in connection with its small aperture as compared with its internal surface area, results in a blackbody of very high effective emissivity. The good heat capacity of this furnace produces very high temperature stability. Monitoring of the temperature of the blackbody is presently determined by a carefully calibrated platinum versus platinum-10 % rhodium thermocouple. Check measurements are made using an optical pyrometer, with an agreement generally of the order of 2–3°C.

Procedures for the use of derived secondary (carbon-filament lamp) standards of total irradiance, at low intensities, have been disseminated (NBS, 1960), and such reference lamps distributed regularly over this period.

During the last few years, NBS has issued standards of (a) spectral radiance (Stair et al., 1960) and (b) spectral irradiance (Stair et al., 1963), applicable to the 0.25–2.5 μm region. In the former instance, these secondary references are tungsten-in-quartz lamps, while in the latter, the tungsten source incorporates iodine as an agent to return tungsten deposited on the inner surface of the quartz envelope to the filament. Consequently, in this cycle, bulb blackening is minimized and thus stronger emission results, particularly in the ultraviolet region. The blackbody employed for the shorter wavelengths is constructed of a high-purity graphite cylindrical enclosure. This graphite blackbody is heated by induction inside a water-cooled coil by a radiofrequency generator operating at 450 kHz, to temperatures approaching 2200°C. Degradation of the graphite at high temperatures is reduced by enclosing the blackbody in an airtight chamber through which dry helium is passed. The concentration of oxygen is further reduced by heating copper coils within the chamber preceding each operation of the blackbody. The effective area of this blackbody is many times greater than the small tungsten enclosures often used in high-temperature work. A much larger opening can thereby be employed, making possible the use of the entire slit of the monochromator used to compare spectrally the secondary lamp specimens against the primary source. The blackbody temperature is determined by means of a pyrometer.

With the object of rendering the wavelength interval of the three types of NBS lamp standards closely similar, the older carbon-filament lamp has been supplemented by a series of calibrated 100-, 500-, and 1000-W tungsten-filament, projector-type lamps, as used in standard photometry (Stair et al., 1967). The working radiometers employed by NBS to compare regularly their primary and working standards, consist of a group of cavity (blackbody) detectors and Eppley laboratory-type thermopiles specially constructed to have uniform response out to wavelengths as long as 20 μm, at least. For further information regarding the newer NBS standards, reference should be made to this volume, Chapters 3, 4, and 8.

1.2.6. National Research Council (Canada) Radiometric Reference

In 1960, Bedford described his standard of total radiation constructed for the purpose of calibrating thermal radiation detectors. This is a low-temperature blackbody cavity source consisting of a modified cone immersed in a stirred liquid bath controlled over the temperature range $+40°$ to $150°C$, supplemented by the thermopile detector, under test, installed in a housing designed to (a) provide a constant temperature surround to this detector, (b) define the aperture of the system, and (c) contain a shutter. In calibrating the detector, the procedure essentially entails measuring its output as exposed, alternately, to the blackbody source, at the selected elevated temperature, and to the shutter at the temperature of the surround. When the thermopile etc. detector views the blackbody, the net radiant flux density W, at the receiver, is given by

$$(1.5) \qquad W_1 = \sigma F_{01} \varepsilon_1 \varepsilon_0 (T_1^4 - T_{01}^4) \quad \text{W cm}^{-2}$$

where σ is the Stefan–Boltzmann constant, ε_1 the emissivity of the blackbody, ε_0 the emissivity of the detector receiver, T_1 the temperature of the blackbody (°K), T_{01} the temperature of the detector, and F_{01} a nondimensional factor involving the geometry of the system.

Equation (1.5) assumes that the detector radiation sensor is spectrally nonselective, so that its emissivity for radiation characteristic of its own temperature is numerically equal to its absorptance for radiation characteristic of the source temperature. Similarly, in this case, where the detector views the shutter, we get

$$(1.6) \qquad W_2 = \sigma F_{02} \varepsilon_2 \varepsilon_0 (T_2^4 - T_{02}^4)$$

Here ε_2 is the shutter emissivity. The Bedford apparatus is designed so that $F_{01} \equiv F_{02}$ and hence subtraction of Eq. (1.6) from Eq. (1.5) yields

$$(1.7) \qquad W_1 - W_2 = \sigma F_{01} \varepsilon_0 [\varepsilon_1 (T_1^4 - T_{01}^4) - \varepsilon_2 (T_2^4 - T_{02}^4)]$$

If V_1 and V_2 are the emfs developed by the thermopile detector when exposed, respectively, to the source and shutter, then the detector sensitivity S' is defined by

$$(1.8) \qquad S' = \varepsilon_0 [(V_1 - V_2)/(W_1 - W_2)]$$

Present-day thermopile models usually have a very closely linear radiation intensity–thermopile emf relationship, in which case Eq. (1.8) simplifies to

$$(1.9) \qquad S_1 = (V_1 - V_2)/\sigma F_{01} \varepsilon_1 (T_1^4 - T_0^4)$$

where T_{01} is replaced by the temperature of the cold junctions T_0 for all practical considerations (e.g., $S' \sim S$ to within 0.1 % in such calibrations described by Bedford).

For his blackbody source, Bedford has derived a value for ε, of 0.998, in accordance with the analyses of cavity emissivity based upon the studies of Gouffé, De Vos, and Edwards. He has adopted the theoretically derived value for the Stefan–Boltzmann constant (i.e., $\sigma = 5.67 \times 10^{-12}$ W cm^{-2} deg^{-4})— see below.

Employing a flushing method (dry oxygen) to remove the effects of atmospheric absorption in the source-detector path, and operating his system vertically with the detector below the source to minimize convective energy transfer effects, he has claimed an overall absolute accuracy of better than 0.5 %. Radiative transfer procedures are accomplished with the aid of a group of calibrated thermopiles. For further information on this radiometric reference, Chapter 6 (this volume) should be consulted.

1.2.7. Investigations Concerning the Spectrum of the "Blackbody"

The following is a summary of a report prepared in connection with the proposal of Max Planck for the Physics Nobel Prize. With regard to this extraction, it should be stated that the content of the report reflects the experimental and theoretical status in 1918. At present, emphasis is placed on the Planck formula chiefly in view of its importance in the development of the quantum theory.

In 1860, Kirchhoff put forward the general law that "the intensity of radiation from a blackbody is dependent solely on the wavelength and the temperature of the radiating body." Stefan found through a critical examination of experimental data on this matter (1879)[1] that the total radiation is proportional to the fourth power of the absolute temperature, viz.,

$$(1.10) \qquad E = \sigma T^4$$

The importance of this law was still better realized when Boltzmann, in 1884, derived it from the second law of thermodynamics and on the basis of the conception of the radiation pressure introduced by Maxwell. This radiation law is now generally referred to as that of Stefan–Boltzmann. It is valid for a perfect blackbody. Such a body is scarcely found in nature, but Lummer and Wien succeeded (1890) in coming very close to it through the construction of a hollow body, surrounded by a bath of given temperature, and with a small hole in the wall, through which the radiation was emitted.

Through very careful measurements by Lummer and Pringsheim (1899) and by Lummer and Kurlbaum (1901), it was shown that the Stefan–Boltzmann law was generally valid between $-180°$ and $+1800°C$. An important

---

[1] The dates in this subsection are given for historical reasons and not as references to classical physics.

question concerned the energy distribution within the spectrum. Langley seems to have been the first who attempted a clarification of this question through experiments (1886). He measured the spectral distribution of the radiation from different sources by means of a bolometer and found that the radiation has a maximum at a certain wavelength, which is shorter with higher source temperatures. Weber and Kövesligethy (1888, 1889) established a simple law, namely:

(1.11) $$\lambda_m \cdot T = \text{const}$$

where λ_m is the wavelength at which the maximum occurs and T the absolute temperature. If the wavelength is expressed in centimeters, the constant was found to have the value 0.294. This law for the shift of the position of maximum wavelength with temperature was later (1894) derived by Wien, in an important theoretical treatise. In 1898, C. Beckmann carried out an investigation by means of the "restrahlung" from fluorite (at 24 μm). Rubens pointed out that the results of Beckman were incompatible with the law suggested by Wien, in 1896, namely:

(1.12) $$E_\lambda = K\lambda^{-5} \exp{-(\beta/\lambda T)}$$

Rubens studied (1900) together with Kurlbaum this rest-radiation from quartz, fluorite, and rock salt ($\lambda = 8.85, 24$, and 51.2 μm) at temperatures from $-18.8°$ to $+1500°$C, with a result unfavorable to the law of Wien. Further, Lummer and Pringsheim performed (1900–1903) a comprehensive experimental investigation concerning the intensity of radiation at various λ-values corresponding to different temperatures.

Following the named theoretical investigation of Wien, Planck (1896) began his very penetrating theoretical study of the radiation of a full blackbody and its dependence on wavelength and temperature. He found, at first on a purely empirical basis (1900), the following formula which seemed to satisfy the experimental results rather well:

(1.13) $$E_\lambda = (c^2 h/\lambda^5)[\exp(ch/K\lambda T) - 1]^{-1}$$

What later brought to Planck the greatest recognition was the agreement of his formula with experimental results (see, for example, Fig. 7.1 this volume), and also the fact that he was able to give clear physical definition to the constants of this formula, c being the velocity of light.

1.3. Precision and Accuracy of Radiometric Devices

In the previous sections of this chapter, *accuracy* has been interpreted as the absolute uncertainty in a measurement; it is the degree of conformity to the true value. On the other hand, *precision* is intimately connected with the response of the measuring instrument or system to changes in the parameter under investigation. The latter can be expanded into its components, viz. reproduction, resolution, and repeatability, one or more of which may be involved in any application.

With regard to the reliability of measurements made with standard radiometers, it should be clear from the aforegoing text that there is no major difficulty, in a national primary physical laboratory, to reproduce its own radiometric reference to within one or two parts in one thousand. Likewise, similar measurement repeatability with working standards, of present-day construction, on a routine basis is quite normal. However, the position with respect to the fixation, universally, of the true absolute reference of radiometry is not so satisfactory. It is doubtful whether this reference has really been established anywhere, to date, to better than 0.5 %. When applications of these various laboratory-maintained radiation scales have to be undertaken, on a transfer basis, for greater generality, a reliability figure of 1 % in the derived secondary references is realistic. The same arguments are valid in the case of the maintenance, in the different participating countries, of the International Pyrheliometric Scale (IPS) and its reproduction locally. For example, over a 10-year period, the IPS has been realized, at Newport, with a constancy, as referred to the Davos (Switzerland) primary standards, of $\pm 0.2\%$ in general. But, as has been pointed out, this does not infer an absolute accuracy of such a figure in the pyrheliometric measurements.

Table III is of interest in these respects. Two sets of calibration results are presented for five laboratory-type thermopiles (selected at random from the Eppley files).

The NBS reference was transferred via the carbon-lamp working standards, at least two transfers being involved. The IPS reference was transferred to the optical bench used for the comparison. Here a thermopile normal incidence

TABLE III. Comparison, at Newport (Eppley Laboratory), of NBS and IPS radiometric references.[a]

Year	Month	Thermopile	Calibrations $\mu V \, \mu W^{-1} \, cm^{-2}$		NBS/IPS
			NBS	IPS	
1966	November	7960	0.258	0.259	0.996
	November	7961	0.261	0.261	1.000
	November	7965	0.230$\underline{5}$	0.229	1.008
1967	March	8191	0.113	0.112	1.009
	April	8162	0.248	0.250	0.992
Mean (ambient temperature 24 $\pm 1°$C)					1.001 $\pm 0.6\%$

[a] After Drummond (1968).

pyrheliometer (with practically the same aperture as the Eppley–Ångström primary group of pyrheliometers employed to standardize it) was the transfer agent. Care was taken to avoid significant errors arising through instrument temperature dependence and linearity of response over the range of radiant flux density encountered. The agreement between the several pairs of calibrations is within $\pm 1\%$ (Drummond, 1968).

Although it has been proposed that the approach to fundamental radiometry, through the source rather than the detector, assures better prospects of attaining the present ideal of 0.1 % in absolute uncertainty of measurement, there are serious difficulties here which are often ignored or not properly eliminated. In the realization of such perfect blackbody radiators, the problems of the correct attributable emissivity and temperature of the cavity arrangement are always present and, perhaps, most debatable of all is the choice of the appropriate value of the Stefan–Boltzmann constant (σ). For example, in Chapter 6 (this volume) are given a number of such values determined by experiment and generally judged to be the most reliable of the many similar attempts. Mostly, the measurements have been carried out at the higher temperatures. It will be noticed that the average scatter about the (unweighted) mean (5.73×10^{-12} W cm^{-2} deg^{-4}) is only about $\pm 0.5\%$. But the corresponding value based, theoretically, upon factors associated with present-day knowledge of atomic structure (see Chapter 6) is more than 1.0 % lower. Also, if we consult the component values available originally to Planck, we see that the derived figure for σ is now about 0.6 % lower than that established 50 years ago.

If we are to improve our certainty of the radiation intensity reference, it will be evident that a combination of both approaches, in the same institutions, appear highly desirable.

1.4. The Extraterrestrial Solar Fluxes

Let us now turn to the primary source of energy—the sun—and consider its most important characteristic, namely this energetic emission, integrally and spectrally, as unmodified by the earth's atmosphere. There is no question that when our technology allows us to make precise solar measurements in free space (see Table IV), use of the unattenuated solar beam should assist materially in the advancement of precision radiometry.

This extraterrestrial flux of solar radiation, integrated over all wavelengths, is generally referred to as the earth solar constant. It is the rate at which energy is received upon unit surface, perpendicular to the sun's beam, in free space, at the earth's mean distance from the sun. A good knowledge of the energy in different regions of the extraterrestrial spectrum is important for the treatment of many geophysical and astrophysical problems. Between

1923 and 1966, ground-based estimates of the total solar irradiance at the outer limit of the terrestrial atmosphere ranged from 1.90 to 2.05 cal cm^{-2} min^{-1} (i.e., 132–143 mW cm^{-2}). Such extrapolated evaluations which have found most support up until now are those of Johnson (1954)—139.4 mW cm^{-2}—and Nicolet (1951)—138.0 mW cm^{-2}, but the new work of Labs and Neckel (1968)—136.5 mW cm^{-2}—deserves attention. The wavelength range containing more than 99 % of the radiation is 0.2–4.0 μm.

There are, however, several important points which must be borne in mind when considering these figures. In the first place, Johnson assigned reliability limits of ± 2 % to his derived solar constant: Nicolet is much more conservative and argued that this uncertainty could be as great as ± 5 %. A further problem is imposed by the limits of current knowledge of the basic radiation reference. It has been indicated that although it is believed that the International Pyrheliometric Scale represents the true radiometric standard to within ± 0.5 %, it is possible that the uncertainty may be as great as ± 1 %. Also, most of the approaches to solar-constant determination have involved, essentially, a study of the classical measurements made during the last half century by Abbot and his co-workers at the Astrophysical Observatory of the Smithsonian Institution in Washington. When these measurements have to be reconciled with solar radiation data derived elsewhere, the question of the radiometric reference becomes even more complicated. Before the advent of the IPS, comparisons between the U.S. working standards (mainly specimens of the Abbot silver-disk pyrheliometer) and those maintained in Europe (Ångström electrical-compensation pyrheliometers in general) were far from satisfactory from the viewpoint of precision radiometry. The details of such comparisons have been discussed earlier in this chapter. Further, it has not always been clear what basic reference was employed by the Smithsonian, especially in view of the several scale changes which have been made. This has led to much confusion in the evaluation of extraterrestrial fluxes by other workers, which was added to by modifications of the corrections to the observed data for solar-constant extrapolation. However, in fairness to the Smithsonian Institution, it must be acknowledged that Abbot and his colleagues were much more concerned with measurement repeatability than with absolute values.

How well the Smithsonian succeeded in its prime objective is borne out in the 30-year series (1923–1952) of annual values of the solar constant where the mean deviation is ± 0.1 %! This "constancy" of the solar constant is further demonstrated in the measurements in the visible spectrum, by Hardy and Giclas (1955); over a five-year period it is indicated that solar variations did not exceed 1.0 %. It is true that rather greater solar variability in the ultraviolet region has been reported by the Smithsonian, but Deirmendjian and Sekera (1956) have demonstrated that this variation may be due, at least in part, to

TABLE IV. Principle values established for the solar constant of radiation.[a]

Year	Authority	Value (cal cm^{-2} min^{-1})	Value (mW cm^{-2})	Radiometric reference
				(a) *Ground-based extrapolations*
1923–1952 (30 years)	Abbot	1.94[b]	135.2	Smithsonian scale, 1932, etc.
1932	Linke	1.94	135.2	
1934–1935	Mulders	1.95	135.9	Blackbody source
1938	Unsöld	1.90	132.4	Smithsonian scale, 1932, etc.
1951	Nicolet	1.98	138.0 ±5%	
1951	Allen	1.97	137.3	?
1954	Johnson	2.00	139.4 ±2%	Smithsonian scale, 1932, etc.
1955	Unsöld	1.96	136.5	
1956	Stair and Johnston	2.05	142.8 ±5%	US NBS scale
1958	Allen	1.99	138.6 ±1%	Smithsonian scale, 1932, etc.
1966	Stair and Ellis (1968)	1.95	135.9 ±3%	US NBS scale
				(b) *Measurements from high-altitude aircraft*
1966–1968	Drummond, Hickey,	1.952	136.0 ±1%[c]	International Pyrheliometric Scale (IPS) (at 83 km from the earth)

	Scholes, and Laue (1967–1968)			
1967	NASA Goddard Space Flight Center (Thakaekara et al., 1969)	1.938	135.0 ±2%	IPS
1967	NASA Ames Research Center (Arvesen, 1969)	1.994	139.0 ±3%	US NBS scale
				(c) *Survey of the literature (but excluding b)*
1968	Labs and Neckel (1968)	1.958	136.5	
				(d) *Measurements from spacecraft*
1969	Jet Propulsion Laboratory (unpublished)	1.942	135.3 ±1%	JPL blackbody—Mariner spacecrafts 6 and 7—Mars (at about 20,000 km from the earth)

[a] After Drummond (1965) and Drummond and Hickey (1968).
[b] Value reported originally, in 1952, was 1.95 but this contains a small scale error (corrected above).

the effects of Mie scattering by dust in the earth's atmosphere. But this does not rule out the possibility of measurable changes, on a time scale of centuries, in the photosphere continuum radiation which accounts for most of the energy. On the other hand, where emission lines originate in the chromosphere and corona (above the photosphere), individual lines show intensity variations which are pronounced in the ultraviolet. It is also possible that small variations occur in the blue region of the spectrum, where Fraunhofer absorption is intense.

Table IV, Section(a), which presents the principal results of the determination of the solar constant from elevated ground locations, is mainly taken from a recent review of the subject (Drummond, 1965). Sections (b)–(d) of the table will be discussed later in this chapter. For the sake of comparison, Table V [mainly Ångström (1962)] has been included. In this table are given values for different aspects of the earth's albedo (i.e., reflectance) which is an associated parameter in the fixation of the thermal balance of our planet (note the much lower value for the planetary albedo established recently, by Vonder Haar and Suomi (1969), from meteorological satellite measurements). The extension of the latter to other planets is shown in Table VI which has been prepared by A. Ångström. The various planetary solar constants can be directly derived from the earth value through application of the inverse square law.

1.4.1. Problems in Solar-Constant Determination (Extrapolation Methods)

Quite apart from the central problem of the fixation of the true standard of radiation and its precise reproduction in solar radiation measurement instrumentation, three other basic problems have to be considered in solar-constant evaluations, especially with regard to the spectral energy distribution concerned. These are as follows:

(a) The spectral response of the equipment (including any imaging systems employed) must be investigated.

(b) The attenuation introduced by the earth's atmosphere must be evaluated and corrected for, if significant over the wavelength interval of interest.

(c) The spectral purity of the measurements must be known precisely, in order that, in the presence of Fraunhofer absorption structure, the resolution can be acceptable for the purpose of the program, and also that comparisons can be made with the results assembled by other investigators.

As indicated, until recently (see next section), all methods of deriving the integral wavelength solar constant and its spectral components have entailed extrapolation of measurements made at the earth's surface, within the wavelength interval over which it is possible to carry out such a program. At the top of the atmosphere, this integrated flux probably represents nearly 95 %

TABLE V. Albedo of the earth and its atmosphere.[a]

Year	Authority	Clouds	Ground and atmosphere	Planetary albedo	Photometric albedo	Basis
1908	Abbot and Fowle	0.65		0.37		General meteorological considerations and radiometric measurements
1919	Aldrich	0.78	0.17	0.43		
1928	Simpson	0.74	0.17	0.43		
1934	Baur and Philips	0.72	0.085	0.415		Detailed meteorological computations and results of radiometric measurements
1948	Fritz	0.47–0.52	0.17	0.347	0.39	Meteorological computations and acceptance of the value 0.39 of Danjon for the photometric albedo
1954	Houghton	0.55	0.135	0.34		Detailed meteorological computations
1962	Ångström I	0.46	0.195	0.33	0.39	Measurements of illumination as dependent on turbidity, general meteorological considerations, and acceptance of Danjon's 0.39 value
1962	Ångström II	0.55	0.195	0.38	0.43	Ångström's derivation of albedo of ground and atmosphere; Houghton's value for the albedo of clouds
1928	Danjon I				0.29	Astrophysical measurements of the brightness of moonlight
1936	Danjon II				0.39	
1969	Vonder Haar and Suomi			0.29		Survey of observations from the first generation of meteorological satellites (1962–1965)

[a] After Ångström (1962) and Vonder Haar and Suomi (1969).

TABLE VI. Albedo (photometric) of the planets.

	Authority			
Planet	Allen	Harris	Kuiper	Mean
Mercury	0.060	0.056	0.058	0.058
Venus	0.61	0.76	0.76	0.71
Mars	0.15	0.16	0.15	0.15
Jupiter	0.41	0.73	0.51	0.55
Saturn	0.42	0.76	0.50	0.56
Uranus	0.45	0.93	0.66	0.68
Neptune	0.54	0.84	0.62	0.67
Pluto	0.16	0.14	0.16	0.15
Earth	0.34	0.36	0.39	0.36
Moon	0.070	0.067	0.072	0.070

of the total representative of all wavelengths. Therefore, additive corrections are necessary to allow for the unobserved components in the ultraviolet and infrared regions, mainly on account of selective absorption by ozone and water vapor, respectively. Our best knowledge to date of the ultraviolet radiation unobserved even on the highest mountains is probably due to the rocket spectrograph ascents carried out, first around 1950, by the U.S. Naval Research Laboratory (Johnson et al., 1954)—see Fig. 1.7. This investigation has continued. The other correction is based upon studies of water vapor absorption within the atmosphere.

It must be borne in mind that in solar-constant evaluations, the extrapolations aimed at correcting for the energy attenuated by the atmosphere, in the spectral region where direct observation is possible, are based upon Bouguer's extinction law. Here, the measured surface irradiance F'_λ is related to the extraterrestrial irradiance F_λ by the expression

$$(1.14) \qquad \log F'_\lambda = \log F_\lambda + m(\log q_\lambda)$$

where m is the optical air mass and q_λ the transmission coefficient of the atmosphere for wavelength λ. Although the basic surface measurements of the solar spectrum at different wavelengths, in this procedure, are generally made at a number of different air mass values (i.e., optical pathlengths), on the clearest days possible under steady atmospheric conditions, an objection will be readily apparent. The law of Bouguer is only strictly valid for monochromatic radiation. In the observing programs of the Smithsonian, the resolution of their spectrobolometers was rather poor. Repetitions during recent years by others (especially Stair and his colleagues (Stair and Johnston, 1956; Stair and Ellis, 1968) at the U.S. National Bureau of Standards],

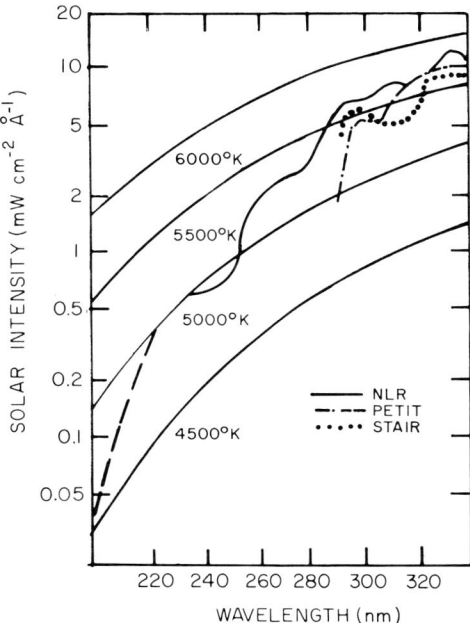

FIG. 1.7. Spectral distribution of the extraterrestrial solar irradiance from 200 to 340 nm [after Johnson et al. (1954)].

took advantage of more modern instrumentation, yielding greatly improved resolving power.

To sum up, the overall procedure employed was to first integrate the relative spectral curve and then to correct for the depleted components in the radiation penetrating to the observing site. Next, the absolute energy reference was established by normalizing the corrected integrated spectrobologram, etc., to the value of total irradiance measured, simultaneously, with a calibrated pyrheliometer. Extrapolation of this curve to zero air mass (i.e., extraterrestrial pathlength) was then accomplished according to Bouguer's law. The final step consisted of integrating and adding the uv and ir corrections for energy not observed at the measurement site (on account mainly of uv absorption by ozone and ir absorption by water vapor). The former of this second pair of corrections was derived (by Johnson and his colleagues) from the U.S. NRL uv rocket spectrograms; the latter correction is based upon studies of water-vapor absorption within the atmosphere (Yates and Taylor, 1960).

However, even with present-day equipment, the presence of Fraunhofer absorption bands makes it very difficult to compare different series of measurements unless the bandwidths covered are identical and similarly resolved.

This difficulty can be partly overcome by integrating the detailed observations of the spectrum over sufficiently wide bands. Such a process reduces the observations to a discontinuous step function representing the energy in different wavebands. If the effects of Fraunhofer absorption can next be removed for each band, it is possible to obtain the irradiance curve as a smooth function, which assists greatly in comparing measurement series and in selecting the most reliable values from the different sources.

In his determination of the solar constant and its spectral distribution, Nicolet adopted this procedure of correcting an assumed smoothed energy curve for the background solar continuum for Fraunhofer absorption. Essentially, he considered (a) a brightness temperature, at the center of the sun's disk, of 7200°K for the region 0.45–0.95 μm, (b) a similar temperature of 5800°K for the region 0.30–0.37 μm, (c) Michard's data for the Fraunhofer absorption fraction of the continuous spectrum, as a function of wavelength, (d) Abbot's surface spectral and integral wavelength measurements, and (e) NRL uv rocket data. It is implicitly assumed in using this method that the correction for Fraunhofer absorption is known with greater accuracy than the relative spectral distribution of the energy. This assumption is reasonably sound for wavelengths around 0.45–0.50 μm when the total absorption, over a 0.01 μm band, rarely exceeds 10 % and very sound for wavelengths greater than 0.60 μm when this absorption rarely exceeds 3 %. At shorter wavelengths, where the Fraunhofer effect may exceed 50 %, this assumption cannot be justified so readily, as Nicolet recognized.

These spectral distributions of the extraterrestrial solar radiation, according to Johnson and Nicolet, are presented in Fig. 1.8.

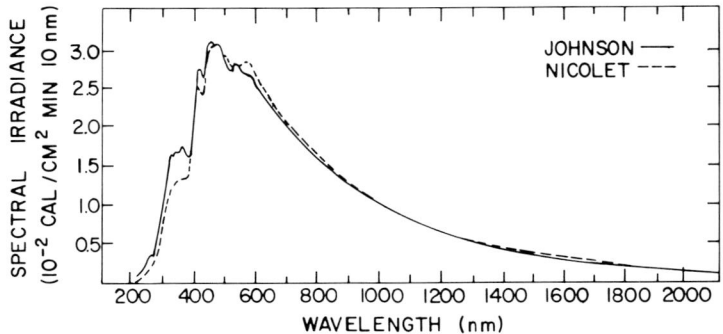

Fig. 1.8. Spectral distribution of the extraterrestrial solar irradiance from 200 to 2000 nm (plotted from data referred to $\Delta\lambda = 10$ nm, according to (a) Johnson and (b) Nicolet [after Drummond (1965)].

1.4.2. Direct Measurement—Eppley and Jet Propulsion Laboratories Research Project

As has been pointed out, an accurate knowledge of the energetic flux of the sun's emission is a necessary requirement in geophysical and climatological studies entailing such central problems as the depletion of the solar beam, through scattering and absorption by atmospheric agencies, and the energy balance of the troposphere and lower stratosphere. Information regarding the solar spectral irradiance above and within the atmosphere is fundamental to any study of the concentrations of absorbing constituents like ozone. Related to questions of the general terrestrial thermal balance is that involving the realization of good radiative equilibrium of spacecraft, particularly in voyages considerably different from that of earth-orbiting satellites.

It is clear that new determinations of the solar constant and its spectral components can now be attempted, directly above the effective terrestrial atmosphere. The techniques of total and filter radiometry have reached such a degree of precision as to encourage their employment in a measurement program of this nature. Therefore, the experiment described here had as its prime object the measurement of the integral wavelength extraterrestrial solar radiation and selected well-defined spectral components of this flux, with the highest possible accuracy in the present state of the art (Drummond *et al.*, 1967a,b; 1968a,b; Laue and Drummond 1968; Drummond and Hickey, 1968). Special attention was paid to investigation of the ultraviolet and low visible regions, particularly in wavelengths where there is considerable disagreement between the evaluations of Johnson and Nicolet (see Fig. 1.9). These are also the regions where good knowledge of the solar emission is necessary in the solution of problems concerned with the thermal equilibrium of spacecraft.

For these reasons, the Jet Propulsion Laboratory (JPL) of the California Institute of Technology invited the Eppley Laboratory, in 1964, to collaborate (under NASA contract funds) in a solar-constant measurement project. The initial plans provided for the operation of specially designed solar-radiation measurement and readout equipment on the USAF/NASA X-15 rocket research aircraft, but instrumentation development difficulties and unavoidable flight program problems delayed the first (and only) such successful attempt until October 17, 1967, over Nevada. However, in the meantime, NASA's Research Center at Cleveland kindly made available their B-57B two-engine jet aircraft (a converted Canberra bomber) in preliminary support of the effort. Six solar-constant measurement flights were made over the Dayton-Columbus area, in Ohio, two in summer conditions (July–August 1966) and four in late winter (March 1967). Since the latter altitudes ranged from 11.5 to 15 km, as compared with the X-15 probe to nearly 85 km, correction for some atmospheric extinction was necessary. A few days after

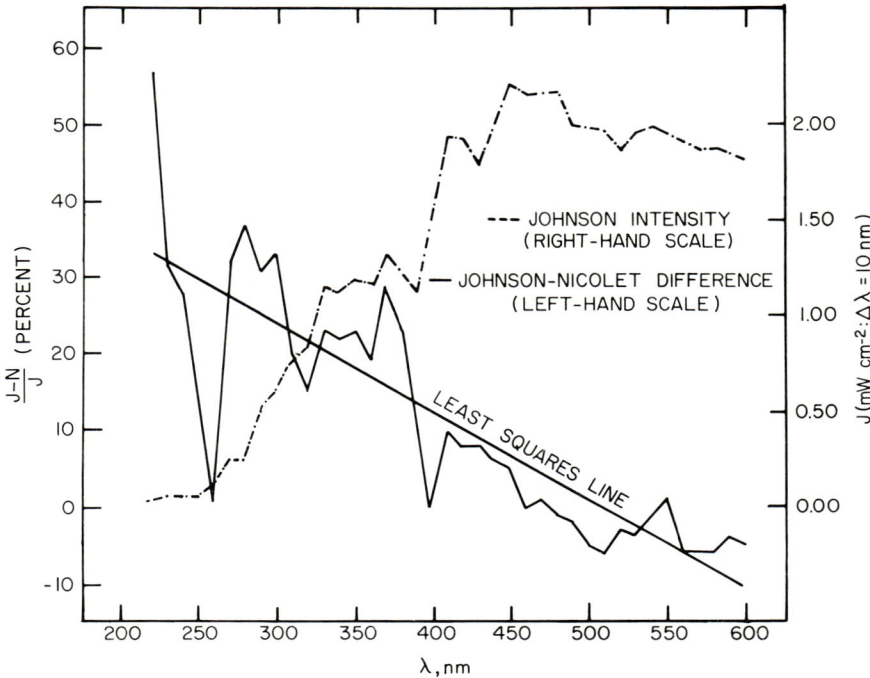

Fig. 1.9. Deviation of the Johnson and Nicolet extraterrestrial curves in the ultraviolet and visible regions [after Drummond and Hickey (1968)].

the X-15 success, the same multichannel radiometer was transferred to the NASA CV-990 (Convair) four-engine jet aircraft based at the Ames Center at Moffett Field, California. One measurement series was obtained over the Pacific, in latitude 11.5°N, at an altitude of 12 km. Then, between June 5 and August 7, 1968, 20 additional CV-990 flights were made from Ames; the second series of 10 was specifically intended for solar observation at 12 to 12.5 km.

In all, over the two years of aircraft measurements, there was participation in 36 flights of which 28 were highly successful. The results from 14 selected flights are summarized later in this section.

1.4.2.1. Instrumentation. The detector was a 12-channel radiometer equipped with fast-response, rugged thermopiles of good stability and also of high sensitivity resulting from the incorporation of lens collectors. Their response was essentially independent of temperature over a wide range (Drummond et al., 1967a,b), although this parameter was continuously monitored during flights. All thermopiles are pressure-sensitive. Hence, channel tests were undertaken from a normal atmospheric ambient down to a pressure of 10^{-3} Torr, which is lower than experienced in the X-15 aircraft. At pressures below 10^{-4} Torr, there is no further change in the sensitivity of this type of detector (Karoli et al., 1960). During the instrumental development period, a comprehensive study was made of the characteristics of narrow bandpass optical filters as related to their production methods. Very significant improvements in such filters resulted; these are described in the concluding chapter of this book. It will suffice to state here that typical radiometer filter wavelength limits (λ nm), mainly adopted in accordance with the curve divergence of Fig. 1.9, were:

Ultraviolet	Visible	Near infrared	
235–344	412–475	1250–2000	
272–322	447–498	$>533, +25°C$;	$>526, -30°C$
298–344	507–589	$>697, +25°C$;	$>687, -30°C$
309–393	595–645		
334–402	645–700		

A selection of the 11 listed narrow bandpass filters was made for the particular type of aircraft flight (e.g., below or above the ozonosphere). The transmission of the narrow bandpass (interference) filters is not significantly influenced by ambient temperature over the range -30 to $+30°C$; the temperature dependence of the two broad bandpass (colored glass) filters is indicated. The narrow bandpass transmittance limits are the 0.01 values; with the broad bandpass filters, the wavelength corresponding to the center of lower cutoff is the reference. (The two glass filters are Schott OG1 and RG8.)

The signal conditioner included a 24-channel commutator with a switching rate operated between 1 channel/3 sec and 16 channels/sec. The commutation of radiometer flux sensors, radiometer temperature, and calibration signals was accomplished by the use of magnetic reed relays driven by chip-type transistor binaries. Because of size and power requirements, a simple pulse circuit and gating logic were used to insure that only one relay was closed at a time. A dc differential amplifier having a gain of approximately $1000 \times$ was employed to amplify the detector signals to a level suitable for recording. This unit was designed to operate over the temperature range -60 to $+100°C$. Tests established the near freedom-from-temperature effects. An important requirement was the minimizing of the output signal zero-shift produced by ambient temperature changes and variation in input voltage. The frequency response of the amplifier was 0–30 Hz to achieve rapid scanning of the input.

1.4.2.2. Aircraft Installation. Figures 1.10–1.12 show the different methods of installation of the radiometer-amplifier, etc. assembly in the three types of high-altitude aircraft involved. The assembly in the aft section of the X-15 port pod (Fig. 1.11) was the most difficult to achieve. Solenoid-controlled, pneumatically operated pistons provided the power for opening the hatch and erecting the filter detector to the preset angle so that, when the X-15 was in level horizontal flight, at its peak altitude, the radiometer could be properly oriented towards the sun. Springs were used to close the hatch upon removal of the actuation pressure. Gaseous nitrogen was stored at 500 psi in three small tanks on one side of the pod. A standard three-way solenoid valve was used to admit the gas into two pistons. A capillary tube was employed to feed the gas to the hatch-operating cylinder. This capillary had three functions: (1) to delay the hatch-opening and detector-erector motion until the door latch had operated; (2) to control the rate of opening, independent of tank pressure and g load; and (3) to control the rate of hatch closing. Readout was on magnetic tape on the aircraft, supplemented by telemetry to an associated ground recording system. Verification of the accuracy of the radiometer orientation was also recorded (the signals being those in two axes from electrically opposed pairs of silicon cells—so adjusted to yield null outputs when the solar pointing was perfect).

Like the X-15 measurements, the radiometer on the B-57B (aircraft unpressurized) flights was exposed to the sun without any intervening window (Fig. 1.10). An optical sight for the pilot ensured proper orientation of the sensors. The signals were tape-printed on an integrating digital voltmeter in the observer's cockpit.

In contrast, the CV-990 aircraft employed is pressurized. This entailed mounting the detectors beneath a quartz window of known transmission. In Fig. 1.12 can be seen the multichannel flight radiometer. Generally, a standard-type strip-chart, recording potentiometer (with preamplification where necessary) was utilized, without recourse to the signal conditioner used in the X-15 and B-57B flight series; channel switching, in this instance, was manual.

1.4.2.3. Environmental Testing. The first flight radiometer and signal conditioning equipment were tested, at JPL, in a horizontal vacuum chamber which could be operated at a pressure of 10^{-5} Torr. A shroud at liquid-nitrogen temperature was used to simulate the "black space environment." Solar energy was simulated by the beam from a 5-kW high-pressure, mercury-xenon arc (mounted in a 24-in. searchlight) irradiating all 12 detectors through a quartz window. The tests consisted of measuring, in turn, the radiometer outputs under vacuum conditions with the chamber walls at room temperature and at liquid-nitrogen temperature. Measurements were also made at STP conditions with and without the chamber window. Radiometer outputs

1. PRECISION RADIOMETRY   37

Fig. 1.10. The Eppley-JPL multichannel radiometer installed on the NASA modified Canberra B-57B jet research aircraft (photograph courtesy of NASA).

Fig. 1.11. Radiometer installation on the USAF/NASA X-15 rocket research aircraft (photograph courtesy of NASA).

Fig. 1.12. Radiometer installation on the NASA Convair CV-990 jet research aircraft; (a) external view, (b) internal view (photographs courtesy of NASA).

were measured with conventional instrumentation, as well as with the signal conditioner. Instrument stability was thus demonstrated.

The program was then extended to complete vibration, shock, temperature, and steady-state acceleration testing. No damage was incurred in the radiometer. However, to avoid repetition of this excessive handling and of the necessary precautions to maintain a dust-, moisture-, and oil-vapor-free environment, it was decided to conduct careful instrument calibrations before and after the initial flights instead of environmental testing (see next section).

Since the modified pod was part of the X-15 aircraft, the environmental testing, in this connection, also involved pilot-safety aspects. In addition to the standard tests, the pod was subjected to a NASA Flight Research Center thermal transient simulation, in which the thermal flight environment is simulated by mounting the pod in a simple wind tunnel. Controlled cooling was accomplished by blowing a mixture of liquid and cooled gaseous nitrogen over the pod. Thermocouples mounted on the Inconel skin of the pod were used to monitor and control the test temperature. The skin temperature was maintained at $-50°C$ to simulate the 1–2 hr prelaunch B-52B flight.[2] At the end of the prelaunch soak, the liquid-nitrogen flow was throttled down, and a 60-sec heating pulse from the controlled infrared lamps was applied to simulate the powered flight aerodynamic heating. At the end of this heating cycle,

[2] The X-15 was normally launched at about 45,000 ft, from a USAF B-52 jet aircraft.

the skin temperature was $+100°C$. During a 90-sec period, corresponding to the X-15 coast, the lamps were swung away from the pod and the hatch mechanism operated for 60 sec. When the hatch was again closed, the lamps were swung into position and the re-entry thermal (approximately 5 W cm^{-2} for 20 sec) applied. The skin temperature reached $+350°C$ on one side and $+250°C$ on the opposite (again simulating true reentry conditions). At the conclusion of the reentry thermal pulse, rapid LN$_2$ cooling was used to reduce the skin temperature to $+100°C$. Because of the planned preset pod roll angle, thermal reentry tests at three different roll angles were performed.

1.4.2.4. Results. With regard to flight measurement reliability, a study of the B-57B raw signal data indicated that, in general, the values for radiometer channels 9 (OG1 filter), 10 (RG8 filter), and 11 (total flux) were within ± 0.3 % of the mean for "on sun" period. After data analysis (e.g., after application of temperature and zero corrections), the mean deviation of corrected signal data for compatable measurements series was better than ± 0.2 %. Scrutiny of the X-15 flight "selected measurement" data gave repeatability figures of the order of ± 0.5 %. From the nature of the CV-990 flights (i.e., pressurized aircraft observing cabin), the procedures were similar to those employed in obtaining ground-based solar measurements.

The only correction applied to the very high altitude (78–83 km) X-15 rocket aircraft measurements was that for reduction to mean solar distance. In the case of the jet aircraft measurements (12–15 km), additional corrections were those for Rayleigh scattering, ozone absorption, and water-vapor absorption. Previous estimates of no significant haze (aerosol) above aircraft level were confirmed, by direct observation, in the July–August 1968 CV-990 flights. This latter type of flight required correction for the presence of the aircraft window. The following values (percentage corrections) for the CV-990 flight of October 22, 1967 are representative but reference should be made to Fig. 1.13.

	Percentage corrections irradiance		
	Total $\lambda > 200$	OG1 > 526	RG8 >687 nm
Earth mean distance	-1.0	-1.0	-1.0
Rayleigh scattering	$+2.2$	$+1.0$	
Ozone absorption	$+2.0$	$+0.7$	
Water-vapor absorption	$+0.3$	$+0.4$	$+0.5$
Unmeasured infrared (beyond aircraft window)	$+3.3$	$+4.6$	$+6.0$

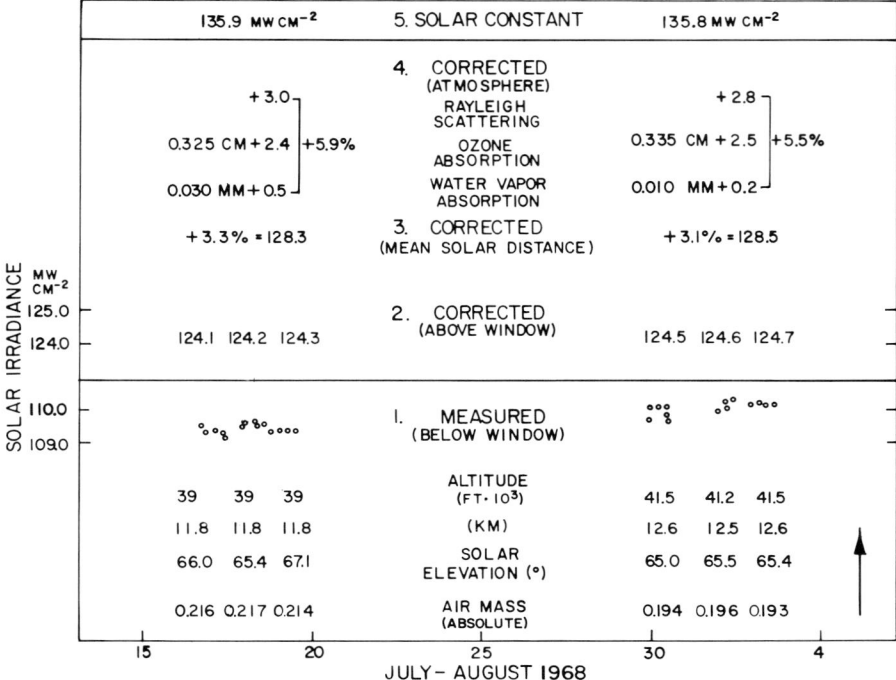

Fig. 1.13. Diagramatic presentation of the Eppley-JPL 1968 NASA CV-990 aircraft solar-constant determinations.

In the case of the B-57B measurements, the last correction is, of course, unnecessary.

Before assessing the principal results, attention should be paid to Fig. 1.14 which shows the various solar radiation measurement locations of the Eppley-JPL experiment, with reference to altitude above MSL. It will be noted that the October 1967 X-15 flight reached the mesopause at about 83 km (there were no others on account of the disbanding of this NASA operation after the tragic crash of the second Eppley-JPL mission a month later).

Table VII is a summary of these measurements. A number of comments are pertinent, viz. (a) the results are mutually highly consistent and therefore confirm the generalizations regarding radiometry precision made in Section 1.3; (b) the agreement of the (atmospheric) corrected lower level jet aircraft measurements with the (uncorrected) rocket aircraft measurements essentially above the terrestrial atmosphere is remarkable; (c) there is no question that the Johnson–Nicolet pyrheliometric solar constant values are too high (about 2–3 % integrally and 6–7 % over the spectral region $\lambda < 600$ nm); and (d)

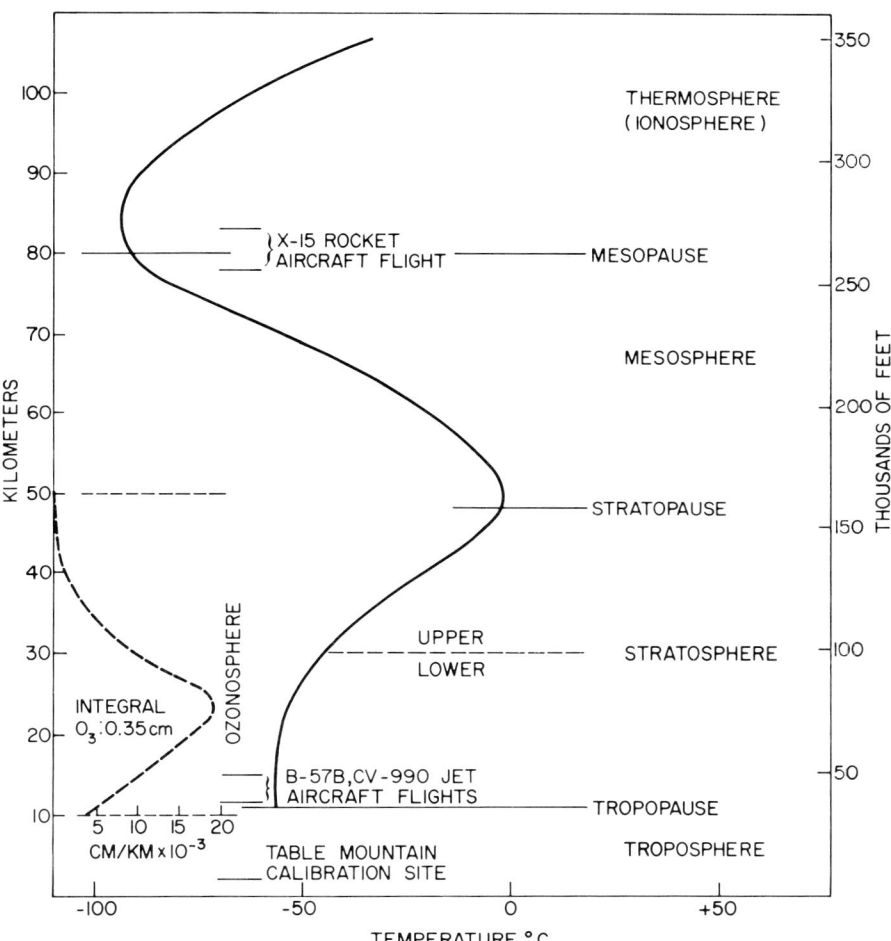

FIG. 1.14. Standard atmospheric vertical profile (U.S. 1962), with common systems of nomenclature, showing the different solar radiation measurement locations of the Eppley-JPL experiment [after Drummond et al. (1968a)].

reference to Table IV(b–d) indicates excellent agreement with the recent (1969) independent JPL determinations from their Mars Mariner 6 and 7 spacecraft missions (absolute cavity detectors), and the Labs-Neckel survey of the literature on the radiation of the solar photosphere (1968).

On this basis, the most probable value of the solar constant of radiation (unweighted mean of the two directly observed fluxes in space and the

TABLE VII. Eppley-JPL solar-constant measurements: summary of principle jet (12–15 km) and rocket (78–83 km) aircraft results.[a]

	Total flux[b] (nm > 200)	Ultraviolet and visible[b] (nm < 607)	Infrared[b] (nm > 607)
1. Measurements			
(a) B-57B jet: July–August 1966 (2 flights: Ohio)	135.8	48.8	87.0
(b) B-57B jet: March 1967 (5 flights: Ohio)	136.0	48.7	87.3
(c) CV-990 jet: October 1967 (1 flight: Pacific 11.5° N)	136.2	48.9	87.3
(d) Mean (weighted)	136.0	48.7$\underline{5}$	87.3
(e) X-15 rocket: October 1967 (1 flight: Nevada)	136.1	48.6	87.5
(f) General Mean [(d) and (e)]	136.05	48.6$\underline{5}$	87.4
2. Ground extrapolations (Johnson's curve)	139.6	52.3	87.3
3. Difference (1 − 2)/2	−2.5$\underline{5}$ %	−7.0 %	+0.1 %

[a]After Drummond *et al.* (1968a).
[b]Given in mW cm^{-2}.

most up-to-date review figure, combining observational (but not the JPL-Eppley efforts) and theoretically derived and treated material), is the average of 1.952 (Eppley-JPL), 1.942 (JPL), and 1.958 (Labs–Neckel); or 1.95 cal cm^{-2} min^{-1} (136 mW cm^{-2}) ±1 % within the experimental limitations of the investigation, including that of the uncertainty of the true reference of thermal radiometry.

1.4.3. Field Calibration of Radiometers for Extraterrestrial Solar Measurements

It will have become apparent that a prime requirement in any precise radiometric measurement program is instrument calibration. This aspect of space radiometry has therefore been presented separately. The illustrative data selected are part of the first effort (where such information is available) in this respect.

Repeated exposure of the radiometer models to powerful tungsten-filament, xenon (compact-arc) and carbon-arc sources was used to establish the repeatability and long-term stability of the total and filtered flux channels. Figure 1.15 shows the radiometers installed for basic calibration, in natural

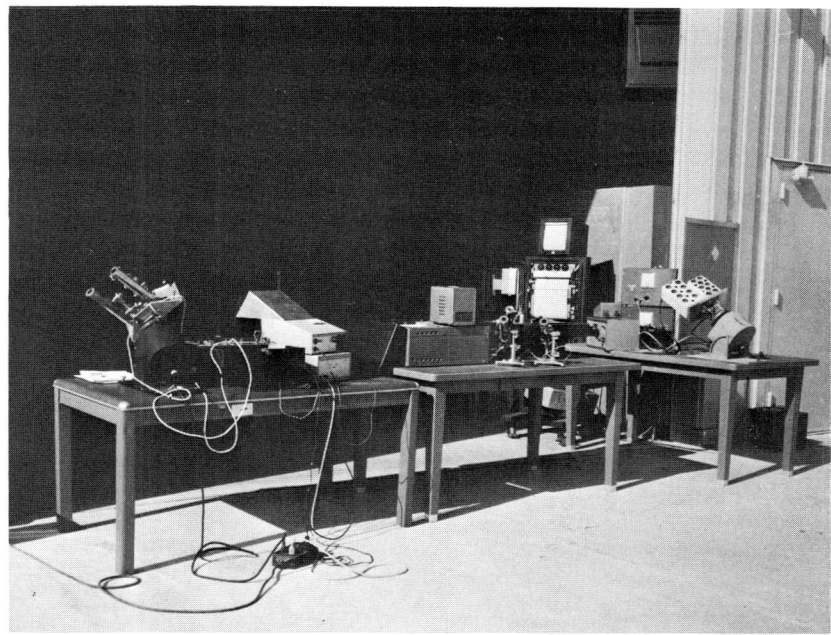

FIG. 1.15. Flight- and surface-model radiometers under calibration at the JPL Table Mountain facility [after Drummond et al. (1967a)].

sunlight, at the JPL Solar Panel Test Facility on Table Mountain, California (34.4°N, 117.7°W; 2.25 km). A group of Eppley–Ångström working-standard electrical compensation pyrheliometers was employed to reproduce the International Pyrheliometric Scale (IPS); these instruments have been compared with the primary pyrheliometric references of the Eppley Laboratory and the Davos Observatory (Switzerland). Reproduction of the IPS, at the test site, has been established to be within one or two parts in one thousand (see Section 1.2.3).

Calibration of the total flux and broad bandpass filter channels was straightforward. The broad filters were removed, and the four channels of each radiometer exposed simultaneously to the solar beam, with the Ångström reference pyrheliometers. Since the standard instruments and the thermopile detectors of the radiometers have responses independent of the wavelength of the incident energy, the only important criterion was that the atmosphere be as transparent as possible to minimize the effects of turbidity in the comparisons. During these evaluations, the atmospheric turbidity (expressed by the Ångström turbidity coefficient) was measured. The main purpose of this measurement was to derive the small correction necessary to equate the 15° radiometer aperture channels to the standard 5° value of the reference instruments (see Section 1.2.3).

With the narrow bandpass filter channels, the procedure was more complicated because it was necessary to reduce the intensity of the calibration energy to a level comparable to that passed by the wavelength selective filters during flight operation. This was accomplished through the use of a series of Inconel-on-quartz neutral density filters (nominal transmittance at 500 nm of 0.05, 0.03, and 0.02). First, these reducing filters were calibrated, in sunlight, using one of the total flux radiometer channels as the detector (due allowance being made for reflection effects within the optical system). Laboratory bench tests have shown that over the ranges of flux density so encountered (0.02–2.0 cal cm^{-2} min^{-1}), the thermopiles in the radiometers are linear in their intensity–output relationship. The bandpass filters were next removed from the radiometers and the channels screened temporarily, with metal covers, to prevent damage to the detectors which might otherwise have occurred on account of the energy concentration by the lenses. Then, one at a time, the channels were exposed to the sun through the calibrated neutral density filters. During this operation, constancy of the sunlight was monitored by one of the total radiation channels and referred to one of the Ångström standards. This method is considered valid because the thermopile receivers are nonwavelength selective. Effects arising from chromatic aberration in the lens channels were investigated.

The final step in establishing the absolute sensitivities of the ten wavelength filtered channels of each radiometer was the application of the appropriate filter factors (referred to in Chapter 8, this volume). An important aspect of the verification of radiometer calibration constancy was the regular comparison program carried out between the surface and flight models, before and after operations.

Figure 1.16 is a flow diagram of the aforementioned procedures. In Table VIII is presented an extract of the radiometer calibration data assembled at the Table Mountain test site. The transfer precision, for primary working standard pyrheliometers to operational instruments, *in the field*, is of the order of $\pm 0.25\%$. This clearly supports the realization of the measurement objectives discussed earlier.

1.5. Instrumental Problems in Precision Measurements of Solar and Terrestrial Radiative Fluxes above the Earth's Surface

1.5.1. Radiation Sensing

After consideration of the major tasks incurred through the realization of the fundamental radiometric reference (here IPS), by transfer operations to

TABLE VIII. Calibration of multichannel radiometers in sunlight at Table Mountain, March 1967.[a]

March 1967 Day		Temperature C°	Number of calibrations	Detector constant (mV mW^{-1} cm^{-2}) Channel			
				9 (OG1 filter)	10 (RG8 filter)	11 (Total)	12 (Total)
Surface model	28	+10	2	—	—	0.0977	0.1014
(for comparison)	29	6	2	0.1022	0.1076	0.0974	0.1010
	30	+12	2	0.1026	0.1076	0.0972	0.1012
Mean (weighted)		+10	6	0.1024	0.1076	0.0974	0.1012
Mean deviation (%)				±0.2	0.0	0.2	0.1
Flight model (1)	28	+10	1	—	—	0.0240	0.0233
(operational)	29	6	5	0.0385	0.0479	0.0240	0.0234
	29	6	1	0.0387	0.0480	0.0240	0.0234
	30	+14	2	0.0384	0.0476	0.0239	0.0233
Mean (weighted)		+8	9	0.0385	0.0478	0.0240	0.02335
Mean deviation (%)				±0.3	0.3	0.2	0.2

[a] After Drummond et al. (1968b).

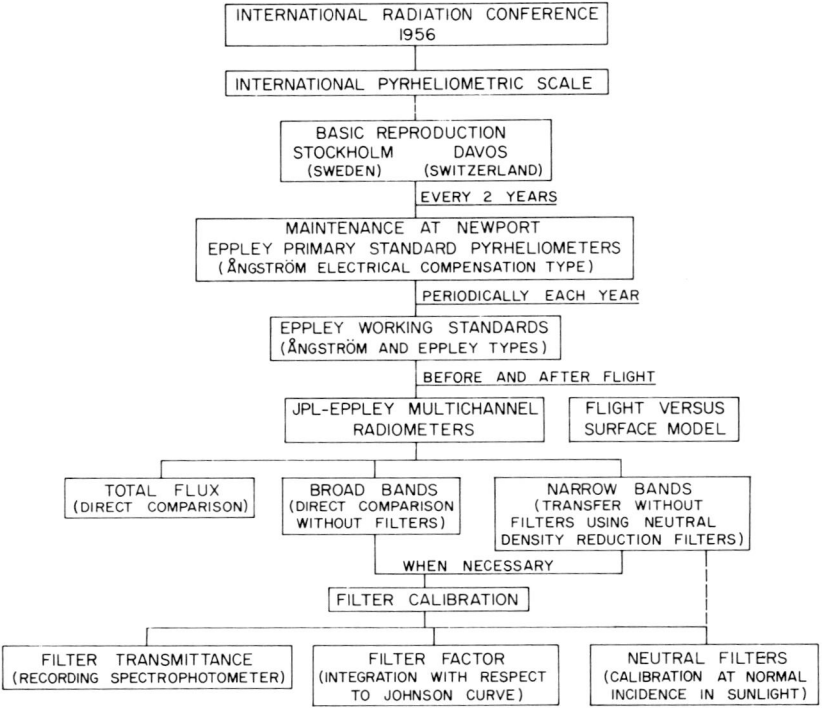

Fig. 1.16. Radiometric calibration procedures [after Drummond et al. (1968b)].

secondary radiometers tested initially in the field and ultimately to be used for solar exposure aloft in aircraft, etc., the next question is that of the choice of the system instrumentation for the particular purpose.

Let us first consider the radiation detector requirements. These are as follows: maximum stability of response, adequate sensitivity and response time, linearity of the intensity–output relationship, independence of the wavelength of the incident radiation, and freedom from effects of ambient temperature and pressure on the establishment of such corrective measures. Of tantamount importance is capability of the detector (and the associated readout system) to withstand all mechanical shocks during aircraft take off, flight, and landing operations, and space vehicle launch, without significant detriment to the instrument performance characteristics determined, at the ground, in preflight and, where possible, in post-flight testing. In-flight calibration, a highly desirable feature, has been accomplished on aircraft. The transfer of this facility to unmanned vehicles is difficult but not impracticable of accomplishment.

An important matter in the application of ground calibrations to operating conditions above the earth's surface is that of standardizing the apertures (i.e., effective fields of view) of the control and secondary radiometers; otherwise the problem arises of different amount of circumsolar sky radiation so accepted during calibration, even on high mountains. Therefore, involved here can be corrective factors for the upper atmosphere or space transfer where this sky component is negligible. Another need for precaution is raised when protective windows (e.g., quartz) are incorporated over the receiving surface of the detector. Although the primary standards generally have no such windows, the secondary radiometers, if so equipped, should exhibit a (known) longer window cutoff characteristic, extending as far as possible into the infrared region, in order that energy at solar wavelengths beyond about 2.5 μm (largely absorbed by water vapor at or near the surface but not so at the dry higher levels) is fully transmitted or properly taken into account. Moreover, it is preferable to employ double rather than single window systems; this reduces reemission effects, the inner one serving as a heat blocking filter whenever the temperature of the outer one is raised sufficiently through absorption to influence the thermal zero of the detector.

When optical aids in the form of lenses have to be employed, such aspects have further to be considered as that of establishing the best position of the lens in the radiometer channel, with respect to minimizing incident angular effects commensurate with maintaining adequate signal amplification and the desired field of view. With the addition of filters (especially those having narrow wavelength bandpasses), which is usually the reason for installing a lens arrangement, chromatic aberration can likewise be important. A lens under a filter, of course, also provides thermal blocking.

Possible solarization of filters by ultraviolet radiation is an essential subject for investigation. Broad-band filters are characterized by a wavelength termed the center of lower cutoff. As the filter factor is the reciprocal of the main-band transmittance, it is a unique value for all sources with negligible amounts of energy beyond 3 μm (the upper effective filter cutoff). This compares with the narrow bandpass filters, when the filter factor usually depends upon the nature of the spectral energy distribution of the source, because of irregularities in such filter transmittance functions (Drummond and Hickey, 1967; Hickey, 1970). Therefore, these wide glass filters provide an excellent means of checking the results of narrow-band filter measurements over a restricted range of the spectrum. When mounted on a radiometer, they yield absolute values both over their own transmission regions and, differentially, between their lower wavelength cutoffs. Comparison of broad- and narrow-band filter measurements involves only a summation process.

1.5.2. Readout Instrumentation

Considering the readout limitations, one notes that a facet of precision radiometry which is often neglected concerns the accuracy of the readout equipment. Because it is possible to record the data to four or five significant figures, the assumption is frequently made that the observed value is to the same number of digits. Under fixed laboratory conditions it may be safe (after the effect of stray noise has been evaluated) to accept the day-to-day or week-to-week calibrations. However, field operations can involve large changes in ambient temperature, high winds with dust or rain, and inadvertent power fluctuations. Another source of trouble can arise, in aircraft at high altitudes, when the air is very dry, resulting in static electrical charges on the readout. In addition to the local environment, electrical interference in general always seems to be more severe in routine operation of the instrumentation than in the preliminary laboratory checkout tests. Constant vigilance and a suspicious nature are basic requirements in minimizing such errors. There is now much information (e.g., Brower, 1968; Abernathy, 1968) concerning the signal-to-noise aspects of instrumentation in the microvolt region.

A 5-mV signal was quite typical in the JPL-Eppley solar-constant measurement program, so that 50 μV represented 1 % of this value. If our objective is to achieve an overall accuracy of this kind in radiometric measurements, then the maximum allowable error in the readout should not exceed 0.1–0.2 % (5–10 μV). It is of interest here to note that the limiting factor of any electrical measurement (Johnson noise), on the basis of a 1000-Ω source at normal room temperature and a 1-sec integrating time or 1-Hz pass band, is 1 nV.

Independent of the readout characteristics, there are two different types of signal amplifiers which can be utilized for radiometric measurements. Each has attributes useful for particular field applications. Proper selection requires consideration of the operating conditions; in particular, the desired data rate, temperature regime, allowable space, power and weight, and the nature of the local interfering signals are relevant.

Chopper amplifiers have been in use for many years. One of their main features is extreme zero stability of the output. The chopping technique allows the use of ac coupled amplifiers and, with appropriate feedback techniques, the result is a very stable amplification system. The heart of the stability of these amplifiers lies in the signal chopping techniques (Shea, 1966; Anderson, 1968). It is a difficult task to produce a chopping device which can modulate and demodulate the signal over a wide temperature range and introduce a minimum of noise into the system. The major disadvantages of the chopper system of amplification concern the requirement for a chopping signal source, the extra weight and power required to drive the chopper, and the filtering necessary to remove the ripple developed during the demodulation. Of necessity, the bandpass of the chopped amplifier is limited to less than 1/4 the chopping frequency, and unless rather involved filtering is utilized, the maximum frequency is more nearly 1/10 the chopping rate. Because of the low-pass filtering, chopper amplifiers exhibit excellent noise-free operation except when the interfering noise has the same frequency as the chopping rate. In this latter case, the interfering signal produces an offset in the output, but so long as the level of interference is constant this in no major problem in radiometric measurements; however, variations in amplitude or phase of the noise result in undesirable variations of the signal. For normal operations, where there is no limitation of space, weight, power, or frequency response, the chopper-amplifier system provides the optimum type of amplification. The phase-locked Brower amplifier represents the latest development in chopped amplifiers.

In cases where the limitations of space, weight, power, and data rate preclude the use of a chopper amplifier, the transistorized differential operational amplifier has proven to be most useful. Indeed, with the advent of microcircuit devices, requiring low power and possessing good thermal stability, differential operational amplifiers are very reliable and economical. Relatively simple straightforward filter techniques are applicable to establish the bandpass of the amplifier. Although the gain stability is not equal to that of the chopper-type amplifiers, the regular use of a calibration signal will minimize this difficulty. Inherent in any transistorized amplifier is the excess random semiconductor noise. In the case of the operational amplifier used for the B-57B (and also the X-15) flights, this amounted to an uncertainty of about 0.015 V in the output signal (maximum 5 V). Ruggedness, stability, and capability of operation under extreme environmental conditions characterize this type of amplifier. A thorough error analysis in the application of transistorized operational amplifiers has been reported (Taskett, 1967).

To conclude, the subject matter of Sections 1.4 (Drummond *et al.*, 1967a,b; 1968a,b) and 1.5 (Drummond and Laue, 1968) has been dealt with in some detail because of the lack of specific information generally available for these aspects, but more important has been the objective of connecting the upper atmospheric and space targets with the better established ground-based knowledge outlined in Sections 1.1 to 1.3.

REFERENCES

Abbot, C. G. (1911). The silver-disk pyrheliometer. *Smithsonian Inst. Misc. Collections* **56**, No. 19, 11 pp.

Abbot, C. G. (1934). The standard scale of solar radiation. *Smithsonian Inst. Misc. Collections* **92**, No. 13, 3 pp.

Abbot, C. G., and Aldrich, L. B. (1913). Smithsonian pyrheliometry revised. *Smithsonian Misc. Inst. Collections* **60**, No. 18, 7 pp.

Abbot, C. G. and Aldrich, L. B. (1932). An improved water-flow pyrheliometer and the standard scale of solar radiation. *Smithsonian Inst. Misc. Collections* **87**, No. 15, 8 pp.

Abbot, C. G., Aldrich, L. B., and Froiland, A. G. (1954). Concerning Smithsonian pyrheliometry. *Smithsonian Inst. Misc. Collections* **123**, No. 5, 4 pp. (See also Aldrich, L. B. (1949). The Abbot silver-disk pyrheliometer. *Smithsonian Inst. Misc. Collections* **111**, No. 14, 12 pp;. and Hoover, W. H., and Froiland, A. G. (1953). Silver-disk pyrheliometry. *Smithsonian Inst. Misc. Collections* **122**, No. 5, 10 pp.)

Abernathy, H. (1968). *Control Engineering* **15**, 92.

Aldrich, L. B. and Abbot, C. G. (1948). Smithsonian pyrheliometry and the standard scale of solar radiation. *Smithsonian Inst. Misc. Collections* **110**, No. 5, 4 pp.

Anderson, B. (1968). *Electro-Technology* **81**, 34.

Ångström, A. (1929). On the atmospheric transmission of sun radiation and on dust in the air. *Geografis. Ann.* **H2**, 156–166.

Ångström, A. (1930). On the atmospheric transmission of sun radation II. *Geografis Ann.* **H2–3**, 130–159.

Ångström, A. K. (1961). Radiation to actinometric receivers in its dependence on aperture conditions. *Tellus* **13**, 425–431.

Ångström, A. (1962). Atmospheric turbidity, global illumination and planetary albedo of the earth. *Tellus* **14**, 435–450.

Ångström, A. K. (1970). This volume, Chapter 9.

Ångström, A. K. and Rodhe, B. (1966). Pyrheliometric measurements with special regard to the circumsolar sky radiation. *Tellus* **18**, 25–33.

Ångström, K. (1893). Eine elektrische Kompensationsmethode zur quantitativen Bestimmung strahlender Wärme. *Nova Acta Reg. Soc. Sci. Ups.* Ser. III, **16** (Translated in *Phys. Rev.* **1**, 1894).

Ångström, K. (1899). The absolute determination of the radiation of heat with the electrical compensation pyrheliometer with examples of the application of this instrument. *Astrophys. J.* **9**, 332–337.

Ångström, K. (1900). Intensité de la radiation solaire à differentes altitudes. *Nova Acta Reg. Soc. Sci. Ups.* Ser. III, 1–46.

Ångström, K. (1903). Energie dans le spectre visible de l'etalon Hefner. *Nova Acta Reg. Soc. Sci. Ups.* Ser. III, 1–12.

"Annals of the Astrophysical Observatory of the Smithsonian Institution," Vols. I (1900), II (1908), III (1913), IV (1922), V (1932), VI (1942), and VII (1954). Washington, D.C.

Arvesen, J. C. (1969). Summary of contribution to 1968 Solar Energy Society Annual Meeting. *Solar Energy* **12**, 401–402.

Barr, E. E. (1970). This volume, Chapter 14.

Bedford, R. E. (1960). A low temperature standard of total radiation. *Can. J. Phys.* **38**, 1256–1278.

Bedford, R. E. (1970). This volume, Chapter 6.

Bedford, R. E., and Ma, C. K. (1968). International comparison of measurements of irradiance. *Metrologia* **4**, 111–116.

Betts, D. B., and Gillham, E. J. (1968). An International comparison of radiometric scales. *Metrologia* **4**, 101–111.

Bossy, L., and Pastiels, R. (1948). Étude des propriétés fondamentales des actinomètres. *Inst. Roy. Meteor. Belgique, Memoirs* **XXIX**, Brussels.

Brower, R. (1968). *Electronics* **41**, 80.

Callendar, H. L. (1910). The radio-balance. A thermoelectric balance for the absolute measurement of radiation, with application to radium and its emanation. *Proc. Phys. Soc. (London)* **23**, 1–34.

Coblentz, W. W. (1915). Measurements on standards of radiation in absolute value. *Bull. Bur. Std.* **11**, 87–97 (issued November 1914).

Coblentz, W. W., and Emerson, W. B. (1916). Studies of instruments for measuring radiant energy in absolute value: An absolute thermopile. *Bull. Bur. Std.* **12**, 503–552.
Coblentz, W. W., and Stair, R. (1933). The present status of the standards of thermal radiation maintained by the Bureau of Standards. *J. Res. Nat. Bur. Std.* **11**, 79–87.
Courvoisier, P. (1963). On the compensation pyrheliometer. *Arch. Meteorol. Geophys. Bioklimatol. Ser. B*: **12**, 422–441.
CSAGI (1957). IGY Instruction Manual, Part VI. Radiation instruments and measurements. *Ann. Intern. Geophys. Yr.* **IV**, 367–466. Pergamon Press, London.
Deirmendjian, D., and Sekera, Z. (1956). Atmospheric turbidity and the transmission of ultraviolet sunlight. *J. Opt. Soc. Am.* **46**, 565–571.
Drummond, A. J. (1956a). A contribution to absolute pyrheliometry *Quart. J. Roy. Meteorol. Soc.* **82**, 481–493.
Drummond, A. J. (1956b). The international scale of radiation. *World Meteorological Organization Bull.* **5**, 75–78.
Drummond, A. J. (1965). The extraterrestrial solar spectrum. *Proc. Inst. Environ. Sci. Am. Soc. Testing Materials Symp. Solar Rad.*, pp. 55–64. Los Angeles, 1965.
Drummond, A. J. (1968). The absolute calibration of thermal radiation detectors. *Instr. Soc. Am. Trans.* **7**, 194–206.
Drummond, A. J., and Hickey, J. R. (1967). Measurement of the total flux and its spectral components in solar simulation systems with special reference to the extraterrestrial radiation. *Solar Energy* **11**, 14–24.
Drummond. A. J., Hickey, J. R , Scholes, W. J., and Laue, E. G. (1967a). The Eppley-JPL solar constant measurement experiment. *Proc. Intern. Astronaut. Fed. Congr. 17th, Madrid* 1966, **2**, 227–236. Gordon and Breach, New York.
Drummond, A. J., Hickey, J. R., Scholes, W. J., and Laue, E. G. (1967b). Multichannel radiometer measurement of solar irradiance. *J. Spacecraft Rockets* **4**, 1200–1206.
Drummond, A. J., and Hickey, J. R. (1968). The Eppley-JPL solar constant measurement program. *Solar Energy* **12**, 217–232.
Drummond, A. J., Hickey, J. R., Scholes, W. J., and Laue, E. G. (1968a). New value for the solar constant of radiation. *Nature* **218**, 259–261.
Drummond, A. J., Hickey, J. R., Scholes, W. J., and Laue, E. G. (1968b). The calibration of multichannel radiometers for application in spacecraft and space simulation programs. *Proc. Intern. Astronaut. Fed. Congr. 18th, Belgrade* 1967, **2**, 407–422. Pergamon Press, New York.
Drummond, A. J., and Laue, E. G. (1968). Instrumental problems in precision measurements of solar radiative fluxes above the Earth's surface. Preprint No. AS176 of contribution to *Intern. Astronaut. Fed Congr. 19th, New York*, 1968. Available from *Am. Inst. Aeronaut. Astron.*, New York.
Eldridge, R. H. (1952). A comparison of substandard pyrheliometers. *Quart. J. Roy. Meteorol. Soc.* **78**, 260–264.
Gillham, E. J. (1959). The measurement of optical radiation. *Research (London)* **12**, 404–411.
Gillham, E. J. (1961). Radiometric standards and measurements. *Nat. Phys. Lab., G. Brit., Notes Appl. Sci.* No. 23, 23 pp.
Gillham, E. J. (1962). Recent investigations in absolute radiometry. *Proc. Roy. Soc. (London) Ser. A* **269**, 249–276.
Gillham, E. J. (1970). This volume, Chapter 2.
Guild, J. (1937). Investigations in absolute radiometry. *Proc. Roy. Soc. (London) Ser A* **161**, 1–38.
Hardie, R. H., and Giclas, H. L. (1955). A search for solar variation. *Astrophys. J.* **122**, 460–465.
Hickey, J. R. (1970). This volume, Chapter 8.

Johnson, F. S. (1954). The solar constant. *J. Meteorol.* **11**, 431–439.
Johnson, F. S., Purcell, J. D., Tousey, R., and Wilson, N. (1954). The ultraviolet spectrum of the sun. *In* "Rocket exploration of the upper atmosphere " (R. L. F. Boyd and M. J. Seaton. eds.), pp. 279–288. Pergamon Press, New York.
Karoli, A. R. (1970). This volume, Chapter 7.
Karoli, A. R., Ångström, A. K., and Drummond, A. J. (1960). Dependence on atmospheric pressure of the response characteristics of thermopile radiant energy detectors. *J. Opt. Soc. Am.* **50**, 758–763.
Kostkowski, H. J., Erminy, D. E., and Hattenburg, A. T. (1970). This volume, Chapter 4.
Kostkowski, H. J., Erminy, D. E., and Hattenburg, A. T. Higher accuracy spectral radiance measurements. *J. Res. Nat. Bur. Std.* (to be published).
Labs, D., and Neckel, H. (1968). The radiation of the solar photosphere from 2000 Å to 100 μ. *Z. Astrophys.* **69**, 1–73.
Laue, E. G., and Drummond, A. J. (1968). Solar constant: First direct measurements. *Science* **161**, 888–891.
Linke, F., and Ulmitz, E. (1940). Messungen der zircumsolaren Himmelsstrahlung. *Meteor. Z.* **57**, 372–380.
Mulders, G. F. W. (1935). On the energy distribution in the continuous spectrum of the sun. *Z. Astrophys.* **11**, 132–144.
National Bureau of Standards (1960). Instructions for using the total radiation standards. National Bureau of Standards leaflet (issued by the Eppley Laboratory), Newport, Rhode Island.
Nicolet, M. (1951). Sur le problème de la constante solaire. *Ann. Astrophys.* **14**, 249–265; (1951). Sur la détermination du flux énergétique du rayonnement extraterrestre du soleil. *Arch. Meteorol. Geophys. Bioklimatol. Ser. B*: **3**, 209–219.
Pastiels, R. (1959). Contribution à l'étude du problème des méthodes actinométrique. *Inst. Roy. Meteor. Belgique, Publ.* Ser. A, No. 11, Brussels.
Shea, F. G. (1966). "Amplifier Handbook." McGraw-Hill, New York.
Stair, R. (1970). This volume, Chapter 3.
Stair, R., and Ellis, H. T. (1968). The solar constant based on new spectral irradiance data from 310 to 530 nanometers. *J. Appl. Meteorol.* **7**, 635–644.
Stair, R., and Johnston, R. G. (1954). Effects of recent knowledge of atomic constants and of humidity on the calibrations of the National Bureau of Standards thermal-radiation standards. *J. Res. Nat. Bur. Std.* **53**, 211–215.
Stair, R., and Johnston, R. G. (1956). Preliminary spectroradiometric measurements of the solar constant. *J. Res. Nat. Bur. Std.* **57**, 205–211.
Stair, R., Johnston, R. G., and Halbach, E. W. (1960). Standard of spectral radiance for the region 0.25 to 2.6 microns. *J. Res. Nat. Bur. Std.* **64A**, 291–296.
Stair, R., Schneider, W. E., and Jackson, J. K. (1963). A new standard of spectral irradiance. *Appl. Opt.* **2**, 1151–1154.
Stair, R., Schneider, W. E., and Fussell, W. B. (1967). The new tungsten-filament lamp standards of total irradiance. *Appl. Opt.* **6**, 101–105.
Tasket, D. (1967). *Electro-Technology* **79**, 40.
Thekaekara, M. P., Kniger, R., and Duncan, C. H. (1969). Solar irradiance measurements from a research aircraft. *Appl. Opt.* **8**, 1713–1732.
Vonder Haar, T. H., and Suomi, V. E. (1969). Satellite observations of the earth's radiation budget. *Science*, **163**, 667–669.
World Meteorological Organization. First and second international comparisons of working standard pyrheliometers. Repts. 1960 and 1964 (mimeographed), Geneva, Switzerland.
Yates, H. W., and Taylor, J. H. (1960). Infrared transmission of the atmosphere. *U.S. Naval Res. Lab. Rept.* 5453, Washington D. C.

—2—
RADIOMETRY FROM THE VIEWPOINT OF THE DETECTOR

E. J. Gillham

2.1. Introduction

The purpose of this article is to describe the principles of radiation measurement, in the ultraviolet, visible, and infrared regions of the spectrum. It is first necessary to inquire what quantities are needed to define the properties of such radiation. This may seem an elementary question, but it is as well to ask it, so that we may be quite clear, when carrying out a radiation measurement, precisely what is being measured and of what use the result will be. We therefore begin by reviewing the elementary concepts involved in a quantitative description of radiation.

Consider a perfectly general field of radiation. The properties of this may vary not only from one point to another, but also with time. This complication can be evaded quite simply by considering a particular point at a particular time. In general, radiation will be passing through this point in all directions. Here again, we simplify by considering a particular direction and arrive finally at the question, how do we describe or specify in quantitative terms the radiation passing through a particular point, in a particular direction at a particular time.

However, to speak of the radiation passing through a point is incorrect, since it is one of the fundamental properties of radiation that the radiant power which can pass through a point is vanishingly small. We must, therefore, consider a small element of area δA about the point (Fig. 2.1a). For obvious reasons, we draw this area perpendicular to the chosen direction, and inquire how much radiant power passes through it in the chosen direction. Here again we must correct the wording of the question, since it is a second fundamental property of radiation that the radiant power which can be conveyed in a perfectly parallel beam is again vanishingly small. We must therefore consider, not only the chosen direction, but an infinity of directions contained within a small solid angle $\delta\Omega$ about it. Now, we can write for the radiant power δP transferred across δA:

$$(2.1) \qquad \delta P = B\, \delta A\, \delta \Omega$$

The radiation traveling through the point in the chosen direction is thus described by this quantity B, which is of the form $B = \partial^2 P/\partial A\, \partial\Omega$ and which does, in fact, tend to a definite limit as δA and $\delta\Omega$ tend to zero. It is important to note that in this definition it is implied that δA is perpendicular to the direction in question, and that if we wish to define B for some other direction of the radiation, then δA must be redrawn perpendicular to it.

We have thus reduced the problem of describing radiation to its most elementary aspect and arrived at a description in terms of a quantity B, which is conformable to the basic properties of radiation. The quantity B must therefore be regarded as the fundamental concept in the quantitative description of radiation. It is a quantity which, in the visible region of the spectrum, we recognize or perceive as brightness,[1] and the accepted term for it when dealing with radiant power is radiance. As to how, in principle, it may be measured, we have only to place at the point in question an aperture of area δA, at some point along the specified direction another aperture subtending the solid angle $\delta\Omega$ at δA, and then, by means of some radiation measuring instrument, measure the radiant power δP transmitted through the two apertures, from which we calculate the radiance as $\delta P/\delta A\, \delta\Omega$.

Radiance has an important property. If we pin our attention on a particular ray or beam of radiation, and follow it along its path through the medium (and this path may include reflections and refractions) then, ignoring any attenuation which the beam may suffer by absorption, reflection, scattering, and so on, it can be shown that, at any point, the radiance B, looking back along the direction of the beam, obeys the relation $B/n^2 = $ constant, where n is the refractive index of the medium at the particular point. In most cases of practical interest, the beam is only accessible to observation in regions where $n = 1$, so that if attention is confined to such regions, we may say that the radiance B is constant along a ray path. The requirement that the beam undergo no attenuation may seem to restrict the practical utility of this property, but in practice such losses are easily taken into account, and this invariance of B often provides a particularly simple way of allowing for the geometry of the radiation field.

If we want to describe the radiation at a point in the field completely, we must give the radiance as a function of direction. If we want to describe the field, we must give this function for every point in the field, except to the

---

[1] Strictly, luminance. The magnitude of the sensation of brightness provoked by looking at a surface having a fixed luminance will vary with visual factors such, for instance, as the pupil diameter of the eye. It has lately been recommended that the word luminosity be used to denote the visual sensation of brightness.

2. RADIOMETRY FROM THE VIEWPOINT OF THE DETECTOR 55

extent that the radiance at one point may be related to that at another by the condition of invariance along a ray path. Add to this a possible variation with time, and the fact that the radiation has a certain spectral distribution and state of polarization; then it is clear that a complete description of the field requires a function of a considerable number of variables.

In such circumstances, it is natural to seek for quantities which will give abbreviated descriptions of the radiation, in the form of averages or sums, which will be of practical interest or utility. One such quantity is irradiance. This refers to the radiant power which crosses an elemental area δA, of prescribed orientation, in all directions within a hemisphere of which the elemental area forms the base, and is defined by the relation $I = \partial P/\partial A$. It is necessary to give the sense of this power flow, and corresponding to the irradiance I at a given point, there will also be an irradiance I' describing the power flow in the opposite direction.

The practical significance of irradiance is that it represents the total radiant power per unit area incident on a plane surface, a quantity which is obviously of importance in problems of radiative heat transfer. It is worth remembering, however, that it is an abbreviated description, and that even in the example just given, the information it provides may be inadequate. Usually, in problems of heat transfer, one is more concerned with the radiation that is absorbed than with that which is incident, and if the absorption factor of the surface varies with the direction of the incident radiation (as in fact it nearly always does), then the absorbed power will not be uniquely related to the irradiance but will also depend on the directional distribution of the radiance.

The irradiance depends on the orientation of δA and will change when this changes. It is interesting to consider the question whether an irradiance can be defined solely from the properties of the radiation field, without first specifying the orientation of δA or, to put it another way, is there, at a given point in a given radiation field, a preferred orientation with respect to which one can define the irradiance.

Let us suppose that at the point in question the radiance is given as a function of direction by $B(\mathbf{r})$, where $\mathbf{r}$ is the unit vector representing direction. We take an element of area of arbitrary orientation specified by the unit vector $\mathbf{n}$ along the normal. It is easily shown from the definition of radiance that the irradiance for this particular orientation is given by

$$I = \int B(\mathbf{r}) \cos \theta \, d\Omega$$

where θ is the angle between $\mathbf{r}$ and $\mathbf{n}$, and the integration is carried out over the hemisphere with δA as base.

We can write this result in a very convenient form by representing the radiance as a vector $\mathbf{B}$ of magnitude B and directed along the beam of radiation to which it refers. We now have $I = \int \mathbf{B} \cdot \mathbf{n}\, \delta\Omega$ or, taking $\mathbf{n}$ outside the integral,

$$I = \mathbf{n} \cdot \int \mathbf{B}\, \delta\Omega$$

Unfortunately, this vector integral still depends on $\mathbf{n}$ since $\mathbf{n}$ sets the limits of integration. We can overcome this difficulty by extending the integration over the whole sphere surrounding this point; in other words, over all values of $\mathbf{B}$, provided we recognize that the result will give, not I, but the net power flow $I - I'$. Thus

$$I - I' = \mathbf{n} \cdot \int \mathbf{B}\, \delta\Omega$$

where this integral is now independent of $\mathbf{n}$ and depends solely on the properties of the radiation at the point in question. Writing for this integral the vector $\mathbf{I}$, we have

$$I - I' = \mathbf{n} \cdot \mathbf{I}$$

from which it is seen that the direction of $\mathbf{I}$ gives the preferred orientation for which the net power flow is a maximum, and that the magnitude of $\mathbf{I}$ gives the magnitude of the corresponding net irradiance.

Of course, we are usually interested in I rather than the net power flow $I - I'$. However, in many practical situations, the radiance at the point of observation is confined to a limited range of directions and is zero elsewhere, so that over an appreciable range of directions of $\mathbf{n}$, I' is zero, and the vector quantity $\mathbf{I}$ gives the ordinary scalar irradiance according to the relation $I = \mathbf{n} \cdot \mathbf{I}$.

Before leaving this subject, it should be mentioned that the concept of irradiance, though appropriate when we are dealing with the radiation falling on a plane surface, may not be so when, on the scale of the radiation field, the receiver is not a plane surface. If, for example, we were interested in the total radiant power incident on a space vehicle of roughly spherical shape, then it would be more useful to give equal weight to the radiance in each direction, instead of applying the weighting factor $\cos\theta$, as in deriving irradiance. A suitable measure of the radiation field would then be the integral $\int B\, d\Omega$ over a sphere, with B regarded in the usual way as a scalar quantity.

So far, nothing has been said about radiation sources, but from the point of view adopted here, which is essentially that of an observer in the radiation field, these present no special difficulties and can be described in terms of the concepts already discussed. If we have a radiating surface, all we need do to describe its properties is to place an observer very close to the surface, and the values of radiance and irradiance which he obtains (on looking at the surface) will describe the properties of the radiator. This approach differs from the usual practice, which is to regard the radiation source as something different from the rest of the field, and to use different terms in describing it. There seems, however, to be no logical justification for this procedure, and so far from enabling one to gain a clearer picture of what is happening, or carry out calculations, it frequently does the opposite.

This can be illustrated by an example. Let us take a small source of radiation of area δA and radiance B and compute the irradiance at some point distant r along the axis. From the point of view adopted here, the solution is easy. The radiance B is what we should measure near the source, and it follows from the constancy of B that it is equally the radiance we observe at the point of observation. The irradiance is thus, by definition, given by $I = B\,d\Omega$ where $d\Omega$ is the solid angle subtended by the source at the receiver, and this result we can write as $I = B\,\delta A/r^2$.

If, however, we think in terms of what the source emits, rather than what an observer sees, then we must start by taking an arbitrary solid angle $\delta\Omega'$, and say that the radiant power emitted into this by the source is $B\,\delta A\,\delta\Omega'$. To get at the irradiance, we must now draw an area $\delta A'$ at the point of observation, such that $\delta A'/r^2 = \delta\Omega'$, from which we deduce that the power incident on $\delta A'$ is $B(\delta A\,\delta A'/r^2)$. To get the irradiance we now divide by $\delta A'$. We have thus had to introduce an element of area $\delta A'$, only to discard it immediately afterwards, and incidentally have missed the particularly simple expression $I = B\,\delta\Omega$ for this irradiance.

Admittedly, in such an example, the advantage of the first method is marginal, but in a more complicated situation, as for instance when the radiator is focused by some optical system, the gain in facility of understanding and computation may be considerable. To all of this, however, there is one important exception. So far, it has been assumed that the source is of known radiance and subtends an appreciable solid angle at the observer. If, on the other hand, we have a radiation source, such as a tungsten-filament lamp, which has a high radiance and a small but ill-defined area, it is obviously more convenient to express its properties in terms of the product $B\,dA$, rather than to give each separately. This is equivalent to defining the radiant power emitted by the source per unit solid angle and is, in fact, the logical justification of the point source concept which has played an important part in photometry.

Before passing on to the next topic, let us examine another application of these ideas by considering a particular type of radiation field, viz., the isotropic field in which the radiance is at every point and for every direction the same. It is well known that the thermal radiation inside an enclosure, the walls of which are everywhere at the same temperature, is of this character. By making a small hole in the enclosure, small enough not to disturb appreciably the field inside, we realize the well-known cavity or blackbody radiator which enables an observer outside the box to see, sensibly undiminished, the same radiance as that inside. This will be the same, therefore, regardless of the direction in which the source is viewed. A radiator having this property is sometimes described as obeying Lambert's law, but this conveys some suggestion of arbitrary behavior, and it is better to regard such a radiator as an isotropic one, an ideal more or less approximated by actual radiators.

A case of practical importance is that of an isotropic radiator in the form of a circular disk; it is useful, for this particular case, to calculate the irradiance at some point along the axis. To do this, we draw a hemisphere about the point of observation of such a radius r as to intersect the circumference of the disk (see Fig. 2.1b). We then have for the irradiance

$$I = \int_{\theta=0}^{\theta=\phi} B \cos\theta \, d\Omega$$

or, writing $d\Omega = dA/r^2$ where dA is an element of area on the hemisphere,

(2.2) $$I = (B/r^2) \int dA \cos\theta$$

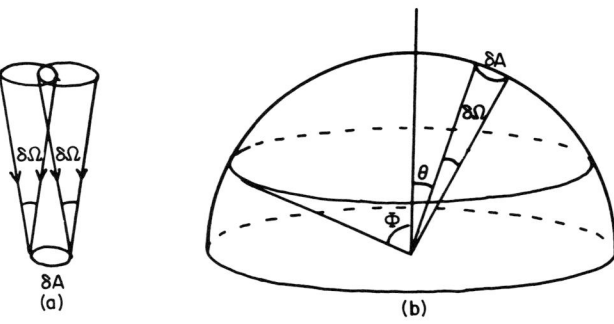

FIG. 2.1. Geometry of a radiation field.

It is easy to see that the integral of $dA \cos\theta$ is equal to the area of the disk or $\pi r^2 \sin^2\phi$, where ϕ is the semi-angle subtended by the disk at the observer. Thus

(2.3) $$I = \pi B \sin^2\phi$$

or, if we write $I_{\pi/2}$ for the irradiance when B extends over a whole hemisphere,

(2.4) $$I = I_{\pi/2} \sin^2\phi$$

This is a particularly convenient form to remember since the output of a radiator is usually expressed in terms of the total hemispherical emission, $I_{\pi/2}$, the Stefan–Boltzmann constant usually being given in terms of this quantity.

2.2. SPECTRAL DISTRIBUTION

We turn now to the question of spectral distribution. In this respect, radiation assumes different forms. It may be monochromatic[2] radiation, or a mixture of monochromatic radiations, or it may be what is known as white radiation and have a continuous spectral distribution, or it may be a mixture of both types. Thus, in describing a general type of spectral distribution, one will give the power in the monochromatic radiation, or lines, in absolute measure, labeled with the appropriate wavelength or frequency; the continuous spectral distribution will be described by a spectral density or concentration which gives the power per unit wavelength or frequency interval, as a function of wavelength or frequency. This description can be applied to any of the quantities such as radiance or irradiance which we have considered. In principle, the measurement of the spectral distribution does not present any new problems—we simply isolate, in turn, each of the spectral components with a monochromator of known transmission and bandwidth and measure the power in it.

The chief difficulty is that this analysis adds, as it were, another dimension to the space required to describe a field of radiation, except in one or two special cases, for example, where the radiation is monochromatic or approximates in spectral distribution to a blackbody radiator at some temperature, in which latter case it can be specified by this temperature. Thus, as in dealing with the spatial properties of radiation, it is natural to look for various abbreviated descriptions of the radiation, this time in the form of a weighted sum over the spectral distribution.

One quantity of this kind has been implicit in all that has been said so far, and that is the total power in the radiation, regardless of its spectral distribution—a quantity usually referred to as the total radiation. Such a quantity has a fundamental significance, in that it is conserved when the radiation is converted into other forms of energy, but it is worth remembering that for practical purposes it has its limitations. As was said when discussing irradiance, we are usually more concerned with the radiation absorbed by a surface than with what is incident on it, and this (for a given surface) will not be uniquely related to the total radiation, regardless of spectral distribution, except in the special and very unusual case for which the absorption factor of the surface is independent of wavelength. This is not to deny the practical importance of total radiation. It is the only quantity one *can* measure when no special assumptions can be made concerning the spectral absorption factor.

---

[2] Strictly speaking, quasi-monochromatic radiation.

One can form other sums or integrals over the spectral distribution by assigning different weights to the radiant power at different wavelengths; the example just given, of the absorption of radiation at a surface, will provide an illustration. Consider the infrared and other lamps used in medical heat-treatment to supply surface heat to the skin. If we had to evaluate the radiant output of lamps used for this purpose, it would be sensible to weight the radiation at each wavelength, $P(\lambda)$, according to the absorption factor $\alpha(\lambda)$ of an average or typical skin at that wavelength, to obtain the sum $\int P(\lambda)\alpha(\lambda)\, d\lambda$ which represents, not the radiation given out, but only that which is usefully employed.

Various other weighted sums appropriate to different purposes can be formed, but it is important to note that this procedure is only valid if the particular action or effect of the radiation that we are concerned with is linearly related to the radiant power. This is to say, the action of each spectral component must be proportional to the radiant power in that component and must also be quite independent of the action produced by any other spectral component that may be present. This condition is satisfied to a very good approximation in photometry, where the luminous content of radiation is measured by using, as the weighting function, the well-known luminous efficiency curve which represents the variation with wavelength of the sensitivity of the eye. It is probably not satisfied when we consider the action of radiation in promoting plant growth.

As a final word on this subject, it should be mentioned that, in principle, there are two ways of measuring such weighted sums. One can measure the spectral distribution and then perform the necessary calculation using the prescribed weighting function. Alternatively, one can, by using filters, construct a radiation-measuring instrument whose sensitivity varies with wavelength according to the weighting function. Such an instrument exposed to the whole radiation will then give a response proportional to the weighted sum and this, with a suitable calibration procedure, may be made to yield its absolute value.

2.3. The Measurement of Radiation—Basic Principles

Having defined the quantities used to describe radiation, let us now consider how these quantities may be measured. In most cases, they are measured with the aid of radiation sources or detectors which have been calibrated at a national standardizing laboratory. However, the intention here is to examine the basic principles of radiation measurement; this can best be done by considering the fundamental radiation measurements which have to be carried out at the standardizing laboratory, in order to provide these calibrations.

2. RADIOMETRY FROM THE VIEWPOINT OF THE DETECTOR

These fundamental measurements are of two kinds. In the first, the power in a beam of radiation is measured directly and, in the second, a blackbody radiator at a known temperature is used as a standard source of predictable spectral or total radiant emission. The two methods are complementary, but here we shall restrict ourselves to the first method, since the second is fully treated elsewhere in this volume. The following description will refer mainly to the methods used at the British National Physical Laboratory (NPL), but these differ in no essential respect from those used at other standardizing laboratories, which would serve as well to exemplify the principles and practices of basic radiometric work.

To measure the radiant power in a beam of radiation directly requires some means of converting the radiation into another, measurable form of energy, with a predictable or measurable conversion efficiency. There are various mechanisms whereby radiation can be converted into other forms of energy over limited regions of the spectrum; for example, there is the photochemical effect used in photography and the various photoelectric effects employed in a wide variety of radiation detectors. Few of these offer much hope for a primary measurement, however, since the efficiency of conversion cannot be measured or, save in the most exceptional circumstances, predicted on theoretical grounds.

There is one process, however, the efficiency of which can be predicted, and that is the conversion of radiation into heat which occurs when radiation is absorbed at the surface of some opaque material. All forms of energy show a natural tendency to become heat, and it is obviously very easy to insure that all the radiation absorbed at the surface is converted into heat; if it were otherwise, if any of the energy appeared in some other form such as fluorescence or electron emission, it would not be difficult to detect. Admittedly, some of the incident radiation will not be absorbed but reflected; in principle, this presents no difficulty since the proportion which is reflected can be determined by purely *relative* measurements of radiation using any suitable uncalibrated radiation detector. We can thus establish the overall conversion efficiency, and all that is then needed is to measure the rate of heat production.

Radiation detectors and measuring instruments based on this concept are known as detectors of the thermal type. They consist, essentially, of a suitably blackened target or receiver, on to which the radiation is directed, together with some means of indicating or measuring the rate at which heat is generated in the target. With the ordinary type of thermal detector, this indication is merely on a relative scale, so that the response is proportional to the radiant power input but gives no absolute measure of it. Nevertheless, such instruments play an essential part in building up a scale of radiation measurement, since they permit what cannot be done by any other means, and that is the comparison of the radiant power in two beams of different wavelengths or spectral distribution. It is not difficult to blacken the surface of the receiver so that the reflection factor is less than 10 or even 5 % over the whole spectral range of interest. If better accuracy than this is needed, it is possible (in theory at any rate) to measure the spectral reflectivity of the receiver. Thus, with the aid of such detectors, one can build up a consistent scale for the measurement of radiation, which will give equal weight to radiant power at any wavelength, and a measure of radiant power in terms of an arbitrary but fixed unit of power.

In order to relate this unit to the ordinary unit of power, the watt, or, in other words, to affix an absolute value to our scale, we now make use of an absolute radiometer, which is simply a special type of thermal detector which can indicate the absolute rate of heat production in the target, in watts, instead of merely giving a relative indication of it. In principle, of course, such an instrument could be used directly for any radiation measurement. In practice, however, its use is generally restricted to the measurement of a few special kinds of radiation, and the results obtained are then extended to other kinds by relative measurements of the type described above. From the point of view of technique and experimental difficulties, the absolute measurements can be regarded as an extension of the relative measurements. It is therefore appropriate to discuss the latter first, and hence we turn now to the radiation detectors of the thermal type on which such measurements depend.

There are various types of thermal detectors, in which different means are employed to detect and indicate the heat produced in the target. In most other respects, however, their mode of operation is similar, and we shall therefore consider only the type most commonly used, the radiation thermopile.

2.4. The Thermopile

Schematically, we may represent this instrument as consisting of a receiver, which might, for example, be a circular disk of thin metal, to which is attached a thermocouple or a number of thermocouples, the cold junctions of which are taken to some comparatively massive object which acts as a heat sink of constant temperature. When the blackened front surface of the receiver is exposed to radiation, its temperature rises more or less quickly to a new equilibrium or steady-state value where the heat losses from the receiver balance the radiant power input. The voltage developed by the thermopile follows a similar course, and the response of the instrument is usually taken to be the steady value attained when thermal equilibrium has been established.

In normal conditions of use, the temperature rise does not exceed a few degrees and very often it may be only a few hundredths or thousandths of 1°C. In these circumstances, Newton's law of cooling usually applies quite accurately and the temperature rise is, therefore, proportional to the radiant power input. The thermoelectric voltage is also proportional to the temperature rise and hence, in turn, to the radiant input. The system is thus a linear one. Representing the heat losses in the form $K\theta$, where θ is the temperature difference between the receiver and the surroundings and K is the thermal conductance from the receiver to the surroundings, we have for the radiant power input $P = K\theta$. If the thermopile has N junctions, each of thermoelectric power ρ, then the output voltage $V = N\rho\,\theta = N\rho\,P/K$, so that the responsivity S_P of the instrument, i.e., the volts developed per unit radiant input, is $S\rho = N\rho/K$.

The speed with which the receiver and thermopile output attain their equilibrium values depends on K, and on the heat capacity of the receiver and of any other parts which change in temperature with the receiver. With many types of thermopile, one can neglect the latter and represent the thermal behavior of the system on an electrical analogy, as a condenser, representing the receiver heat capacity, in parallel with a resistance representing the heat conductance K. Thus, on the sudden application of a radiant power input, the receiver temperature and the response will rise in an exponential manner with a time constant of C/K. For some types of thermopile there may be, in addition, various distributed heat capacities and resistances, so that this simple representation with two lumped circuit elements will fail, and it will not be possible to define a single time constant or even give an unambiguous figure for the responsivity.

It is worthwhile considering some of the factors which determine the performance of a thermopile and the minimum radiant power which it can detect. This is not an academic problem—in many radiation measurements it is necessary to work near the limit of detector performance, and it is useful to know what the limit is and how it can be improved.

One point must first be mentioned. So far, the responsivity has been defined in terms of radiant power input, on the assumption that the thermopile is used to measure the radiant power in a beam which can all be directed on to the receiver. There is another type of measurement in which the radiation extends over a greater area than the receiver, and what is measured is irradiance rather than radiant power. Thus, in addition to the responsivity S_P, we may define another responsivity S_I in terms of the voltage developed per unit irradiance. The relation between the two is obviously $S_\mathrm{I} = AS_\mathrm{P}$ where A is the area of the receiver.

The second point to make is that the efficiency of a thermopile does not depend solely on its responsivity, which merely indicates the open-circuit emf developed, but also depends on its internal resistance. The real criterion, in fact, is the electrical power which the thermopile can deliver to the electrical amplifying or indicating system used with it. Nowadays, with various means available for amplifying small voltages and currents, this requirement may not seem very important insofar as the overall sensitivity of a radiation detecting system is concerned. However, when we consider the fundamental limit set to detection by the Johnson or resistance noise in the thermopile, we see that it is indeed the right criterion, since the noise voltage, as given by the well-known Nyquist formula, is $V^2 = 4kTR\,\Delta f$. Thus, what the output of the thermopile has to contend with is a fixed noise power, V^2/R, rather than noise voltage, and the greater the power developed by the thermopile, per unit of input radiant power, the better will be its performance. This quantity is given by S^2/R, which thus serves as the real figure-of-merit for a thermopile.

We can derive some simple relationships for this quantity by considering a thermopile of a particular type, and then imagining that a number of these are grouped together to form a composite thermopile. The available power is clearly the same, whether they are connected in series or in parallel, so we will assume that they are connected in series.

Let us first consider the value of S^2/R where S is the power responsivity S_P. If S_P and R are the values for the single thermopile, then the corresponding values S_P' and R' for a composite thermopile of n units are $S_\mathrm{P} = S_\mathrm{P}'$ (since each element only receives $(1/n)$th the power, but there are n in series) and $R' = nR$. Thus, we have

$$(2.5) \qquad \frac{S_\mathrm{P}'^2}{R'} = \frac{1}{n}\frac{S_\mathrm{P}^2}{R}$$

If we consider S_I for a field of radiation extending over the whole composite thermopile, then $S_\mathrm{I}' = nS_\mathrm{I}$ and $R' = nR$ so that, in this case,

$$(2.6) \qquad \frac{S_\mathrm{I}'^2}{R'} = n\frac{S_\mathrm{I}^2}{R}$$

We can regard this comparison between the composite thermopile and the simple one as a comparison between two thermopiles of similar construction and efficiency but different receiving areas, so we see that for a thermopile of a particular type the figure-of-merit in terms of S_P is inversely proportional to the receiver area, whereas for S_I it is proportional to the area.

This suggests what is indeed the case, that there is an essential difference in technique between measuring the total power in a radiation beam of limited extent and measuring irradiance in an extended field of radiation. In the first instance, we use a thermopile with as small an area as the radiation can be concentrated on; in the second, we use as large an area as can be covered by the radiation.

It is worth analyzing the behavior of a thermopile a little further to see what other factors besides receiver area affect the figure-of-merit. It is convenient to do this in terms of the power responsivity S_P, if for no other reason than that in measuring radiant power, the power available is usually much less than for measuring irradiance, and the figure-of-merit consequently assumes greater importance.

We need only consider a thermopile with a single thermocouple, since the number of junctions has no effect on the optimum figure-of-merit. For the responsivity of such a thermopile, we have $S_P = \rho/K$. Because part of the heat conductance is due to conduction along the thermocouple wires to the heat sink, we can reduce K and hence increase S by making the wires longer or thinner. This, however, will increase the resistance of the thermopile and may, therefore, lead to a poorer instead of better value for the figure-of-merit. Thus, there is an optimum value for the heat conductance of the wires which is, in fact, equal to the heat conductance K_0 and represents the heat losses from the receiver by every other means, i.e., by radiation, air conduction, etc. If we make the simplifying assumption that the electrical and thermal conductivities of the two materials used for the thermocouple are the same, σ and K say, then it turns out, on making the calculation, that when this optimum condition obtains, the figure-of-merit is given by

$$(2.7) \qquad \frac{S_P{}^2}{R} = \frac{1}{16 K_0} \frac{\rho^2 \sigma}{K}$$

We note first that if K_0 is proportional to the area of the receiver, as is usually the case, then we have the inverse dependence on receiver area, which we obtained previously by another argument. The second factor depends solely on the properties of the thermocouple materials; it is a figure-of-merit for them. This might be expected to vary enormously from one material to another, but there is a natural connection between electrical and thermal conductivity (expressed for pure metals by the Wiedeman–Franz constant) and, unfortunately, there is a tendency for materials with a high thermoelectric power to have an abnormally low value for σ/K. As a result, the figure-of-merit for ordinary materials, such as manganin and constantan, is only about 1/10 that of the best one which can be achieved with semiconductor materials of elaborate composition.

2.5. Use of Radiation Thermopiles—Measurement of Total Radiation

Having dealt with the basic properties of the radiation thermopile, we turn now to its use in a typical radiation measurement. Let us start by considering a measurement of total radiation, e.g., the kind of wideband radiation such as that emitted by a tungsten-filament lamp or a blackbody radiator.

From the definition of total radiation, it would seem that for such a measurement we would require a detector with absolutely constant sensitivity through the spectrum, in order to give equal weight to the radiation at different wavelengths, and it might be wondered why we start with a measurement which presents this initial difficulty of principle, rather than with the conceptually simpler problem of measuring monochromatic radiation. The answer is that broadband radiation is always available in much greater quantities than monochromatic radiation, and the measurement of it is correspondingly simpler.

As radiation of this kind is usually available in large quantities and spread over large areas, the measurements made on it are usually measurements of irradiance. Figure 2.2 shows a typical apparatus for measuring the irradiance produced by a tungsten-filament lamp. To get some idea of the magnitudes involved, let us assume that the lamp is a projector-type lamp of 1 kW rating, and that the irradiance is being measured at a distance of 1 m. Then, assuming that all the electrical power input is dissipated as radiation distributed isotropically about the lamp, we get a rough value for the irradiance as

$$(2.8) \qquad I = \frac{10^3}{4\pi \cdot 10^4} \sim 10^{-2} \quad \text{W cm}^{-2}$$

As for the spectral distribution, this will extend from the blue end of the visible spectrum out to a wavelength of 3 or 4 μm, with an approximately Planckian distribution having a maximum at about 1 μm.

The type of thermopile commonly used for this type of measurement, at the NPL, is the Moll thermopile (manufactured by Kipp) shown in Fig. 2.3

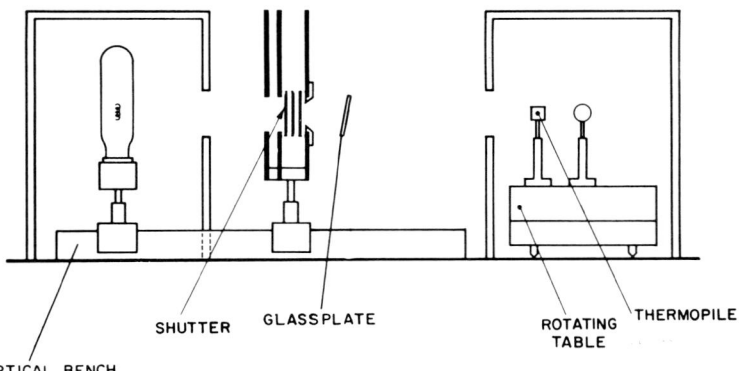

Fig. 2.2. Apparatus for measuring irradiance.

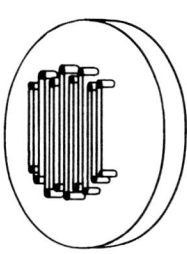 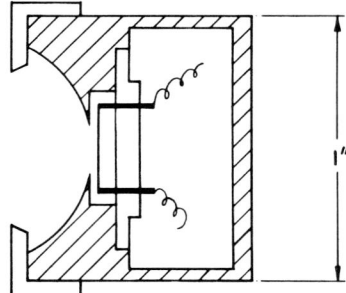

Fig. 2.3. Moll-type thermopile.

(Moll, 1925). It differs from the ideal representation of a thermopile, as previously described, in that the thermopile itself performs the function of the receiver, being in the form of a number of thin flat strips rather than wires, which are arranged one above the other to form an almost continuous (blackened) surface, serving as the receiver. Each strip consists of two halves, one manganin and the other constantan, welded together at the middle and soldered at each end to an electrically insulated copper pin which is in good thermal contact with the body of the instrument and acts as a cold junction. About 20 of these strips, connected in series in a zig-zag fashion, constitute the thermopile. The array of strips covers an area of about 8×6 mm; the area exposed to radiation is limited by a circular diaphragm, of 6 mm diameter, which serves to screen the cold junctions from the incident radiation.

The response of this thermopile to radiation is quite rapid and reaches a perfectly stable value in 2 or 3 sec after the exposure to radiation commences. The response then remains quite steady, and there is no trace of the slow drift downwards exhibited by some thermopiles as the result of heating up of the cold junctions. The response is quite accurately proportional to irradiance, at least up to irradiances of about 0.1 W cm^{-2} and the responsivity in terms of irradiance is about 60 mV W^{-1} cm^2. The resistance of the thermopile is usually about 30 Ω.

It is clear when we consider the construction of this thermopile that the sensitivity will vary considerably along the length of each strip, being a maximum in the center (where the hot junction is located) and falling practically to zero at the ends where the strip is in contact with the cold junctions. The value given above for the responsivity is an average one which holds when the thermopile is exposed to a uniform irradiance. If this condition is not observed, then the mean or effective sensitivity will change very considerably. It might be thought that this would be a serious drawback but, in fact, with this type of measurement, it is not difficult to insure that the radiation is sufficiently uniform for errors of this kind to be negligible.

As an illustration of a type of thermopile which does not suffer from this defect, we may cite the circular type of thermopile manufactured by the Eppley Laboratory (Coblentz, 1914) which has a separate target with a wire thermopile of 6 or 8 junctions attached to it like spokes on the hub of a wheel. With this arrangement, which approximates closely to our schematic representation of a thermopile, the sensitivity is far more uniform. A typical figure for the responsivity of such an instrument, using Bi-Ag as the thermocouple materials, is 120 mV W^{-1} cm^2 for a receiving area of about 0.75 cm^2 and a resistance of about 12 Ω. The corresponding figure-of-merit is about 10 times greater than that of the Moll-type thermopile, a result attributable in part to the greater receiving area and in part to the superior properties of the Bi-Ag thermocouple.

The radiation thermopile operated in air, without a protecting window, is sensitive to drafts and sudden fluctuations of atmospheric pressure such as occur in boisterous weather or when doors are slammed. However, a window has a limited spectral range of transmission and often shows some selective absorption or scattering within this range, so that it is often necessary to dispense with one and tolerate this inconvenience. A thermopile used in this way has a sensitivity which extends over the whole range of wavelengths encompassed by thermal radiation and, in particular, is sensitive to the room temperature radiation emitted by its surroundings.

An immediate consequence of this is that if the thermopile differs in temperature from its surroundings, then it will, in effect, be looking at a blackbody radiator and will be exposed to an irradiance of $\sigma(T_1^4 - T_0^4)$ times some geometrical factor depending on its angle of view. Making the usual approximation of writing this as $4\sigma T_0^3 \Delta T$, which is equal to about 5×10^{-4} W cm^{-2} $°C^{-1}$ at ordinary room temperature, and assuming that the field-of-view is a cone of semiangle ϕ, we find that this amounts to about a $5 \times 10^{-4} \sin^2 \phi$ W cm^{-2} $°C^{-1}$ difference between the thermopile and its surroundings. For the Moll thermopile shown in Fig. 2.3, the semiangle is about 30°, giving an irradiance of about 10^{-4} W cm^{-2} $°C^{-1}$ difference, which means, in the particular radiation measurement we have been considering, a change of 1% per °C.

To avoid an error of this kind would seem to demand fairly close attention to the temperature of the thermopile. This can be avoided by using a shutter (such as is shown in Fig. 2.2) to obscure the radiation from the lamp and by taking as the measure of the irradiance produced by the lamp, the difference between the thermopile readings obtained with the shutter open and closed. The shutter, or rather the aperture across which it moves, subtends a comparatively small solid angle at the receiver, so that by far the largest part of the background radiation which the thermopile receives remains unaffected by the operation of the shutter, and therefore makes no contribution to the differential response. By the same token, the thermal radiation emitted by the shutter (if it happens to be at a different temperature from the thermopile) will usually have an insignificant effect. However, to minimize this possibility, it is desirable to keep the angular subtense of the aperture through which the shutter is viewed as small as practicable, consistent with the thermopile being able to view the whole of the source, and to prevent any great increase in the temperature of the shutter, such as might result from the proximity of the radiation source. The shutter in this apparatus is, therefore, of a multileaf construction, and the surface facing the lamp is left unpainted to reduce the absorption of radiation. The additional screens on the lamp side of the shutter serve to protect the screen facing the thermopile from the lamp radiation. If it did warm up, it would not affect the differential reading, but it would probably produce an inconveniently large zero reading.

As already mentioned, the sensitivity of the Moll-type thermopile is about $60 \text{ mV W}^{-1} \text{ cm}^2$, so that with an irradiance of 0.01 W cm^{-2}, the voltage which has to be measured is about 0.5 mV, out of a resistance of 20 or 30 Ω. For routine work, such a voltage is conveniently amplified with a chopper-type dc amplifier and indicated with a digital voltmeter. For standardization work, it is often preferable to retain old-fashioned methods and to use a manually-operated potentiometer, together with a galvanometer or, where necessary, a galvanometer amplifier, as the null detector. It is worth noting that since the required result is obtained as the difference between two readings, the electrical measurements are free from the troubles of parasitic emfs which beset the absolute measurement of small voltages.

The chief bugbear in measurements of this kind is the fluctuation caused by drafts and atmospheric pressure disturbances, to which reference has previously been made. Drafts can be excluded by means of suitable screens, although it is well to insure that the temperature of the thermopile surroundings is as uniform as possible in order to prevent convection currents. The pressure fluctuations are less easily dealt with. In bad conditions, they may produce voltage fluctuations corresponding to about $20 \ \mu\text{W cm}^{-2}$ and in good conditions about $2 \ \mu\text{W cm}^{-2}$. One way of mitigating their effect is to connect in opposition to the thermopile another one of similar type, which is carefully screened from the radiation being measured. This may reduce the fluctuations, by a factor of 10, to a value of $0.2 \ \mu\text{W cm}^{-2}$ in good conditions, and so permit (with a reasonable number of readings) the measurement of an irradiance of $10 \ \mu\text{W cm}^{-2}$ with a precision of better than 1%.

2.6. Establishment of the Radiation Scale

Let us now consider how these and similar methods for measuring radiation with a thermopile may be employed in building up a radiometric scale. To recapitulate and amplify what was stated earlier, the sequence of events is this. Using an absolute radiometer we make an absolute measurement of radiation on a particular type of radiation suited to the purpose. In fact, for reasons which will appear later, this radiation is that from a tungsten-filament lamp filtered through 2 cm of water. This measurement enables us to calibrate a Moll-type thermopile, or rather a set of such thermopiles, for this type of radiation at a particular level of irradiance. These thermopiles then serve as secondary or working standards of irradiance to perpetuate the scale in the interval which elapses before the next absolute measurement. In addition, recognizing that our thermopiles may exhibit selectivity, in other words, that their responsivities to other types of radiation may be significantly different, we measure this selectivity, and so derive, from our primary calibration, responsivity values for other types of radiation.

Before describing how the selectivity is measured, something else should first be stated about the use of these thermopiles as secondary or working standards of irradiance. The feasibility of using them in this way is wholly dependent on their stability, i.e., the degree to which they retain an unchanged responsivity over a period of months or years. Fortunately, a well-constructed thermopile, however unfavorably it may compare in some regions of the spectrum with other types of radiation detectors, with regard to sensitivity and convenience of use, is pre-eminent in this respect. Specially so are the types already described, those with a comparatively large receiving area, operated in air at atmospheric pressure, and used primarily for the measurement of irradiance. Such a thermopile, if properly handled, will remain stable in responsivity to within $\pm 0.25 \%$ over long periods.

Returning to the measurement of selectivity, it has already been indicated in principle how this can be done. The sensitivity of a thermopile only changes with the spectral quality of the radiation because its absorption factor changes. This absorption factor can be determined by measuring the reflection factor. However, a thermopile is usually coated with a matte blackening of some kind, so that the reflected radiation is reflected diffusely in all directions. The measurement of such diffusely reflected, and therefore weak, radiation presents considerable practical difficulties, particularly in the infrared where no highly sensitive detectors are available.

At the NPL, therefore, we have adopted the method of constructing a special type of radiation thermopile in which the receiver is of the cavity or blackbody type and, in consequence, has an absorption factor which is sensibly equal to unity for all types of radiation in which we are interested

(Gillham, 1962). This thermopile is, in fact, nonselective, and we can, therefore, measure the selectivities of our standard thermopiles by comparing them with this instrument, using different types of radiation.

One of the instruments of this type used at the NPL has a receiver consisting of a tube of 0.5 mm thick aluminum, 10 mm in diameter and 15 mm long, closed at one end by a flat base, and painted internally with a matte black paint. The radiation entering the receiver is restricted by a diaphragm of 6 mm diameter, so that it clears the side walls and falls only on the base. If it is assumed that the small amount of radiation reflected from the black-painted base is reflected diffusely, with a Lambertian or isotropic distribution, then the fraction of it which escapes from the receiver will be $\sin^2 \theta$, where θ is the semiangle subtended at the base by the open mouth. The mouth has an aperture of about 7 mm diameter, so that $\sin^2 \theta$ is about 1/20th. The rest of the reflected radiation will be intercepted by the walls of the receiver and almost completely absorbed. Thus, since the reflection factor of the black paint is less than about 5 % for all wavelengths, the effective absorption factor of the receiver is within 0.25 % of unity for all wavelengths.

This, in itself does not insure that the sensitivity will be constant to the same degree. It is also essential that the reflected radiation caught by the walls be as effective in producing a response as the radiation absorbed at the base. In other words, the sensitivity must be uniform over the whole interior surface of the receiver. This is achieved partly by means of a judicious distribution of the thermojunctions, over the exterior surface of the receiver, and partly by making the walls of the receiver thick enough so that the whole of the receiver is at a reasonably uniform temperature.

The large surface area of the receiver makes the instrument insensitive, and the thick walls make it slow acting. The responsivity is only about 12 mV W^{-1} cm^2; about 2 min exposure is required to obtain a perfectly steady reading. For these reasons, the instrument can really only be used for the measurement of broadband radiation, where a fairly high level of irradiance is procurable.

Thus, to examine the selectivity of our standard thermopiles by comparison with this cavity detector, we normally use four different types of broadband radiation. The first of these, for obvious reasons, is similar to that used in the absolute measurements. The other radiation sources employed are a tungsten-filament lamp with no filter except a thin glass plate to exclude the long wavelength thermal radiation from the bulb, a nichrome-wire radiator operated at about 1000°C, and a black-painted surface heated to about 200°C.

These measurements, when carried out on our thermopiles, reveal a spread in the responsivity for the different sources which is usually less than 3 or 4 %. The results may therefore be interpolated to give the responsivities of the thermopiles for intermediate types of radiation, with an error which probably does not exceed about 0.5 %.

2.7. Measurement of Monochromatic Radiation

We turn now to the measurement of monochromatic radiation. Here, the amount of radiant power which has to be measured is usually very small, generally of the order of 10 μW or less, and it is therefore necessary to concentrate all the radiation available on the detector; the thermopiles used for this work are therefore of the type used for measuring radiant power rather than irradiance. From the expression for the figure-of-merit for such thermopiles, $S^2/R = (1/16K_0)(\rho^2\sigma/K)$, the intuitively obvious will be seen; in designing a suitable instrument, it is desirable to minimize K_0 by making the area of the receiver as small as possible, and by making the heat conductance per unit area as small as possible, for example, by operating the instrument in vacuo. In addition, the thermocouple should be constructed of materials for which the factor $\rho^2\sigma/K$ is as high as possible. A typical thermopile of this kind is the Schwarz type (manufactured by Hilger and Watt) which uses semiconductor materials for the thermojunctions.

The radiation is usually provided by a monochromator with an exit slit of, say, 5 or 10 mm length. In order that the receiving area of the thermopile may be made as small as possible, it is common practice to produce a demagnified image of the slit, usually with mirror optics. The limit to this procedure is set by the permissible degree of convergence in the beam incident on the detector. However, for the kind of accurate measurement required in standardization work, this technique has its disadvantages. It is difficult with mirror optics to concentrate the radiation on to a small receiver without losing an unpredictable fraction of the radiation by vignetting, optical aberrations, scattering, and so on. It is, therefore, our usual practice to employ a Schwarz-type thermopile with a receiving area of 9 $\times$ 0.5 mm on to which the radiation from the exit slit from a normal monochromator can be focused at approximately unit magnification.

A technique frequently used in routine measurements is to chop the radiation at a frequency of about 10 Hz and amplify the resulting ac output of the thermopile, a method which combines the usual shutter techniques with the advantages of ac amplification. This approach is very suitable for automatic recording purposes but unfortunately it requires the use of fast-acting thermopiles, the receivers of which have to be blackened to a minimum thickness in order to keep down the heat capacity. As a result, such thermopiles are usually far more selective than the slower acting ones.

Thus, the thermopile we normally use has a response time of a second or so, and its output is measured by a manual potentiometric method employing a galvanometer or chopper-type dc amplifier as the indicating instrument. The measurements are made in the usual way, with the shutter operated in front of the entrance slit of the monochromator in order to eliminate the effect of the room-temperature radiation which it emits. The thermopile is installed in an enclosure which provides a thermally stable environment, but even so it is often advantageous to employ the compensated form of this instrument, which consists of two identical receivers connected in opposition both viewing the same environment.

The responsivity of a typical thermopile of this kind is about 20 VW^{-1} out of a resistance of 100 Ω, which gives a figure-of-merit very close to that calculated for the semiconductor materials employed, on the assumption that the only heat losses (apart from those through the thermocouple) are by radiation from the receiver. The Johnson or resistance noise in 100 Ω for a 1 Hz bandwidth is of the order of 10^{-9} V, corresponding to a fluctuation in incident radiant power of 5×10^{-11} W; the minimum detectable power, using the instrumentation outlined above, is not greatly in excess of this value.

A thermopile which is to be used for the direct and accurate measurement of radiant power ought to have a perfectly uniform sensitivity over its receiving area. If it has not, then its response will depend not only on the incident radiant power, but on the way this power is distributed over the surface; and it is very difficult, particularly with a small receiver, to insure that this distribution conforms to a prescribed pattern.

Unfortunately, this requirement is very rarely satisfied by a high-sensitivity thermopile of the type just described. For example, the 9×0.5 mm receiver actually consists of three separate elements one above the other, and the responsivities of these elements may differ by 10 or 20 % from each other. It would therefore be very difficult to calibrate such a thermopile, and then use it for the direct measurement of the radiant power from a monochromator, since this would require the spatial distribution of the radiation to be precisely the same in the calibration as in the subsequent measurements.

What can be done with such a thermopile, however, is to use it for comparing the radiant power at different wavelengths, for this does not require that the distribution be known or conform to a prescribed pattern, but merely that it not change with wavelength. This condition can usually be realized, provided certain precautions are observed. One must be certain, for example, that the spatial distribution of radiance in the radiation source which illuminates the entrance slit does not vary with wavelength. The exit slit of the monochromator must either be kept at a constant height and width, or, if these have to be changed, the effect must be measured and proper allowance made.

As to the use of such relative measurements, we may take as an example the calibration of a photoelectric detector for spectral responsivity. It is often found convenient to measure first the relative spectral responsivity, by comparing at different wavelengths the response of the detector with that of the thermopile, when the radiation from the exit slit of the monochromator is directed on to each in turn. This, of course, is equivalent to the relative measurement of radiant power just discussed. All that then remains to be done is to measure the absolute responsivity at one particular wavelength in the range covered by the relative measurements. This can usually be effected at a wavelength where a large amount of monochromatic radiation is obtainable, using a thermopile which is less sensitive than the type used for relative measurements but better adapted to the absolute measurement of radiant power.

So far nothing has been said about the selectivity of the thermopiles used for the relative spectral measurements. This could be examined in the same way as with the working standards of irradiance, using broadband radiation of various kinds. For accurate work, however, it is necessary to measure the responsivity to monochromatic radiation as a function of wavelength. This is done at the NPL by a method similar to that used for broadband selectivity measurements, but with a cavity-type detector of more suitable design.

In this detector the receiver consists of an open rectangular box made from copper foil, the aperture in the box being about 9×1 mm, and therefore of a suitable size and shape to collect the radiation from the monochromator exit slit (Fig. 2.4). To the rear face of this is cemented a pair of strip thermocouples similar to those used in the Moll thermopile. The interior of the

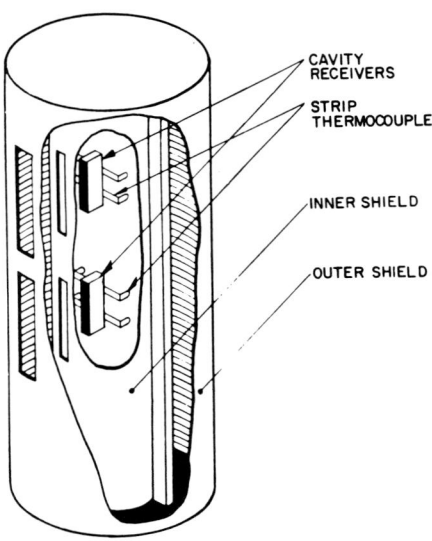

Fig. 2.4. Cavity thermopile for spectral selectivity measurements.

receiver is painted black, and a second similar receiver mounted underneath is used as a compensating element.

This instrument has an appreciably higher responsivity, in terms of volts/watt, than the other type and permits the measurement of spectral selectivity over the wavelength range from about 0.5 μm in the visible out to about 20 μm in the infrared. The increased responsivity has been obtained at the cost of some decrease in blackness, the cavity being appreciably more open than is usually the case. However, auxiliary measurements can be made on the black paint with which the interior is coated, by measuring the spectral selectivity of an ordinary surface thermopile blackened with the same paint; and thus, by what amounts to a process of successive approximation, the effective absorption factor of the cavity can be determined quite accurately.

2.8. The Absolute Measurement of Radiant Power

Having completed the relative measurement, the need sometimes arises (as already mentioned) to make an absolute measurement of radiant power using monochromatic radiation of appreciable intensity. This requires a thermopile with a highly uniform sensitivity across its receiving area. A thermopile used at the NPL which nearly satisfies this requirement is shown in Fig. 2.5. The receiver consists of an aluminum disk 1 cm in diameter and 0.1 mm thick, to the under surface of which is cemented an array of 24 thermojunctions in a uniformly spaced pattern covering the greater part of the disk. These thermojunctions are fabricated from 40 swg. nichrome and constantan wire and are insulated electrically from the receiver and from the

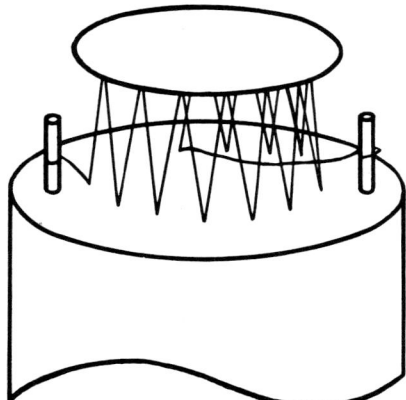

FIG. 2.5. Thermopile for radiant power measurement.

base, which is also of aluminum, by an insulating layer formed by anodic oxidation.

The variation of sensitivity, as examined by moving a small focused spot of radiation over the surface of the receiver, appears to be no greater than about 1 %, so that by using this thermopile in conditions not too far removed from those employed in calibrating it, the errors due to nonuniformity over the receiver area can be kept considerably less than this. The good uniformity results partly from the use of a comparatively thick receiver and partly from the uniform distribution of the thermocouples. The price paid for this is a large thermal capacity for the receiver, as the result of which about 0.5 min is required for a steady reading. The responsivity, on the other hand, is quite high, being about 250 mV W^{-1} out of 20 Ω, which gives a figure-of-merit not far removed from the maximum for this size of receiver and the thermocouple materials used.

As might be expected, this instrument is employed mainly in the visible and near infrared regions of the spectrum, where a number of monochromatic radiations are readily obtainable. It is therefore mounted in a hermetically sealed enclosure provided with a fused silica window. By thus eliminating the disturbing effects of the atmosphere, it is possible to measure a radiant power of 1 μW with 1 % precision. As for the calibration, this is performed by placing a diaphragm of about 6 mm diameter and accurately known area in front of the instrument, and then measuring its sensitivity to irradiance by direct comparison with the working standards of irradiance.

For measurements at even lower levels of radiant power, there is available an evacuated Schwarz-type thermopile with a circular target of about 2 mm diameter. This can be supplied as a nonstandard item with a specially thick target, in which case the sensitivity is uniform to about 5 %. This thermopile improves the limit of measurement by a factor of 10 or more.

2.9. The Absolute Measurement of Radiation

We turn now to the absolute measurements of irradiance by which the NPL working standards of irradiance are calibrated, and which therefore form the starting point for the sequence of calibrations described above. These absolute measurements, it will be recalled, are carried out with an absolute radiometer on the radiation produced by a tungsten-filament lamp plus a water filter.

An absolute radiometer is simply a thermal detector provided with some means of measuring the heat produced in the receiver absolutely. This could be accomplished by a number of calorimetric techniques, but the method almost invariably employed is an electrical substitution one in which the response of the detector to radiation is compared with the response produced by a known amount of electrical power supplied to the receiver. Thus, an absolute radiometer consists essentially of a thermal detector in which the receiver is provided with an electrical heating element or other means of supplying a measurable amount of electrical power.

One of the absolute radiometers used at the NPL is shown in the Fig. 2.6 (see also Gillham, 1962). It consists of two circular aluminum disks, each 44 mm in diameter and 6 mm thick, which are mounted facing each other and supported by somewhat larger disks of thin mica, which are cemented to the outer faces of the aluminum disks and clamped round the periphery to the body of the instrument. The gap between the two disks is bridged by a thermopile of nichrome and constantan wire which runs zigzag between the two disks, so that the junctions (12 for each disk) cover the area of the disk in a pattern of uniform spacing. The junctions are cemented to the disks with an adhesive and insulated from it by an anodic oxide coating on the aluminum.

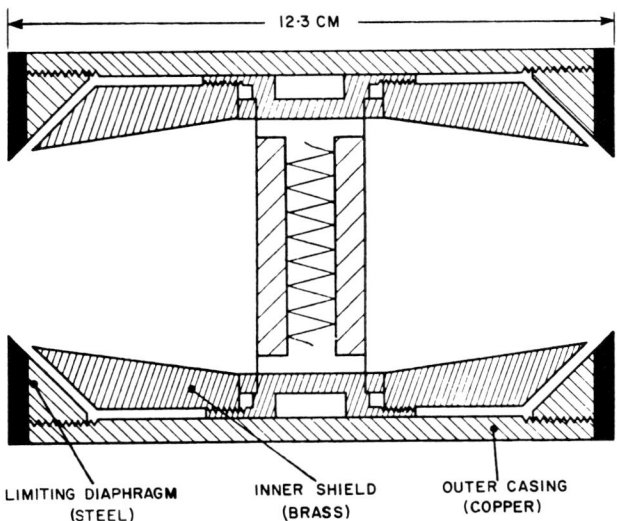

FIG. 2.6. NPL absolute radiometer.

The receivers, mounted inside a housing, constitute a radiation thermopile of the compensated type. The arrangement is a little different from the usual one for a compensated thermopile, where the cold junctions of the receiver and of the identical compensating element are each taken separately to a common heat sink. Here, we can regard the plane bisecting the thermopile wires as a virtual heat sink, at least insofar as the temperature difference between the two disks and the resulting output voltage is concerned. The advantage of this method is that it enables the achievement of an almost perfect degree of compensation. The instrument is perfectly symmetrical and almost completely unaffected by slow changes in the temperature of the body or of the surroundings. In use, of course, only one receiver is exposed to the radiation being measured. With such a thick receiver, the response is naturally very slow; on the other hand, the sensitivity is very uniform which is a desirable feature (as will be seen later).

The electrical heater is on the outer face of the mica disk which supports the receiver and is in the form of a thin gold film deposited by vacuum evaporation in the pattern shown in Fig. 2.7. This resistance grid covers a circular area slightly smaller than the underlying aluminum disk and is provided at each end with three electrodes (of considerably thicker gold film) which extend to the edge of the mica disk, where connection is made to them by means of wires soldered on with indium. The two outer electrodes serve as current leads and are connected, for a reason to be given later, so that each carries half the current through the grid. The inner electrode is a potential lead for measuring the potential difference across the grid. The outer surface of the mica carrying this resistance grid is coated with a matte black paint over the area corresponding to the underlying aluminum.

The instrument is used for measuring irradiance in a uniform beam which extends over the whole of the instrument, and it is therefore provided with a limiting diaphragm at the front of the housing, which confines the radiation falling on the receiver to roughly the same area as that covered by the resistance grid.

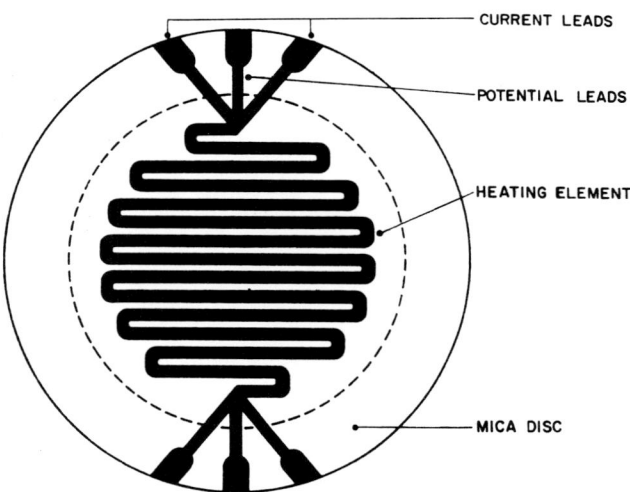

Fig. 2.7. NPL absolute radiometer—heating grid.

2. RADIOMETRY FROM THE VIEWPOINT OF THE DETECTOR

Thus, in principle, measurements are made by supplying the resistance element with electrical power which is adjusted until it produces the same heating effect on the receiver and hence the same thermopile response as the radiation which is to be measured. The electrical power, as given by the product of the heater current and the voltage across the potential leads, is then equal to the absorbed radiant power, from which the *incident* radiant power may be deduced when the absorptivity of the paint has been determined experimentally from a reflectivity measurement. The irradiance in the plane of the limiting aperture can then be calculated, knowing the precise area of the aperture.

The thermal time constant of the radiometer is about 6 min, and it is thus impracticable to secure equality between electrical and radiant heating simply by switching from one to the other and observing the steady-state response. The procedure adopted, therefore, is to adjust the two types of heating to be approximately equal by a preliminary trial, and then to take the radiometer through a number of cycles in which periods of exposure to radiation of 100 sec duration alternate with equal periods in which the receiver is heated electrically. This results in a sawtooth waveform, the amplitude of which gives a relative indication of the difference between the two inputs, say $E-R$. A subsidiary measurement in which periods of heating alternate with periods of no heating gives another sawtooth of amplitude proportional to E, from which (by dividing) we get $1 - (R/E)$ and hence the exact ratio between R and E.

To obtain the reflectivity of the receiver surface, we measure the spectral reflectivity in an apparatus which uses a sensitive photoelectric detector to obtain the reflected radiation. This is one of the reasons why the use of the radiometer is restricted to the particular type of radiation mentioned; the reflectivity measurements need only be made in a region of the spectrum where such sensitive detectors are available. In the instrument used for measuring the spectral reflectivity, the radiation is incident normally on the surface, and the radiation reflected in all directions is sampled by an integrating sphere, so that the reflectivity measured is that appropriate for determining the absorption factor. The mean value of reflectivity for the total radiation employed in the absolute measurement is then calculated by weighting the spectral reflectivity curve according to an estimated spectral distribution for this total radiation. Since the reflectivity is only about 2 % and does not vary greatly over the spectrum, this estimate need be only an approximate one.

It is interesting to examine some of the possible sources of error in the use of this absolute radiometer, since they throw light on the more general problem of absolute measurement and the design of absolute radiometers. One source of error is the heat generated by the passage of current through the current leads to the heating element. Some of this heat will be communicated to the receiver, and the total electrical power supplied to it will, therefore, be slightly greater than that indicated by the current and voltage measurements. However, this additional heating effect can be readily determined simply by passing the same current in through one lead and out through its neighbor, so that none passes through the heating element. The magnitude of this correction is about 1 %.

Apart from this and other easily allowed-for errors, the accuracy of the radiometer rests primarily upon the equivalence between the electrical and radiant heating of the receiver, a factor which is not easy to examine experimentally, and which has to be assured mainly by suitable design. Clearly, exact equivalence can be obtained either by making the spatial distribution of electrical heating the same as that of the radiant heating, or by insuring that the receiver has a completely uniform sensitivity, so that no matter where in it the power is dissipated, the response of the thermopile is the same. In practice, both conditions are aimed at, since neither can be realized completely. The electrical heating element gives uniform heating over approximately the same area as that exposed to radiation; the sensitivity is uniform over the surface of the receiver to better than 1 %.

There is, however, a difference between the two modes of heating in that the absorption of radiation takes place in an absorbing layer of black paint of appreciable thickness. If this layer has a significant thermal resistance between the plane where the absorption of radiation effectively takes place and the inner surface in contact with the heater, then the outer surface of the black will be hotter, and the heat loss from it slightly greater when the receiver is heated by radiation than when it is supplied with the same amount of heat electrically. It can be shown that if the thermal resistance of the black, per unit area, is r, and the thermal resistance which represents the heat losses per unit area from the surface of the black is R, then the result will be an undervaluing of the radiation by a fractional amount r/R. For a receiver operated in air, and losing heat by air conduction, convection, and radiation, R is of the order of $1000°C\ W^{-1}\ cm^2$. It is not unusual for an ordinary black coating to have a thermal resistance per unit area of $5°C\ W^{-1}\ cm^2$, with the result that the thermal resistance error may be as much as 0.5 %.

The magnitude of the thermal resistance has been determined experimentally by supplying a known power per unit area to the black with a radiation beam of known irradiance, and measuring the temperature rise of the black surface by a method of total radiation pyrometry. The accuracy of this measurement, as might be expected, is not very high and it is, therefore, desirable to keep the thermal resistance of the black as small as possible in order that the correction applied should be a minimum.

It is natural to inquire what is the absolute accuracy of these measurements, when all the sources of error described above have been eliminated. The best indication of the accuracy of the NPL absolute measurements is that given by a comparison between the results obtained with different absolute radiometers. In addition to the radiometer described above, the NPL uses two other types, based on the same principles but differing considerably in other respects (Gillham, 1962). One is based on a thermopile similar to the 1-cm diameter circular target thermopile used for measuring radiant power. The heating element is again carried on a mica disk cemented to the front face of the target. It is not coated with a matte black paint, but with a thin absorbing film which reflects specularly and has negligible thermal resistance. The other radiometer uses a cavity receiver similar to that in the instrument described for broadband selectivity measurements. The results obtained with the three different types of radiometer show a total spread of about 0.25 %. In view of the considerable differences between the three instruments, it is felt that the absolute accuracy of the measurements made with them must be close to this figure.

References

Coblentz, W. W. (1914). Various modifications of thermopiles having a continuous absorbing surface. *Bull. Bur. Std.* **11**, 131–176 (see also Eppley Lab. Bull. No. 3, 1964).

Gillham, E. J. (1962). Recent investigations in absolute radiometry. *Proc. Roy. Soc. (London)* **A269**, 249–276.

Moll, W. J. H. (1925). The Moll and Burger thermopile as sold by Kipp and Zonen, Delft, Holland. *Phil. Mag.* **50**, 624–635.

For general reference, see also:

Betts, D. B., and Gillham, E. J. (1968). An international comparison of radiometric scales. *Metrologia* **4**, 101–111.

Callendar, H. L. (1910). The radio-balance. A thermoelectric balance for the absolute measurement of radiation, with application to radium and its emanation. *Proc. Phys. Soc. (London)* **23**, 1–34.

Gillham, E. J. (1959). The measurement of optical radiation. *Research (London)* **12**, 404–411.

Gillham, E. J. (1961). Radiometric standards and measurements. *Nat. Phys. Lab., Gt. Brit. Notes Appl. Sci.* No. 23, 23 pp.

Gillham, E. J. (1964). Further work on a radiometric method of perpetuating the unit of light. *Proc. Roy. Soc. (London)* **A278**, 137–145.

Guild, J. (1937). Investigations in absolute radiometry. *Proc. Roy. Soc. (London)* **A161**, 1–38.

Preston, J. S. (1961). Photometric standards and the unit of light. *Nat. Phys. Lab., Gt. Brit., Notes Appl. Sci.* No. 24, 32 pp.

Preston, J. S. (1963). A radiometric method of perpetuating the unit of light. *Proc. Roy. Soc. (London)* **A272**, 133–145.

— 3 —
SOURCES AS RADIOMETRIC STANDARDS

R. Stair

3.1. INTRODUCTION

The accurate measurement of thermal radiation requires the use of either a standard source or a standard detector. Most early work in this field was centered around meteorological problems and was based upon two independent approaches initiated by K. Ångström in Sweden and by C. G. Abbot in the U.S.A. Both of these approaches resulted in the use of a standardized detector as the working standard. Later investigations in continental Europe, England, and Canada have followed closely the lines first established by Ångström through the use of some type of electrical compensation to match the heating by the radiation from the source (in particular the sun) under study. For further information, the relevant references given in Chapter 1 of this volume should be consulted.

As explained, in detail (Drummond, 1970), the fundamental principle of the Ångström electrical compensation pyrheliometer is based upon equal heating of two similar blackened metal strips, the one by radiation from the sun, the other by carefully metered electrical energy. In principle, the idea is excellent; however, in practice the difficulties are great. Here, the prime problem involved is the assurance that any receiving surface is heated or responds thermoelectrically identically for both electrical and radiant energy. Accurate evaluation of the variations in the effects of conductivity, reflectivity, and emissivity of the blackened surfaces for electrical and radiant energy presents a challenge to the investigator.

In America, Abbot and his associates developed a waterflow calorimeter at the Smithsonian Institution. This instrument is based upon the principle that a definite amount of energy is required to raise the temperature of a specific volume of water by a known amount. Again, while the principle of this instrumentation is excellent, its application is very difficult—probably even more so than in the case of the Ångström pyrheliometer. As a result, most Smithsonian meteorological measurements have been based upon a simpler (secondary) standard, the silver disk pyrheliometer, which consists essentially of a massive blackened and insulated disk of silver onto which the solar rays impinge. The rates of heating and cooling of this disk are determined by, alternately, exposing and shielding it from the solar rays, meanwhile carefully reading its temperature at specified intervals. More complete descriptions of the silver disk and the electrical compensation pyrheliometers have been given by Drummond (1969).

Since these types of instrument cannot be considered truly absolute, it is not surprising that through the years differences in their results have occurred. It is not the purpose of this contribution to try to assess these differences or to endeavor to reconcile them. This has been attempted quite satisfactorily elsewhere and the international Pyrheliometric Scale thus derived was introduced universally in 1956 (Drummond, 1961, 1970).

In parallel with the outdoor meteorological needs, there have been the requirements, indoors, for standard detectors, particularly by the various national physical laboratories concerned with precision thermal radiometry. These developments are also discussed in this volume, mainly in Chapters 1 and 2. However, at the U.S. National Bureau of Standards (NBS), a series of blackbody radiators has been adopted, for many years, as the pertinent radiometric reference, the transfer to secondary instrumentation being effected through the employment of calibrated lamp sources.

3.2. Standards of Total Irradiance

For other than meteorological measurements, until about 50 years ago, the only standards in existence in this area were crude oil lamps or candles. It was in 1913 that Coblentz (1914) recognized the importance of a convenient standard of thermal radiation as an essential requirement in scientific laboratories engaged in researches involving the use of radiant energy. He accordingly set up the carbon-filament lamp as a practical working standard covering a limited range of total irradiance. Then, as today, the blackbody was considered the reference standard, but its use was available only to the few primary scientific laboratories. The carbon-filament lamp standard has received wide acceptance in scientific research, not only in the U.S.A. but throughout the world. (See Chapter 8, this volume, for illustrations of the original and the present types of this lamp.) It has been, and remains, extremely useful, but within recent years has been recognized as being inadequate to cover many new uses for which higher or lower irradiances, or in which spectral energy distributions, are required. Accordingly, work at the NBS has proceeded towards setting up new secondary standards of total irradiance, spectral radiance, and spectral irradiance. The old carbon-filament standards were calibrated in terms of the emittance of a blackbody as given by the Stefan–Boltzmann law; the new standards, as will be explained later, were calibrated in terms of the radiance of a blackbody as given by Planck's law.

The Stefan–Boltzmann law relates the emittance M (total radiant flux from a unit area of a blackbody) to the absolute temperature by the relationship

(3.1) $$M = \sigma T^4$$

The precise value of the Stefan–Boltzmann constant σ can be derived from other physical constants from the relationship

(3.2) $$\sigma = \frac{2\pi^5 k^4}{15 c^2 h^3}$$

where k is the Boltzmann constant (1.38054×10^{-16} erg deg^{-1}K), c is the velocity of light (2.997925×10^{10} cm sec^{-1}), and h is Planck's constant of action (6.6256×10^{-27} erg sec; then $\sigma = 5.6697 \times 10^{-12}$ W cm^{-2} deg^{-4}K.

When the NBS standard of total irradiance was set up, in 1913, a value for σ of 5.70×10^{-12} W cm^{-2} deg^{-4}K was accepted as best representing the extensive experimental radiometric work previously performed with a blackbody. The precision possible in experimental radiometry has not permitted the determination of this value more accurately. However, the extensive research on the physical constants, through the years, of many investigators has established the value of this constant to possibly five significant figures as shown. It is to be observed that this new value differs but little from that employed by Coblentz in 1913 (namely 5.7). It is interesting, however, to note that later experimental values of σ are higher than those obtained before 1913. These data (see Chapter 6, this volume), principally obtained between 1916 and 1933, scatter around a mean of 5.767. A value recently obtained by E. J. Gillham[1] of 5.77 furnishes an additional indication that the experimental value of this constant may well be near this value. On the other hand, the theoretical value of σ, as noted above and based upon the relationship,

(3.3) $$\sigma = \frac{2\pi^5 c^2 h}{15 c_2^4} \quad \text{or} \quad \frac{2\pi^5 k^4}{15 c^2 h^3}$$

results in a value of 5.669, only slightly below the value originally employed by Coblentz. Thus, the best value of σ for radiometric calibrations still remains to be determined.

Because of recent needs for standards with a higher accuracy and a wider range of total irradiance, preferably at a higher color temperature than was available with the carbon-filament standards, work has been done toward

---

[1] Private communication.

setting up new lamp standards. The new standards (Stair et al., 1967) are tungsten-filament lamps of three sizes (100-, 500-, and 1000-W) operated at color temperatures between 2700 and 2850°K. Their calibrations are based upon the spectral radiance L_λ of a blackbody.

$$(3.4) \qquad L_\lambda = \frac{c_1 \lambda^{-5}}{\exp(c_2/\lambda T) - 1}$$

where $c_1 = 1.19096 \times 10^{-12}$ W cm^2 (sr)$^{-1}$, $c_2 = 1.43879$ cm deg, λ is the wavelength in centimeters, and T is the temperature in degrees K (TKTS). These values of constants c_1 and c_2 are consistent to within 0.2 % with the value of 5.67 for σ.

In order to eliminate problems which arise because of significant water-vapor and CO_2 absorption beyond 4 μm, a quartz plate, which does not transmit beyond 4 μm and calibrated for spectral transmittance, was interposed between the blackbody and the receiver. Atmospheric absorption of the energy from the lamp standard is almost insignificant, first because of the operation of the lamp at a relatively high temperature and, second, because the glass envelope is shielded except for a narrow area [1 in. (2.5 cm)] of the bulb in front of the filament. Figure 3.1 shows the experimental set-up employed in comparing the irradiance from the blackbody with that from a 100-W tungsten-filament lamp. The blackbody is set at a distance of about 33 cm— the lamp at about 1.3 m. Calibrated, water-cooled apertures determine the blackbody irradiances at the set temperatures which are controlled to ± 0.2 °K over extended periods of time. The thermopile (or thermocouple) is rotated to alternately face the blackbody and the lamp under study. The thermoelectric outputs are evaluated through the use of a nanovoltmeter and strip-chart recorder. To keep the two thermoelectric voltages comparable when

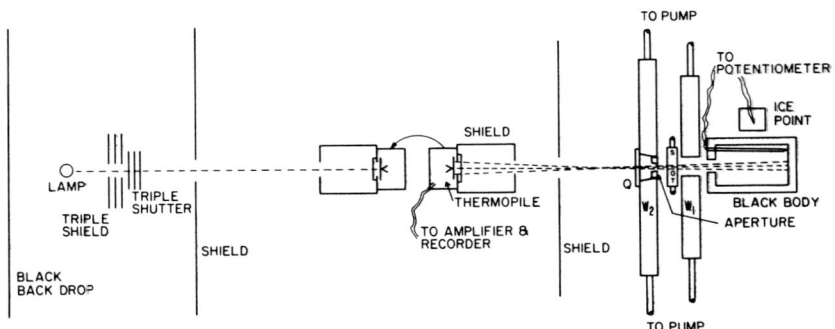

Fig. 3.1. Layout of blackbody, thermopile, lamp shutter, shields, and aperture, as employed in comparison of lamps with blackbody.

calibrating the 100-, 500-, and 1000-W lamps, the blackbody temperature is altered, the distances of the lamp varied or an attenuating chopper is set up between the lamp and detector as required. The blackbody, which was operated up to 1400°K, is constructed of a casting of an alloy of 80% nickel and 20% chromium and has a 3-in. (7.5 cm) outside diameter, 6-in. (15 cm) length, and a wall thickness of ½in. (1.3 cm). The low reflectivity of this oxidized metal, coupled with the small aperture [3/8 in. diameter (9 mm)] as compared with the internal surface area, results in a blackbody effective emissivity (when applying the deVos method of determining cavity emissivity) of 0.999. The high heat capacity of the associated furnace gives the blackbody a very high thermal stability. The automatic oven control keeps the blackbody temperature within $\pm 0.2°$K over long periods of time. Most of the measurements were made with a cavity-type detector (see Fig. 3.2).

Figure 3.3 illustrates the relative spectral irradiance of a blackbody operated at 1300°K, the temperature employed in some of the NBS work. Also, there is indicated on the same scale the spectral irradiance from a blackbody at 300°K (approximately the temperature of the water-cooled shutter) which may be considered zero on this scale. The spectral transmittance of the quartz plate determines the long-wave cutoff of the blackbody irradiance. Hence, the effects of the water-vapor absorption at 6 μm and longer wavelengths are eliminated, as is the case with much of the CO_2 absorption at 4.2 μm. Only the H_2O bands at shorter wavelengths need be considered. On this chart, the water-vapor absorption for the 33-cm pathlength between the blackbody aperture and thermopile is indicated on the basis of an amount equivalent to 0.0001 precipitable cm NTP. This amount of absorption by water vapor is based upon data recently published by Wyatt et al. (1964),

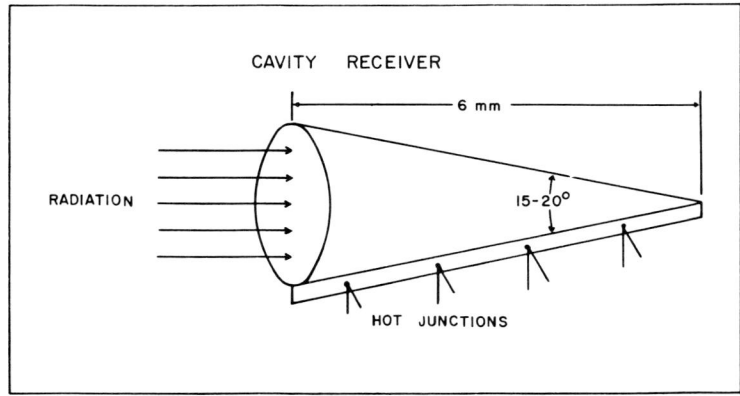

FIG. 3.2. Conical blackbody detector consisting of a formed gold-foil cone blackened inside and having several thermojunctions attached along a single fold.

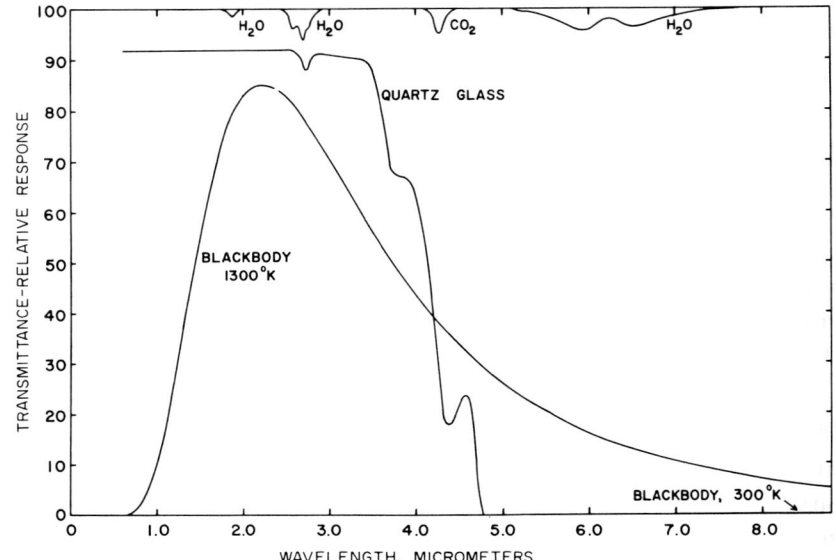

Fig. 3.3. Spectral emission of a 1300°K blackbody; transmittance of a quartz plate; and transmittances of a 33-cm path for water vapor and carbon dioxide.

while the CO_2 values are based upon the data by the same co-workers (Stull et al., 1964). The combined absorptions amount to approximately 0.5%.

The water-vapor content of the atmosphere was determined from measurements of the temperature and relative humidity of the laboratory during the course of the measurements which ranged closely around 75°F and 70%.

In Table I are shown data obtained on the three original carbon-filament lamp standards set up by Coblentz. As noted earlier, his data (column 5) were obtained by using the Stefan–Boltzmann law and making no corrections for water-vapor absorption. Our data were based upon the Planck law of

TABLE I. New and original calibrations on three primary carbon-filament lamps (μW cm^{-2} at 1.00 m).

Lamp No.	1965 values			1913 values	Difference (%)
	Trial 1	Trial 2	Mean		
C-1	358.9	361.4	360.15	359.6	+0.15
C-2	330.9	330.0	330.45	333.2	−0.83
C-3	367.9	367.5	367.7	368.0	−0.08
Average					−0.25

radiation with a correction of +0.5% being made for water-vapor absorption. His data incorporated a value for σ of 5.7 while a value of about 5.67, as pointed out above, is consistent with constants which we used in Planck's equation.

In Table II, data on total irradiance are given for three groups of tungsten-filament lamps (100-, 500-, and 1000-W) when operated at 0.75, 3.60, and 7.70 A, respectively. These values were not corrected for any absorption of radiant flux by water vapor between the lamp and radiometer. Any correction in the original data is small. However, it is possible that some correction should be made if the lamps are to be used in a humid atmosphere (Stair and Johnston, 1954).

Since the new standards operate at much higher temperatures than the carbon-filament lamps, the peak of the spectral energy curve is shifted toward the shorter wavelengths; consequently, they conform more closely with the spectral curves of the NBS standards of luminous intensity, spectral radiance, and spectral irradiance. The lamps employed are commercial projection-type, tubular-bulb lamps having C-13 type coiled filaments and may be operated on either ac or dc. The useful life for the 100-W lamps (at least 50 hr for 1% change) falls somewhat short of that previously available through the use of the carbon-filament lamps. A similar lamp life may be

TABLE II. Total irradiances from three groups of tungsten-filament lamp reference standards.

Lamp No.	(V)	Irradiance at 1 m (μW cm^{-2})
100-W lamp standards operated at 0.75 A		
7741	97.57	570.2
7745	96.56	583.9
7746	94.94	549.5
7749	94.04	542.8
500-W lamp standards operated at 3.60 A		
1	90.18	2963
3	90.00	3007
4	89.51	3072
5	89.24	2967
1000-W lamp standards operated at 7.70 A		
1	100.94	7234
2	101.04	7398
3	101.16	7350
4	101.03	7238

expected in the case of the 500- and 1000-W lamp standards since their operating color temperature (when operated at 3.60 and 7.70 A, respectively) lies between 2800 and 2850°K. Otherwise, their characteristics are similar, with outputs being about five and one-half and thirteen times those for the 100-W lamp standards.

An estimate of the accuracy of the values assigned to the lamps is based on the following factors: (1) The measurement of the blackbody temperature, through the use of thermocouples, is determined to within 0.5° at 1300°K based on the International Practical Temperature Scale (IPTS). However, in order to obtain the best blackbody radiance values, the thermocouple calibrations were corrected to the Thermodynamic Kelvin Temperature Scale (TKTS) which, at 1300°K, required an addition of 1.4° to the IPTS temperature. The standard deviation uncertainty of the correction is about 0.25°. These uncertainties in the blackbody temperature produce uncertainties of ± 0.2 and $\pm 0.1\%$, respectively, in the radiance values. (2) The distance measurements are known to within 0.5 mm, thereby producing an uncertainty of $\pm 0.2\%$. (3) The transmittance of the quartz plate is known with an uncertainty of $\pm 0.4\%$. (4) The area of the blackbody aperture produces an uncertainty of $\pm 0.1\%$. (5) The precision of measurement in terms of a standard deviation is $\pm 0.3\%$. (6) The uncertainty of the blackbody emissivity is 0.1 %.

Combining the above factors in quadrature results in an estimated standard deviation of about 0.6% in the values of total irradiance assigned to the lamps.

Care in the operation of these lamps should provide any radiometric laboratory with a precise source of known total irradiance, useful in a multitude of studies in this area (Coblentz, 1914; Stair et al., 1965a; Stair et al., 1967).

3.3. Standards of Spectral Radiance

Much research has been carried out on tungsten lamps, in particular regarding the spectral emissivity of tungsten. In many laboratories, the determination of the spectral distribution of radiant energy from a tungsten-filament lamp, for employment as a standard of spectral radiance or irradiance, has been obtained through making use of the published values of the emissivity of tungsten and the observed color temperature or brightness temperature of the filament (Stair and Smith, 1943; Grum, 1963). An early model of such a lamp for use as a standard of spectral irradiance, in the ultraviolet, is illustrated in Fig. 3.4. Later models were constructed with ribbon filaments. The calculation of the spectral emission from such lamps was based upon a doubtful assumption that all samples of tungsten are identical in emissivity. No account was taken of the effects of impurities present, or of the size or shape of the filament, or of its mechanical or crystalline structure. All these

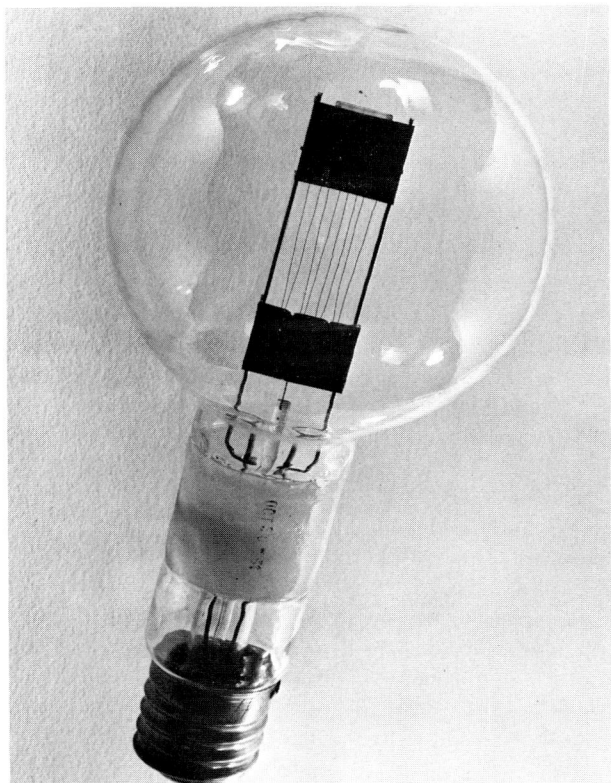

Fig. 3.4. Tungsten-in-quartz lamp.

properties affect markedly the true spectral and total emissivity. Furthermore, it has been found that interreflections within the lamp envelope affect the total spectral radiation from a particular tungsten filament. In order to obtain the nearest correct spectral radiance or irradiance of a lamp. it becomes necessary that the particular lamp, as set up for use, be calibrated against a blackbody.

3.3.1. Apparatus and Method

Before discussing the apparatus and method employed in comparing a group of tungsten ribbon-strip lamps (Stair *et al.* 1960) with blackbodies at various temperatures, a discussion is in order relative to the type of lamp chosen for use as a standard of spectral radiance.

Investigations by Worthing (1917, 1926) many years ago resulted in the accumulation of considerable information relating to the radiation characteristics of tungsten, in particular regarding the effects of the angle (polarization, etc.) from which the filament is viewed. These results proved the necessity of using a flat-ribbon filament to insure a reproducible source. A lamp was designed, at the NBS, some time ago which incorporated the foregoing qualities and was accepted for commercial production by the General Electric Company, as their type GE 30A/T24/3. There is a later modification (see Chapter 8, this volume for an illustration of this type). This lamp was chosen for use as the new standard of spectral radiance. It has a mogul bipost base and a nominal rating of 30 A at 3.5 V.[2] Radiant energy is emitted from the flat strip filament through a $1\frac{1}{4}$ in. fused silica window placed parallel to, and at a distance of about 3 to 4 in. from, the plane of the filament. This separation (necessitated by a graded seal) assists greatly in reducing the deposit of metallic tungsten on the window as the lamp ages.

Three standards of spectral radiance have been established covering the wavelength regions 0.25–0.75, 0.5–2.6, and 0.25–2.6 μm, respectively. The first two were obtained, independently, through the use of two blackbodies having temperatures of about 2200 to 2600°K and 1200 to 1400°K, respectively.

The principal apparatus employed in comparing the radiance of the lamp with that of the blackbody was set up, as shown in Fig. 3.5, to cover the spectral region from 0.7 to 2.6 μm. The lamp and blackbody were mounted, side by side, on an optical bench (constructed from a lathe bed and table) so they could alternately be placed at the focal point of the auxiliary optical system. This blackbody was operated up to 1400°K; it was constructed of a casting of an alloy of nickel and chromium, as described above (Section 3.2). The experimental work consisted of alternately allowing the radiation from the blackbody (set at a specific temperature) and that from a strip lamp (set at a fixed current) to enter a spectroradiometer, after imaging by the same optical system, and then measuring the relative radiances of the two sources at selected wavelengths. A PbS cell, supplemented by an RCA 7102 photomultiplier, was employed as detector. The output was amplified through the use of a 510-Hz tuned amplifier and read on a vacuum-tube voltmeter.

In Fig. 3.6 is shown the spectroradiometer mounted on the optical bench with the auxiliary optics rigidly secured on the front of the monochromator. This arrangement was employed for the spectral region 0.25–0.75 μm, wherein a high-temperature graphite blackbody was used. It was possible to move the spectroradiometer, so that the lamp and the blackbody would be alternately at the correct object distance of the auxiliary optics. These

[2] 6 V in the earlier design.

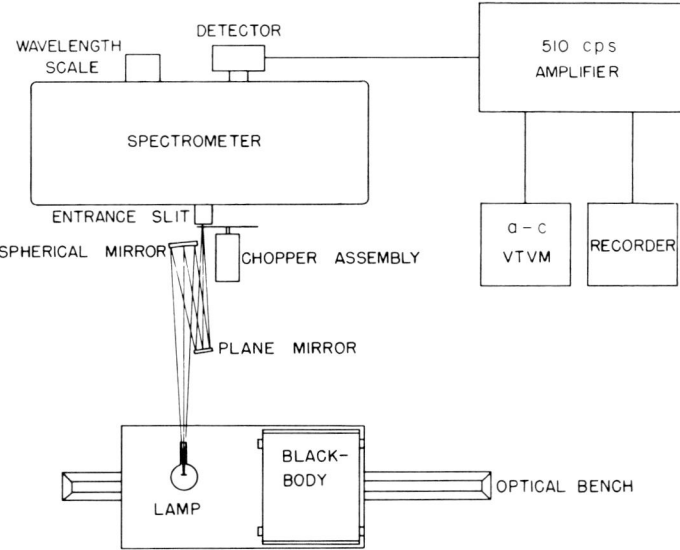

Fig. 3.5. Instrumental setup of blackbody, monochromator, lamp, and associated equipment for establishing the lamp standards of spectral radiance for the spectral region of 0.7 to 2.6 μm.

consisted of a plane mirror and a spherical mirror (as above) having a 71-cm radius of curvature. The placing of the lamp and the blackbody, alternately, at the same optical position insured equal light paths and the use of identical optics.

Leiss double quartz prism monochromators with relative aperture ranges of f/6.4 to f/7.2 were employed in each case.

The blackbody employed for the high-temperature work was constructed of high purity graphite. It is illustrated in Fig. 3.7 and consists of a cylindrical enclosure $4\frac{1}{2}$ in. (11 cm) long and $1\frac{1}{2}$ in. (4 cm) in diameter, having walls 3/16 in. (0.5 cm) thick. The exit of the tube has a 3/8 in. (1 cm) opening shielded by a conical graphite end-piece 3/4 in. (2 cm) long. The very low reflectivity of the graphite (rough machined surface) and the relatively small aperture (as compared to the total internal surface) results in a blackbody of high effective emissivity. It is heated by induction inside a water-cooled coil, by a radiofrequency generator operating at a frequency near 450 kHz. The blackbody tube is insulated by firmly packed boron nitride powder inside a high temperature porcelain tube (closed at one end) 4 in. (10 cm) in outside diamter and 6 in. (15 cm) in length. An alundum ceramic tube, placed midway between the graphite core and the high temperature porcelain tube, increased the mechanical stability of the unit. Deterioration of the

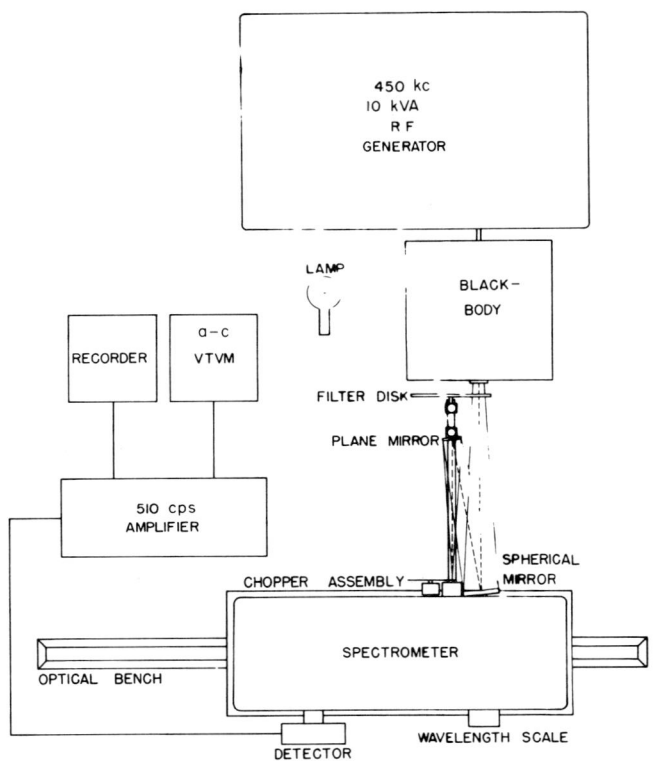

Fig. 3.6. Instrumental set-up of blackbody, monochromator, lamp, and associated equipment for establishing the lamp standards of spectral radiance for the spectral region of 0.25 to 0.75 μm.

graphite, at high temperature, was reduced by enclosing the blackbody unit in an airtight chamber through which dry helium was passed. The concentration of oxygen was further reduced by heating copper coils within the chamber preceding each operation of the blackbody.

In summary, the spectral radiances of two groups of strip lamps were evaluated in terms of those of the two blackbodies, at specfic temperatures, as defined by the Planck relationship. Although set up independently, the two standards are in close agreement over the common spectral range of 0.5 to 0.75 μm. The third standard of spectral radiance simply combines the two spectral ranges, by means of a single strip lamp, operated at a single current, with calibration covering the full range from 0.25 to 2.6 μm.

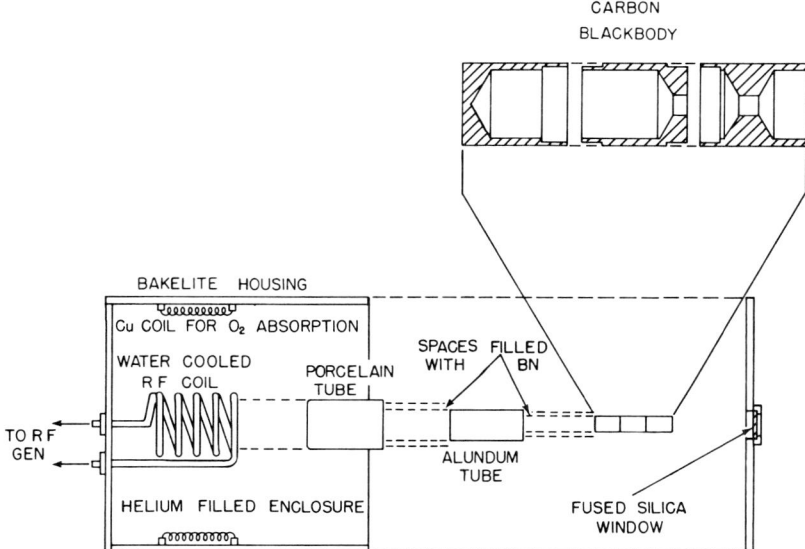

FIG. 3.7. Graphite high-temperature blackbody.

3.3.2. Use of the Standards of Spectral Radiance

The auxiliary optics employed with this standard may be composed of two units such as those employed in the original calibrations, namely a plane mirror and a spherical mirror (each aluminized on the front surface). If the spherical mirror is placed at a distance from the lamp filament equal to its radius of curvature, and the plane mirror set about one-third to two-fifths this distance from the spherical mirror and facing it (at an angle of 10° or less), an image of the filament equal in size to that of the filament itself may be focused upon the spectrometer slit. Little distortion of the filament image occurs, provided good optical surfaces are employed and all reflection angles are kept to less than 10°.

No diaphragm or other shielding is required in the use of these standards, except for a shield to prevent direct radiation from the lamp (not falling on the concave mirror) from entering the spectrometer, since in their use an optical image of the filament is focused upon the spectrometer slit.

In order to calibrate a spectroradiometer with this standard lamp, a knowledge of the spectral reflective characteristics of the mirror surfaces is required to evaluate the radiant energy properly at the spectrometer slit. A good aluminized surface should have a spectral reflectivity in excess of 87 % throughout the region 0.5–2.6 μm, increasing slightly with wavelength except for a slight dip in the region 0.8–1.0 μm. In practice, the proper reflectance losses can best be determined through the use of a third mirror (a second plane one) which may be temporarily incorporated into the optical set-up, from time to time, as shown in Fig. 3.8.

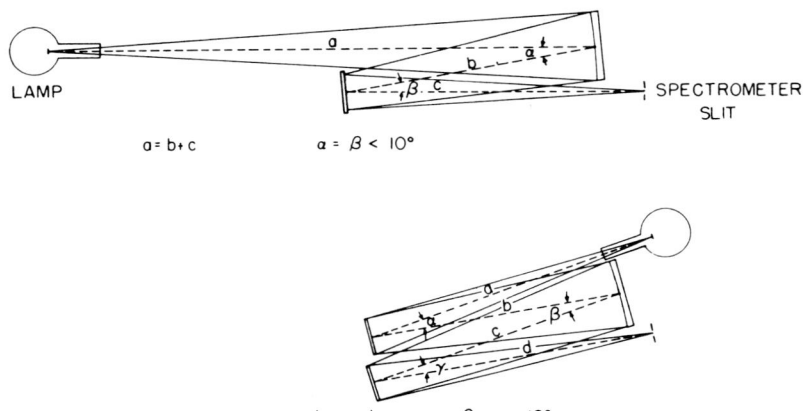

Fig. 3.8. Auxiliary optics when using a lamp standard of spectral radiance in the calibration of a spectroradiometer and for determining the spectral reflectance of the aluminized mirrors employed.

In experiments where sources of radiation are being compared, no knowledge of the spectral reflectance of the auxiliary mirrors, the spectrometer transmission characteristics or the spectral sensitivity of the detector is required. Furthermore, when the same auxiliary optics are employed, no measure need be made of the spectrometer slit widths, or slit areas, provided the slit is fully and uniformly filled in both cases.

Operation of these standards, in our laboratory has been on alternating current (Condell and Byrne, 1957) to reduce filament crystallizing effects which may occur when the operation is on direct current (unless the direction of the current flow is reversed from time to time). To reduce line voltage, a stepdown transformer (1-kVA capacity) having a ratio of 10 to 1 or a 50-A variable transformer may be employed. Then, to realize fine control, a variable transformer (10-A capacity) is wired into the circuit to control the input of the heavy-duty stepdown transformer. For still finer control, a second variable transformer is employed with a radio filament transformer, to add a small voltage (0–2.5 V) to the primary voltage fed into the stepdown transformer. This method is very effective in accurately controlling the larger lamp currents. The heavy-duty (1 kVA) stepdown transformer is preferred to that of the 50-A variable transformer, since the latter is subject to contact damage when operated for long intervals of time at high current values.

These lamp standards are expensive laboratory equipment and it is suggested that they be operated with great care. A working standard should be calibrated, and the standard should be used to check the working standard as required.

The method of calibration against a blackbody is direct and leaves no question relating to filament temperature or tungsten emissivity. The radiance of the lamp is equated to that of the blackbody. Probably the principal source of uncertainty in the results depends upon the accuracy of determination of the blackbody temperatures.

3.4. Standards of Spectral Irradiance

The standards of spectral radiance have found wide use in the calibration of spectroradiometric and other equipment in which a small area, like a slit, is to be irradiated. However, the use of the standard is limited by the small area which can be irradiated, by the low irradiance which the standard provides in the ultraviolet, and by the auxiliary optics required. For many types of spectroradiometric calibration, a standard of spectral radiance has been found to be very useful and will no doubt so continue. However, in many cases, a standard of spectral irradiance is needed. To fulfill this need and to provide a source of higher irradiance in the ultraviolet, a new standard of spectral irradiance has been developed (Stair et al., 1963).

Before proceeding with the establishment of a standard of spectral irradiance for the region 0.25–2.6 μm, based upon the radiance of a blackbody at a known temperature, a study was made of possible lamp designs. First thought was given to the specially constructed tungsten-in-quartz lamp, previously employed in this laboratory as a standard of spectral irradiance based upon color temperature and the published spectral emissivity of tungsten. It was recognized that this lamp had three principal deficiences, viz., its low operating temperature, poor optical quality and bulkiness of the quartz envelope, and relatively large filament area. Only very low ultraviolet irradiances could be realized at a spectrometer slit. Also, since each tungsten filament differs somewhat in emissivity as a function of its specific shape and surface condition, the computed values of spectral irradiance were uncertain by an indefinite amount.

Attention was given next to lamps available commercially. The 6.6 A/T4Q ICL 200-W quartz-iodine lamp (see Fig. 3.9) was examined and found to have acceptable characteristics for use as a standard. It is of robust construction and has a small quartz envelope of relatively good optical quality, so that the intensity usually varies only slightly over a considerable solid

Fig. 3.9. A 200-W quartz-iodine lamp standard of spectral irradiance.

angle centered normal to the axis of the lamp. The filament is a compact coiled coil with overall dimensions approximating $\frac{1}{8} \times \frac{1}{2}$ in. The small size of the lamp envelope (about $\frac{1}{2} \times 2$ in.), together with the small area of the filament, permits placing the lamp within a few centimeters of the slit of a spectrometer. Since this lamp is being set up as a standard of spectral irradiance, to be employed without auxiliary optics, relatively high irradiance at a slit can be realized simply by placing the source close to it.

Because of its high operating temperature, the quartz-iodine lamp emits a relatively large percentage of ultraviolet radiation (see Fig. 3.10). The high temperature is made possible through the unique chemical action of the iodine

3. SOURCES AS RADIOMETRIC STANDARDS

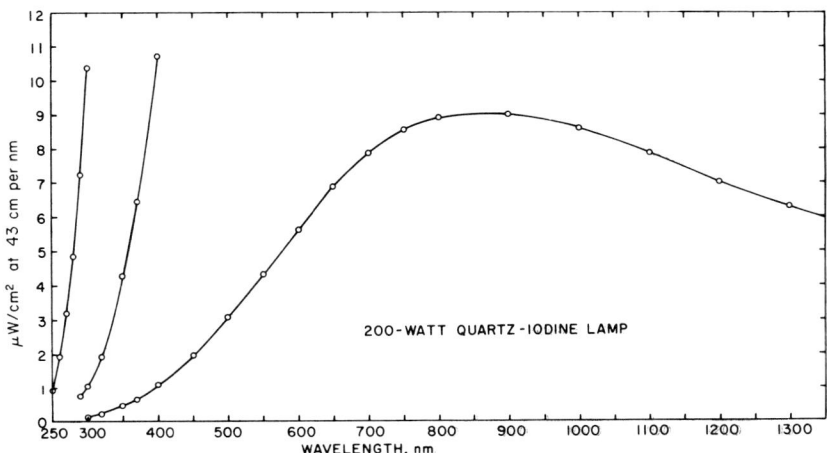

Fig. 3.10. Spectral irradiance from a 200-W quartz-iodine lamp standard.

vapor (Zubler and Mosby, 1959; Allen and Paugh, 1959) which results in the return of evaporated tungsten from the bulb back to the lamp filament, thereby keeping the envelope clean and prolonging the useful life of the lamp. The design life of this lamp when operated at 6.6 A is 500 h. For calibration as a standard, the current is set at 6.50 A which usually gives a filament temperature around 3000°K.

A similar 1000-W lamp (see Chapter 8, this volume, for illustration) is now employed since its output (Table III) is approximately five times that of the 200-W lamp. It is slightly larger, being enclosed in a quartz envelope of dimensions approximately $\frac{1}{2} \times 3$ in.

The experimental work connected with the introduction of the standard of spectral irradiance offered an opportunity to check the agreement between the existing standards of spectral radiance, total irradiance and luminous intensity. Although consideration was given to setting up this new standard directly against a blackbody, a number of difficulties involved in that procedure resulted in its being set up through comparisons with the standards of spectral radiance, supplemented by measurements against the standards of total irradiance and of luminous intensity, all three of which had been established through direct comparisons with the radiances from blackbodies at specific temperatures. The results indicated close agreement between these standards, in some cases to 1 % and, certainly, in all cases to within a few percent. Hence, all NBS standards in this area are based upon the radiance of the Planckian blackbody.

TABLE III. Spectral irradiance of 1000-Watt tungsten-filament quartz-iodine lamps in μW/cm^2-nm at a distance of 50 cm when operated at 8.30 A.

λ (nm)	QM-11	QM-12	QM-13	Mean
250	0.0189	0.0220	0.0207	0.0205
260	0.0340	0.0389	0.0367	0.0365
270	0.0582	0.0650	0.0619	0.0617
280	0.0934	0.103	0.0984	0.0983
290	0.141	0.155	0.148	0.148
300	0.201	0.221	0.212	0.211
320	0.382	0.416	0.402	0.400
350	0.874	0.937	0.914	0.833
370	1.34	1.43	1.40	1.39
400	2.32	2.46	2.41	2.40
450	4.51	4.76	4.68	4.65
500	7.50	7.76	7.65	7.64
550	10.8	11.2	11.0	11.0
600	14.2	14.7	14.4	14.4
650	17.5	18.1	17.8	17.8
700	20.5	21.0	20.9	20.8
750	22.5	23.1	22.9	22.8
800	23.8	24.4	24.2	24.1
900	24.6	25.2	25.1	25.0
1000	24.0	24.6	24.5	24.4
1100	22.4	23.0	23.0	22.8
1200	20.4	21.0	21.0	20.8
1300	18.4	18.9	18.9	18.7
1400	16.5	16.9	16.9	16.8
1500	14.6	14.9	15.0	14.8
1600	12.9	13.1	13.2	13.1
1700	11.3	11.4	11.5	11.4
1800	9.80	9.90	9.98	9.89
1900	8.49	8.59	8.62	8.57
2000	7.33	7.42	7.45	7.40
2100	6.39	6.50	6.50	6.46
2200	5.69	5.72	5.75	5.72
2300	5.04	5.10	5.14	5.09
2400	4.56	4.60	4.64	4.60
2500	4.18	4.19	4.26	4.21

3.4.1. Use of the Standards of Spectral Irradiance

The quartz-iodine lamp is a useful working standard for use in spectral irradiance measurements within the region 0.25–2.6 μm. The methods of calibration are based upon indirect comparisons with the radiances of blackbodies and, thus, do not involve evaluation of filament temperature or tungsten emissivity. A principal uncertainty results from difficulties involved in accurate blackbody high-temperature evaluation.

These standards require no auxiliary optics. If any are employed, proper corrections must be made for their optical characteristics. The lamp is simply placed at a measured distance from the detector or spectrometer slit. However, if the standard and test source differ significantly in geometrical shape, it must be ascertained that the instrument transmittance and detector response are not adversely affected thereby; otherwise large errors may result. Spectrometers usually vary significantly in transmittance over their apertures and with wavelength. Many detectors are highly variable in sensitivity, over their surface area (Stair et al., 1965a), and may require diffusion of radiant flux over their surface to insure accurate radiant energy evaluations. In general, it may be stated that in order to compare unlike sources accurately, an integrating sphere or other averaging device must be employed at the entrance slit of the spectroradiometer.

These standards are normally operated on ac and it is recommended that, for longer life, they be so operated. To reduce the line voltage, a 20-A variable autotransformer may be employed for coarse control. For fine control, a second (5-A) variable transformer may be used in conjunction with a radio filament transformer, to add a small variable voltage to the primary of the 20-A unit. It has been found that this method is very effective for accurate control of the lamp current, for either the 200- or the 1000-W lamp.

In Fig. 3.11 are shown some methods for checking lamp spectral emission data. When differences are taken between successive wavelength values of spectral irradiance, any single error in the tabulation stands out as a break or an inflection in the resulting curve. Likewise, when ratios of the lamp spectral irradiances versus similar point values on some other curve (for example, another lamp or blackbody source) are plotted, other characteristic smooth curves are obtained. Again, breaks or inflections may indicate errors.

3.5. Low-Intensity Standards of Spectral Irradiance

To serve the interests of workers in certain fields of extremely low irradiance, such as is present in the air glow and in phosphorescent, fluorescent, biochemical, and associated areas, an extremely low-intensity standard of spectral irradiance (Stair et al., 1965b) has been set up (Fig. 3.12). This standard,

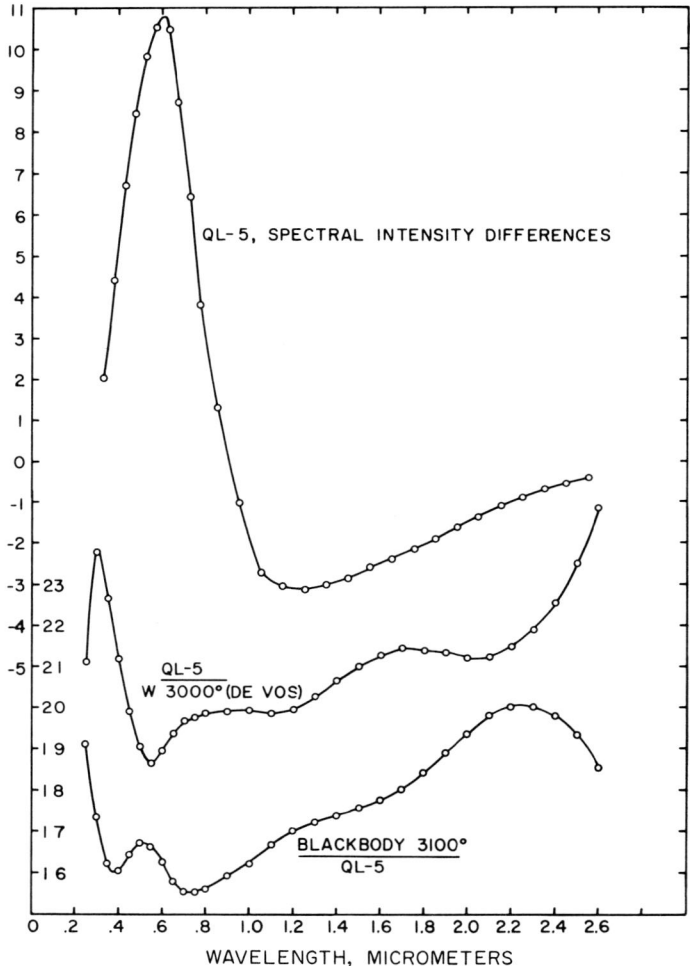

FIG. 3.11. Curves illustrating a method of checking spectral irradiance data.

in effect, combines the NBS standard of spectral radiance with a small aperture which serves as a point source of radiant flux of known spectral intensity. In combination with one or more convex spherical mirrors, employed in the method described by Engstrom (1955), it supplies an irradiance of the order of 10^{-6} down to or below 10^{-16} W cm^{-2} at nanometer wavelength intervals.

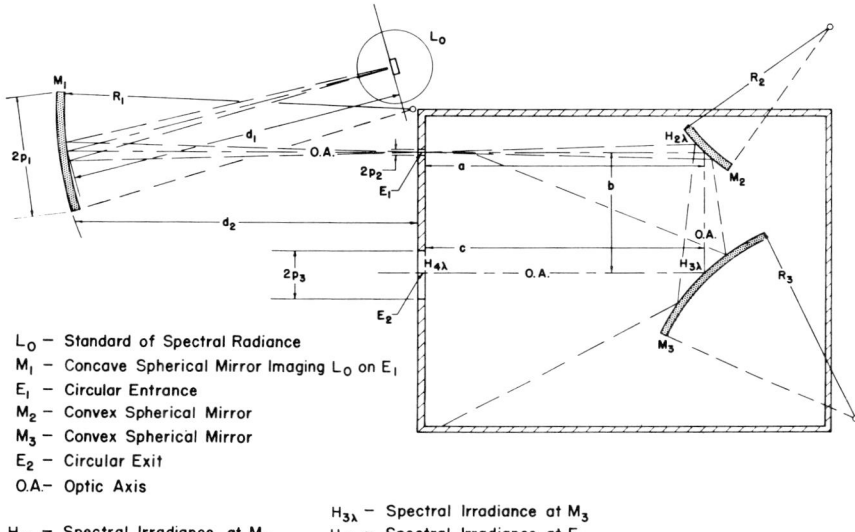

L_0 — Standard of Spectral Radiance
M_1 — Concave Spherical Mirror Imaging L_0 on E_1
E_1 — Circular Entrance
M_2 — Convex Spherical Mirror
M_3 — Convex Spherical Mirror
E_2 — Circular Exit
O.A.— Optic Axis
$H_{2\lambda}$ — Spectral Irradiance at M_2
$H_{3\lambda}$ — Spectral Irradiance at M_3
$H_{4\lambda}$ — Spectral Irradiance at E_2

FIG. 3.12. Schematic diagram of the optical system of the low-intensity standard.

Although the usual standard of spectral radiance, coupled with the use of miniature diaphragms, might be employed in obtaining a standard for extremely low values of spectral irradiance, the interference-fringe effects resulting therefrom, together with the necessity of working at great distances from the standard, rule out its use in this area without the use of auxiliary optics of the type shown in Fig. 3.12.

This standard requires great care in its construction and handling. Its accuracy, furthermore, depends greatly upon these factors. Although the mathematics connecting the irradiance at the exit slit opening with the radiance emitted by the lamp are rather complicated, the substitution of numerical values into the resulting equations is rather simply accomplished. No attempt will be made here to go into the details of these equations as the reader should find the necessary information in the original paper (Stair et al., 1965b).

3.6. Concluding Remarks

In conclusion and summary, some comments are now appropriate, illustrating the use of some of the standards and their calibration, and providing additional pertinent information in connection with radiometry in this area.

First a word about detectors. No commercial detector, either thermal or photoelectric, except possibly the cavity or blackbody element, may be considered as *absolutely neutral in spectral response over any extended spectral range (unless experimentally so proven)*. Such detectors must be used with this consideration in mind. The cavity, or blackbody, unit may be considered as approximately neutral if properly coated and used; that is, with the radiant flux confined only to the lower portion of the conical cavity. Properly coated flat surfaces may have a *nearly neutral response over extended ranges*. An example is Parsons' coating black (Fig. 3.13) which, however, usually results in a slow detector. Near neutral detectors, for selected spectral regions, are available as gold-black and lamp-black coated surfaces (Fig. 3.13).[3] These detectors may vary in efficiency with receiver blackening thickness and method of application. With regard to the new standards of total irradiance, at NBS, much of the work was done with a cavity-type detector. Data taken through the use of a heavily coated lamp-black element were in close agreement. Some data obtained with a gold-black unit were not in agreement—probably because of improper aperture adjustment.

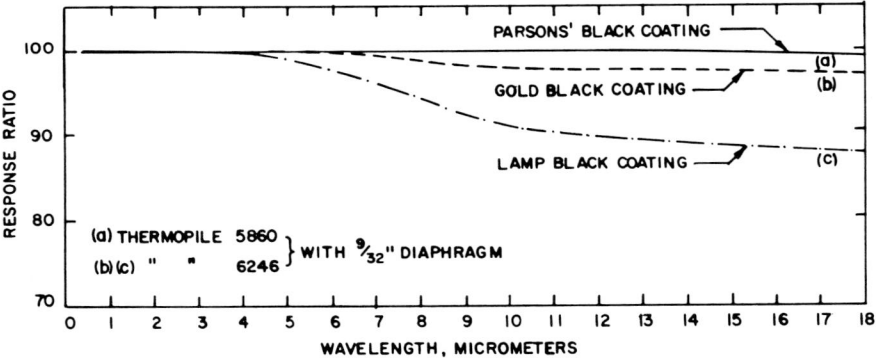

FIG. 3.13. Spectral sensitivity of Eppley thermopiles (Nos. 5860 and 6246) coated with Parsons' black, gold black, and lamp black versus a NBS cavity detector (No. 3), as measured with dc instrumentation.

In measurements of spectral irradiance, differences in source size (or aperture) require that some type of integration be employed to insure that the conventional spectroradiometer or detector evaluates two sources properly. The spectrometer produces a spectrum which is distorted in a number of ways as a result of polarization, selective absorption, instrumental mechanical defects, scattered light, etc., so that the exit beam emerges in an uneven pattern on a detector which, most likely, will have a nonuniformly sensitive surface. The result may be improper evaluation throughout the spectrum.

[3] 3M paint has also been shown to be spectrally flat over extended wavelengths—Ed.

To ensure against erroneous result on account of these factors, either a diffusing block or an integrating sphere has been positioned before the entrance slit, or other opening, of the spectrometer or radiometer. The curves in Fig. 3.14 show what relative responses may be obtained in the near infrared, when employing blocks or spheres coated with certain diffusers, with a 1000-W quartz-iodine lamp, as source, and a PbS cell as detector. Those in Fig. 3.15 illustrate similar data for the ultraviolet, wherein the detector is an ultraviolet-sensitive photomultiplier. In both cases, a Carl Leiss spectrometer was employed; the instrumentation arrangement is depicted in Fig. 3.16.

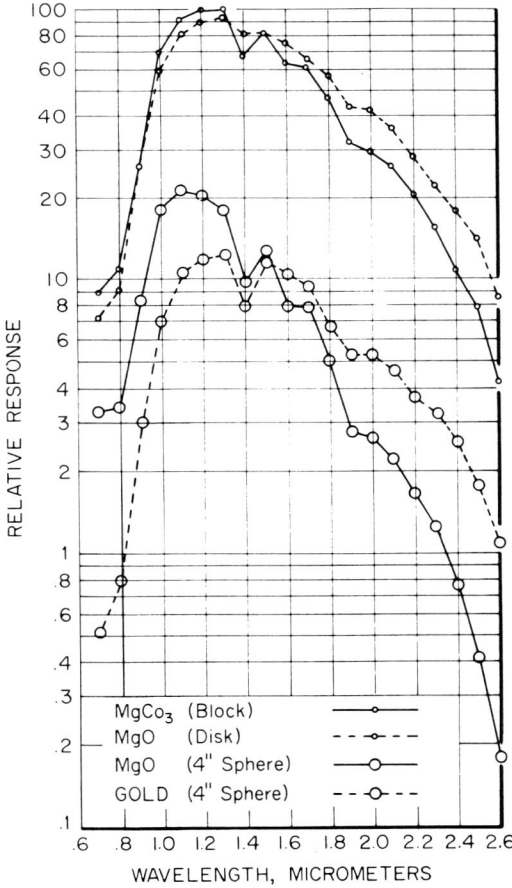

FIG. 3.14. Relative response in the infrared spectrum of four diffusers when used in combination with a Carl Leiss spectroradiometer employing a PbS cell as detector and a 1000-W quartz-iodine lamp as source.

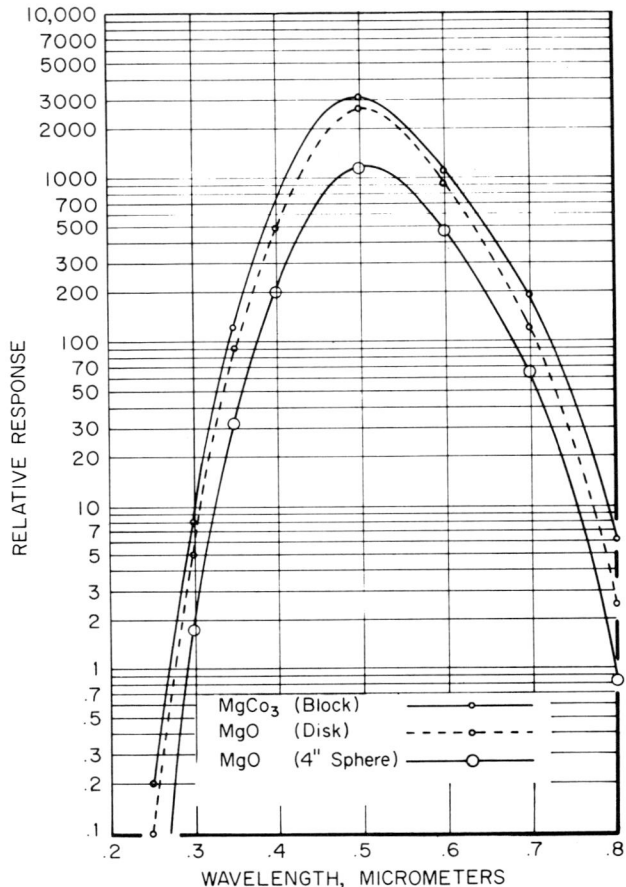

FIG. 3.15. Relative response in the ultraviolet and visible spectrum of three diffusers when used in combination with a Carl Leiss spectroradiometer employing an EMI type 6256B photomultiplier as detector and a 1000-W quartz-iodine lamp as source.

Finally, a photoelectric spectroradiometer is shown, in Fig. 3.17, as employed in an experiment to compare 1000-W standards of spectral irradiance in the infrared spectral region. The diffusing sphere insures that the irradiance from each source falls upon the detector in exactly the same manner—same wavelength, direction, and pattern of energy distribution on the detector surface. Wavelength setting is accomplished through rotation of the filter wheel, which can be manually or automatically operated, either through the use of a motor and geneva-drive mechanism or a TV antenna rotor (Stair et al., 1965c). The signal may be chopped at 510 Hz and amplified by a tuned

3. SOURCES AS RADIOMETRIC STANDARDS 107

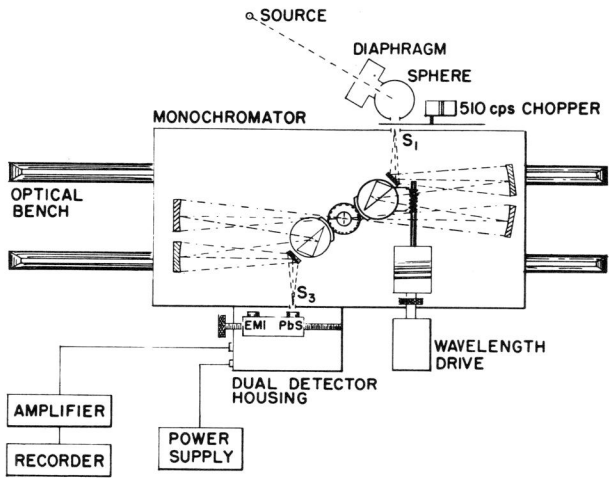

FIG. 3.16. Optical layout of the Carl Leiss monochromator and block diagram of complete double prism spectroradiometer as employed in comparing spectral irradiances.

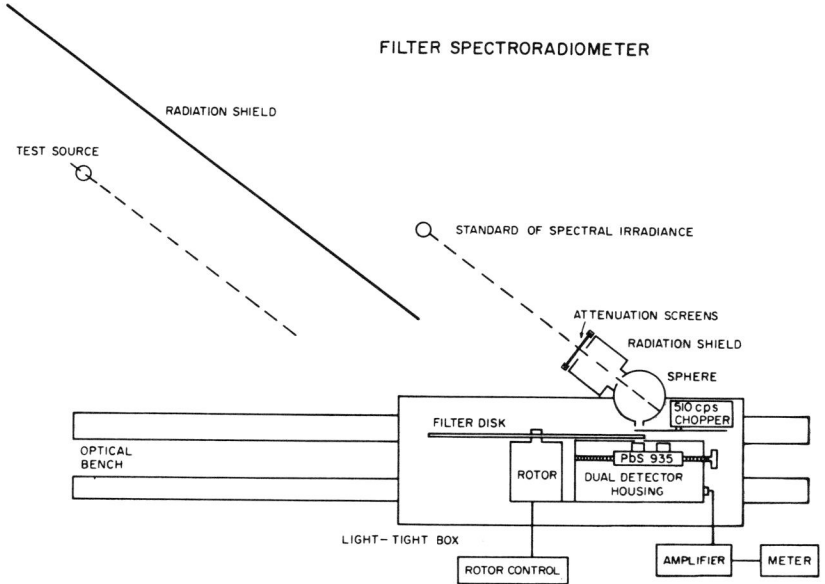

FIG. 3.17. Block diagram of photoelectric filter spectroradiometer. The filter disk and rotor motor, the two photoelectric detectors, the chopper mechanism, and the integrating sphere, with its radiation shield, are in a light tight box which is mounted on an optical bench permitting rapid movement between the standard and test sources.

amplifier (Stair and Johnston, 1956) whose output is fed into a vacuum-tube voltmeter. For use in the laboratory, new and improved Brower lock-in amplifiers are ordinarily employed, with the signal read directly from the amplifier meter or else fed into a strip-chart recorder.

REFERENCES

Allen, C. J., and Paugh, R. L. (1959). Applications of the quartz lighting lamp. *Illum. Eng* **54**, 714.
Coblentz, W. W. (1915). Measurements on standards of radiation in absolute value. *Bull. Bur. Std.* **11**, 87–97.
Condell, W. J., and Byrne, F. T. (1957). Spectral intensity of ac operated tungsten lamps. *J. Opt. Soc. Am.* **47**, 1135–1136.
Drummond, A. J. (1961). Current developments in pyrheliometric techniques. *Solar Energy* **5**, 19–23.
Drummond, A. J. (1970). This volume, Chapter 1.
Engstrom, R. W. (1955). Luminous microflux standard. *Rev. Sci. Instr.*, **26**, 622.
Grum, F. (1963). Modification of a Beckman spectrophotometer for direct measurements of spectral energy distribution of light sources. *Appl. Opt.* **2**, 237–245.
Stair, R., and Smith, W. O. (1943). A tungsten-in-quartz lamp and its applications in photoelectric radiometry. *J. Res. Nat. Bur. Std.* **30**, 449–459.
Stair, R., and Johnston, R. G. (1954). Effects of recent knowledge of atomic constants and of humidity on the calibrations of the National Bureau of Standards thermal-radiation standards. *J. Res. Nat. Bur. Std.* **53**, 211–215.
Stair, R., and Johnston, R. G. (1956). Preliminary spectroradiometric measurements of the solar constant. *J. Res. Nat. Bur Std.* **57**, 205–211.
Stair, R., Johnston, R. G., and Halbach, E. W. (1960). Standard of spectral radiance for the region of 0.25 to 2.6 microns. *J. Res. Nat. Bur. Std.* **64A**, 291—296.
Stair, R., Schneider, W. E., and Jackson, J. K. (1963). A new standard of spectral irradiance. *Appl. Opt.* **2**, 1151–1154.
Stair, R., Schneider, W. E., Waters, W. R., and Jackson, J. K. (1965a). Some factors affecting the sensitivity and spectral response of thermoelectric (radiometric) detectors. *Appl. Opt.* **4**, 703–710.
Stair, R., Fussell, W. B., and Schneider, W. E. (1965b). A standard for extremely low values of spectral irradiance. *Appl. Opt.* **4**, 85—89.
Stair, R., Schneider, W. E., Waters, W. R., Jackson, J. K., and Brown, R. E. (1965c). Some developments in improved methods for the measurement of the spectral irradiances of solar simulators. NASA Contractor Rept. CR–201.
Stair, R., Schneider, W. E., and Fussell, W. B. (1967). The new tungsten-filament lamp standards of total irradiance. *Appl. Opt.* **6**, 101–105.
Stull, V. R., Wyatt, P. J., and Plass, G. N. (1964). The infrared transmittance of carbon dioxide. *Appl. Opt.* **3**, 243–254.
Worthing, A. G. (1917). The true temperature scale of tungsten and its emissive powers at incandescent temperatures. *Phys. Rev.* **10**, 377–390.
Worthing, A. G. (1926). Deviation from Lambert's law and polarization of light emitted by incandescent tungsten, tantalum and molybdenum and changes in the optical constants of tungsten with temperature. *J. Opt. Soc. Am. and Rev. Sci. Instr.* **13**, 635–647.

Wyatt, P. J., Stull V. R., and Plass, G. N. (1964). The infrared transmittance of water vapor. *Appl. Opt.* **3**, 229–241.

Zubler, E. G., and Mosby, F. A. (1959). An iodine incandescent lamp with virtually 100 per cent lumen maintenance. *Illum. Eng.* **54**, 734.

— 4 —

HIGH-ACCURACY SPECTRAL RADIANCE CALIBRATION OF TUNGSTEN-STRIP LAMPS*

H. J. Kostkowski, D. E. Erminy, and A. T. Hattenburg

4.1. Introduction

The capability of determining spectral radiance of tungsten-strip lamps with an uncertainty of less than 1 % has recently been reported (Kostkowski et al., 1965, work to be published). This chapter is a summary of this work, particularly those aspects that might be generally useful in radiometry.

The approach that was used in the spectral radiance calibration was the classical method of comparing the unknown to a blackbody. This method was undertaken because it is the basis of radiation pyrometry where, at least at one wavelength and relative to the International Practical Temperature Scale, considerable success has been achieved (Lee, 1966).

It is generally accepted that the spectral radiance of a blackbody is given by the Planck radiation equation

(4.1) $$L_{b\lambda} = C_1 \lambda^{-5}/[\exp(C_2/\lambda T) - 1]$$

Experimental verification of this equation has not been realized to better than about 5 %. Nevertheless, because of its strong theoretical base, it has been assumed that the above relation is exact. Therefore, a spectral comparison of a source with a blackbody will result in the determination of the spectral radiance of the source with an accuracy which depends on the accuracy of C_1, C_2, λ, and T; the quality of the blackbody; and the accuracy of the comparison.

The initial goal in our effort was a total uncertainty of 1 % or less. Therefore, an attempt was made to limit individual sources of error in terms of spectral radiance to about 0.1 %. The question arises as to how well the above constants are known and how well the other parameters would have to be determined to achieve this criterion.

---

* The research associated with this chapter (which is an official contribution of the U.S. National Bureau of Standards and not subject to copyright) was supported in part by the National Aeronautics and Space Administration.

4.2. Accuracy Requirements

4.2.1. Accuracy of C_1

The first radiation constant C_1 is related to the velocity of light c and the Planck constant h by the equation

(4.2) $$C_1 = 2c^2 h$$

The Committee on Fundamental Constants of the National Academy of Science–National Research Council has recommended (1963) the value $C_1 = 1.19096 \times 10^{-12}$ W cm^2 (sr)$^{-1}$ and estimates an uncertainty in C_1 in terms of one standard deviation of 0.0027 %. Thus, C_1 is reported to be known about 40 times more accurately than required.

4.2.2. Accuracy of C_2

The second radiation constant C_2 is related to the atomic constants by the equation

(4.3) $$C_2 = hc/k$$

where k is the Boltzmann constant. The above mentioned committee has also recommended a value of $C_2 = 1.43879$ cm deg with an uncertainty in terms of one standard deviation of 0.0042 %. This corresponds to an error of 0.045 % for the spectral radiance in Eq. (4.1) at a wavelength of 550 nm and 2500°K, a typical strip-lamp brightness temperature. Thus, C_2 is also known sufficiently well for the 0.1 % criterion.

4.2.3. Wavelength Accuracy

The wavelength accuracy required in comparing a blackbody to a strip lamp can easily be obtained from the equation

(4.4) $$\frac{T_1 - T_2}{T_1 T_2} = \frac{\lambda_2 - \lambda_1}{\lambda_1}\left(\frac{1}{T_1} - \frac{1}{T_c}\right) \simeq \frac{\Delta T}{T_1^2}$$

which is commonly used in optical pyrometry (Kostkowski and Lee, 1962) to relate the brightness temperature T_1 at a wavelength λ_1 to the brightness temperature T_2 at another wavelength λ_2, where T_c is the color temperature as defined in the reference given above. Using Wien's equation as an approximation to Planck's equation, i.e., dropping the -1 in Eq. (4.1) and differentiating with respect to T, one obtains

(4.5) $$\frac{dL_{b\lambda}}{L_{b\lambda}} = \frac{C_2}{\lambda T}\frac{dT}{T}$$

Combining this with Eq. (4.4)

$$(4.6) \qquad \frac{\Delta L_{b\lambda}}{L_{b\lambda}} \cong \frac{C_2}{\lambda_2}\left(\frac{1}{T_1} - \frac{1}{T_c}\right)\Delta\lambda$$

Assuming $\Delta L/L = 0.001$, $T_1 = 2500°K$, $\lambda = 550$ nm, and using $T_c \simeq 2860°K$ (obtained for tungsten from the " Handbook of Physics and Chemistry "), one obtains $\Delta\lambda \simeq 0.5$ nm. This then is the wavelength uncertainty permitted at 550 nm in spectrally comparing the blackbody to a tungsten-strip lamp.

4.2.4. Accuracy of Blackbody Temperature

The uncertainty permitted for the temperature of the blackbody can be obtained from Eq. (4.5). Solving for ΔT for $\lambda = 550$ nm, $\Delta L/L = 0.001$, and $T = 2500°K$, results in $\Delta T = 0.24°$.

4.2.5. Blackbody Quality

To satisfy our criterion, the blackbody must have an emissivity of about 0.999 and a stability of 0.1 %. A rough estimate of the permissible temperature nonuniformity can be obtained as follows. Assume that the emissivity of the cavity material is 0.9 and the walls of the cavity have either a temperature T or $T - \Delta T$, with each irradiating the wall target area equally as illustrated in Fig. 4.1. To a first approximation

$$L_\lambda(_{target}^{wall}) \cong 0.9 L_{b\lambda}(T) + 0.1 \times 0.5 L_{b\lambda}(T) + 0.1 \times 0.5 L_{b\lambda}(T - \Delta T)$$

$$\cong 0.95 L_{b\lambda}(T) + 0.05[L_{b\lambda}(T) - \frac{\Delta L_{b\lambda}}{\Delta T}\Delta T]$$

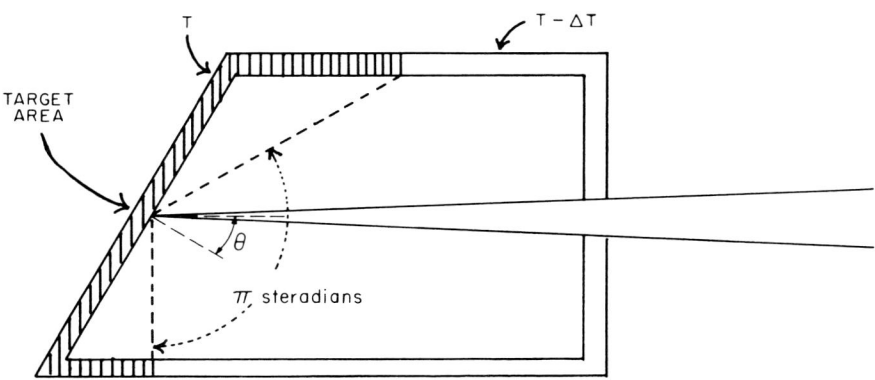

FIG. 4.1. Cavity with nonuniform wall temperature.

and using Eq. (4.3)

$$L_\lambda(^{wall}_{target}) \cong 0.95 L_{b\lambda}(T) + 0.05 \left[L_{b\lambda}(T) - \frac{C_2}{\lambda T^2} L_{b\lambda}(T) \Delta T \right]$$

$$\cong L_{b\lambda}(T) \left[1 - 0.05 \frac{C_2}{\lambda T^2} \Delta T \right]$$

(4.7) $$\frac{L_\lambda(^{wall}_{target})}{L_{b\lambda}(T)} = \varepsilon_{\text{effective}} = 1 - 0.05 \frac{C_2}{\lambda T^2} \Delta T$$

assuming $\varepsilon_{\text{effective}} = 0.999$, $T = 2500°K$, and $\lambda = 550$ nm, $\Delta T \simeq 5°$.

4.2.6. Accuracy of Comparison

The accuracy of the blackbody-strip lamp comparison must be 0.999 and L_λ should not drift by more than 0.1 % in the length of time required for the calibration and its application.

These then are the general requirements. The remainder of the chapter will consider the design and techniques utilized in trying to realize these requirements and the results that have been achieved.

4.3. THE BLACKBODY

4.3.1. Description

The blackbody used for this work consisted of a graphite tube resistively heated in an argon atmosphere. The major features of the blackbody and control system are shown in Figs. 4.2 and 4.3, respectively. With this control system, it has been possible to limit the instability and drift to about 0.1 % over a 30- to 45-min period. This is sufficient time in which to determine the blackbody temperature, make a comparison at some wavelength, and again check the temperature.

4.3.2. Temperature Uniformity

Limiting the temperature nonuniformity to about 5° was achieved by modifying the shape of the graphite tube until, when sighting with a visual optical pyrometer at the cavity hole on the side or at the various holes along the axis of the tube from the top, this condition was confirmed.

4.3.3. Emissivity

The emissivity of the blackbody cavity depends on the geometry and dimensions of the cavity and on the partial reflectivity (deVos, 1954) of the material, particularly the wall opposite the cavity sighting hole. To a first approximation, which is usually quite good in high quality cavities,

(4.8) $$\varepsilon = 1 - r\Omega$$

where r is the partial reflectivity and Ω is the solid angle originating at the wall target and subtended by the cavity opening.

The partial reflectivity of the wall target, relative to an incident and reflected angle θ (as shown in Fig. 4.1), was minimized by having the wall shaped like a coarse thread. Laboratory measurements of the room temperature partial reflectivity, which could easily be performed with a visual pyrometer, resulted in a value of 0.02 (sr)$^{-1}$. Using Eq. (4.8), an $\varepsilon = 0.999$, and $r = 0.02$ (sr)$^{-1}$, the required Ω is 0.033 sr. Thus, since $\Omega \simeq \pi R^2/D^2$ and the depth of the cavity D is about 1 mm, the radius of the cavity hole R required to realize an emissivity of 0.999 is 1 mm. This is the size of the cavity opening adopted and usually used.

A more direct measurement was employed to confirm the 0.999 emissivity. The method consists of making changes in Ω and determining the relative change of spectral radiance at two widely spaced wavelengths, say 325 and 650 nm. A convenient equation (Kostkowski, 1966) to use in predicting the change is

(4.9) $$\frac{L_{\lambda'}^{(2)}}{L_{\lambda'}^{(1)}} = \left(\frac{\varepsilon_1(\lambda)\tau_1(\lambda)}{\varepsilon_2(\lambda)\tau_2(\lambda)}\right)^{\lambda/\lambda'} \bigg/ \left(\frac{\varepsilon_1(\lambda')\tau_1(\lambda')}{\varepsilon_2(\lambda')\tau_2(\lambda')}\right)$$

Equation (4.9) gives the ratio of spectral radiance of source 1 to that of source 2 at a wavelength λ', when the two sources are adjusted to have the same spectral radiance at wavelength λ. The ε's are the spectral emissivities of the emitters and the τ's are transmittances of any windows surrounding the emitters. If there are no windows and if we assume that the sources are gray bodies, Eq. (4.9) becomes

(4.10) $$\frac{L_{\lambda'}^{(2)}}{L_{\lambda'}^{(1)}} = \left(\frac{\varepsilon_1}{\varepsilon_2}\right)^{\lambda/\lambda'} \bigg/ \left(\frac{\varepsilon_1}{\varepsilon_2}\right) = \left(\frac{\varepsilon_1}{\varepsilon_2}\right)^{(\lambda-\lambda')/\lambda'}$$

Thus, if it is assumed that the graphite cavity is at least a gray body, the ratio of the spectral radiances at λ' of two such cavities, with exiting angles of Ω_1 and Ω_2, is

$$\frac{L_{\lambda'}^{(2)}}{L_{\lambda'}^{(1)}} = \left(\frac{1-r\Omega_1}{1-r\Omega_2}\right)^{(\lambda-\lambda')/\lambda'}$$

provided that the power input is adjusted so that their spectral radiances are equal at wavelength λ. Determination of this ratio for measured values of Ω_1 and Ω_2 thus permits a determination of the partial reflectivity. These measurements were performed, and within the total precision available (Std $\sim 0.3\%$) the partial reflectivity was determined to be 0.02 ± 0.02 $(\text{sr})^{-1}$. An $r = 0.04$ $(\text{sr})^{-1}$, together with a 1-mm radius opening, would result in a cavity emissivity of 0.9987.

4.3.4. Absorption

Gas absorption is another factor that could affect the blackbody quality of the radiation. The radiation leaving the cavity hole could be partially absorbed by gases in the vicinity of the hole but at a lower temperature than the cavity walls. The major gases of concern are C, C_2, and C_3, it being determined from the literature that the concentration of C_4, C_5, and the higher polyatomics would be significantly less (Drowart *et al.*, 1959). Other molecules such as O_2, N_2, and CN should not be a problem as a result of the argon flushing.

From the energy level diagram of atomic carbon (Herzberg, 1944) and the selection rules governing atomic transitions, one would not expect any carbon lines to appear in absorption within our spectral region of 200 to 850 nm and temperatures of 3000°K or below. Nevertheless, a spectral search was made for the strong 247.8 nm line and the forbidden line, at 462.1 nm, using narrow slits. No evidence could be found of the presence of either line.

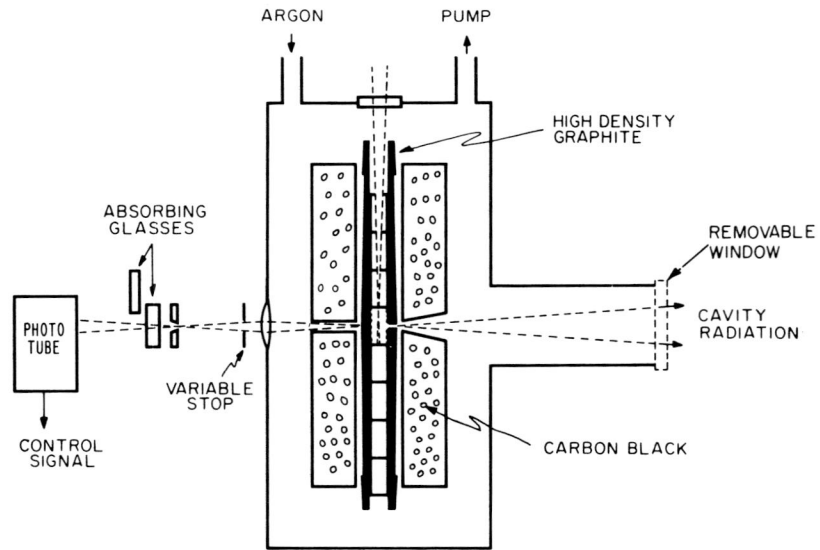

Fig. 4.2. High-temperature blackbody.

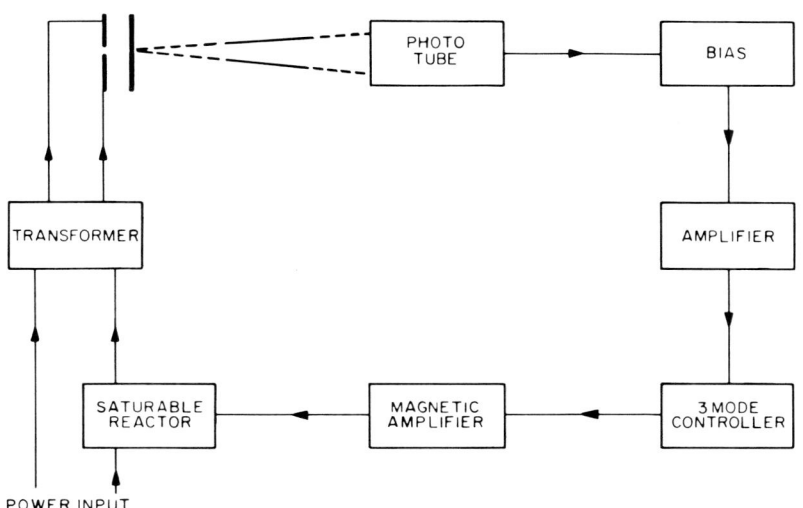

FIG. 4.3. High-temperature blackbody control system.

The molecule C_2 has very strong spectral bands in the blue and green referred to as the Swan bands (Herzberg, 1950; Clementi, 1960). If a significant portion of the cavity radiation was being absorbed by this molecule, absorption lines characteristic of this band would be detectable in a spectral scan of the cavity radiation. No such absorption spectra could be detected.

Little is known about the energy levels of spectra of C_3. However, there is a broad, largely continuous band in the neighborhood of 400 nm that has been attributed to C_3 (Kreis and Broida, 1956). Such a band, especially if it were present to the extent of 1 or 2 % in absorption, would not be easily noted. The following sensitive technique was evolved for determining the possible presence of this band.

The relative emissivity of a tungsten ribbon, at least over 20 or 30 nm, appears to be well known. For example, the relative emissivity between 430 and 400 nm is the same within about 0.2 % whether calculated from deVos' (1954) or Larrabee's (1959) published values. Therefore, if a strip lamp is adjusted to have the same spectral radiance at 430 nm as the cavity, the ratio of spectral radiances at 400 nm can be calculated from Eq. (4.6) and should agree with an experimentally determined ratio if the cavity radiation at 400 nm has not been partially absorbed. The calculations and the experimental measurements were performed and the two differed by 0.2 % with a total uncertainty of 0.3 %. Thus, it is unlikely that C_3 absorption at 400 nm is a significant factor in affecting the blackbody quality.

The only absorption lines present in the cavity radiation throughout the spectral range 200–850 nm were those of sodium, at 589.0 and 589.6 nm, and of potassium at 766.5 nm. These are probably produced by impurities in the graphite. Though rather weak, the blackbody cavity should not be used near these wavelengths.

As a result of the various calculations and experimental measurements described, it is believed that the graphite high-temperature blackbody largely satisfies the initial requirements.

4.4. The Spectroradiometer

4.4.1. Design Criteria

The uncertainty of an NBS calibration of a commercial visual optical pyrometer at 2500°K is about 5° relative to the International Practical Temperature Scale, IPTS (Kostkowski and Lee, 1962). The smallest uncertainty that has been reported for a calibration at 2500°K is 0.6°, and this is for the specially designed NBS photoelectric pyrometer (Lee, 1966). Thus, it appears that the measurement of the blackbody temperature will be the major uncertainty in the spectral radiance determination, and that the instrumentation should be designed to minimize this uncertainty. As a result, the following measures were taken.

(1) Use a photoelectric pyrometer for the temperature measurement.

(2) Design the photoelectric pyrometer as an integral part of the spectroradiometer utilizing similar optics, detectors, target areas, etc. This should minimize any errors in transferring the temperature scale to the blackbody.

(3) Calibrate the pyrometer portion of the spectroradiometer from basic principles, i.e., a primary calibration.

4.4.2. Description

Figure 4.4 shows a block diagram of the spectroradiometer that has been developed. At 654.57 nm (the wavelength of a convenient spectral line in thorium) it is used as a photoelectric pyrometer. A tungsten-strip, vacuum pyrometer lamp, selected for maximum stability, is used as the equivalent of the small lamp in an ordinary pyrometer. Absorbing glasses are employed for temperatures above about 1320°C, as in an ordinary pyrometer. The pyrometer lamp is compared to the blackbody by alternately positioning the two sources on the entrance slit of the monochromator.

In front of the entrance slit is one of two metal slit masks, limiting the slit height to either 0.2 or 0.8 mm. The masks are engraved with two perpendicular

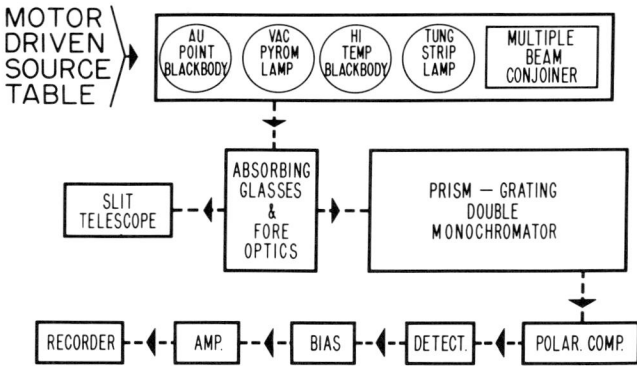

FIG. 4.4. Spectroradiometer.

scales having 0.1 mm divisions. The vacuum pyrometer lamp and other sources are placed in double gimbal mounts providing 6 degrees of freedom with micrometer-type motion. With the aid of the slit telescope, it is thus possible to position or to investigate the effects of changing position or orientation of the sources, to about 0.02 mm or a few minutes of arc. The gradients present in some lamps make this capability highly desirable.

A double monochromator is used to ensure that scattered radiation from wavelengths other than that of interest will be sufficiently small, so as not to introduce a significant uncertainty in comparing sources having different spectral distributions. This is equivalent to keeping the wings of the slit function small. The prism-grating rather than a prism-prism type double monochromator was selected in order to have greater dispersion and greater wavelength reproducibility.

The electronics shown in Fig. 4.4 is that with which, through pyrometry, we have had considerable success in high-precision ($<0.1\%$), low-level current ($<10^{-8}$ A) measurements. The radiation is not chopped and a dc amplifier is used. With the biasing system employed, 0.1 in. on a recorder chart is equivalent to 0.2% spectral radiance. Generally, we are able by integrating over 1–2 in. on the recorder chart to obtain a precision in terms of a standard deviation of 0.05 to 0.15%, depending on the wavelength and temperature, when comparing a lamp to the blackbody. Since the two radiations are adjusted to be almost identical, i.e., we have essentially a null-type instrument, no errors arise due to nonlinearity of the photomultiplier or electronics. The photomultiplier has an S20 response and a quartz window. When it is used at a current level of about 10^{-8} A and is in operation for about 45 min, fatigue produces at most a very small linear change with time which has a negligible effect on the measurements ($<0.05\%$).

4.5. Temperature Calibration

Calibrating the pyrometer portion of the spectroradiometer from basic principles (a so-called primary calibration) means performing experiments that reflect the definition of the IPTS. The IPTS above the melting point of gold is defined (Stimson, 1961) in terms of the ratio of two blackbody radiances, one being at the temperature of equilibrium between liquid and solid gold (gold point) and the other at the unknown temperature T. Mathematically,

$$(4.11) \qquad \frac{L_{b\lambda}(t)}{L_{b\lambda}(t_{Au})} = \frac{\exp[C_2/\lambda(t_{Au} + T_0)] - 1}{\exp[C_2/\lambda(t + T_0)] - 1} = \mathscr{R}$$

where $t_{Au} \equiv 1063°C$, $T_0 \equiv 273{,}15°$, and $C_2 \equiv 1.438$ cm deg are defined values.

4.5.1. Low-Range Calibration

The low-range calibration consists of determining electrical currents for the vacuum pyrometer lamp, which correspond to spectral radiance ratios of the lamp relative to the gold point. These ratios, and the wavelengths associated with them, can then be used to calculate blackbody or brightness temperatures from Eq. (4.11). The brightness temperature is usually limited to about 1300 to 1350°C in order to reduce lamp drift.

The starting point for such a low-range calibration is a gold-point blackbody. The one used is shown in Fig. 4.5 and was developed for calibrating the NBS photoelectric pyrometer (Lee, 1966). Various tests and calculations previously reported (loc. cit.) indicate that the emitted radiation in the vicinity of 654.5 nm is within 0.02 % of that from a perfect blackbody at the temperature of freezing gold. The gold-point radiance could be transferred to the vacuum pyrometer lamp; i.e., a ratio of one, with a standard deviation uncertainty of about 0.05 %. However, in this transfer, a correction is necessary due to the effect of scattered radiation (originating outside of the target area) when comparing two sources of significantly different sizes. The gold-point blackbody has an effective radiating diameter of 25 mm and the strip lamp a similar area of about 1.3 × 30 mm. The magnitude of this effect was determined by performing the complementary experiment. A hole in an illuminated diffuse surface was imaged on the slit of the spectroradiometer. For any particular condition of the entrance optics, the radiation detected depends on the size and shape of the illuminated surface. This was changed by placing black velvet paper in front of the illuminated surface simulating the gold-point blackbody and the strip lamp. The correction varied from about 0.05 to 0.3 % in spectral radiance, depending primarily on the age and condition of the coating on the first mirror of the spectroradiometer optical system.

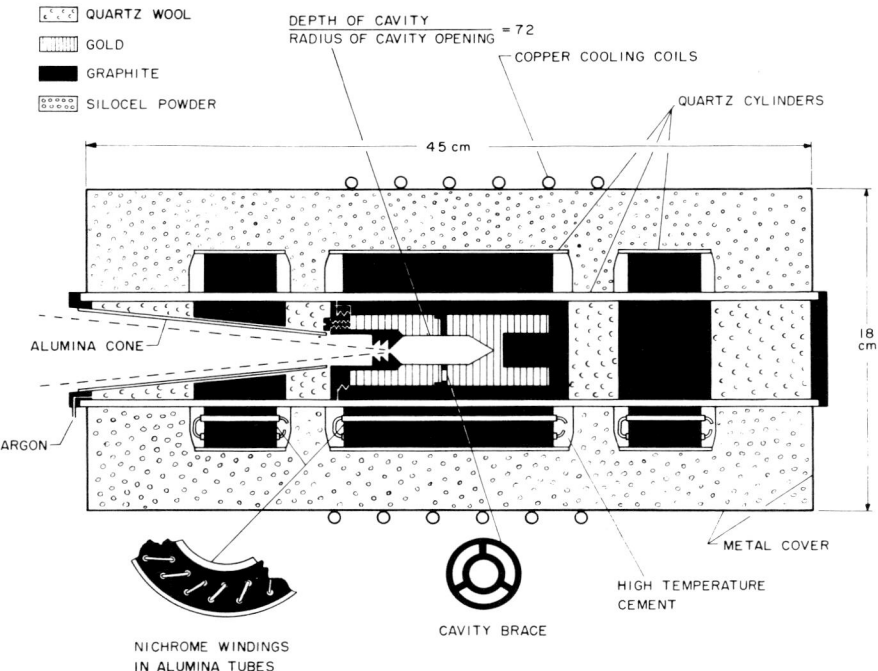

Fig. 4.5. Gold-point blackbody.

The wavelength in Eq. (4.11) is required to be known to at least 2.5 Å (0.25 nm) if 0.12° (0.1 % in L_λ) accuracy is desired at about 1300°C. From the theory of optical pyrometry (Kostkowski and Lee, 1962), the wavelength in Eq. (4.11) is the mean effective wavelength between temperatures t_{Au} and t, designated $\lambda_{t_{\text{Au}}-t}$, where

(4.12) $$\frac{1}{\lambda_{t_{\text{Au}}-t}} \cong \frac{1}{2}\left(\frac{1}{\lambda_t} + \frac{1}{\lambda_{t_{\text{Au}}}}\right)$$

and

(4.13) $$\frac{1}{\lambda_t} = \frac{\int [L_{b\lambda}(t)\sigma(\lambda_0 \lambda) R(\lambda)/\lambda]\, d\lambda}{\int L_{b\lambda}(t)\sigma(\lambda_0 \lambda) R(\lambda)\, d\lambda}$$

The wavelength λ_t is called the effective wavelength, $\sigma(\lambda_0 \lambda)$ is the slit or apparatus function, and $R(\lambda)$ the relative spectral response of the entire system. The slit function was experimentally determined by spectrally scanning over a narrow emission line. The relative response $R(\lambda)$ was obtained by spectrally scanning the positive crater of a low-current graphite arc, the spectral radiance of which is largely equivalent to a 3800°K blackbody, from about 640 to 670 nm with narrow slits. The relative response is then the spectroradiometer output divided by the product of the arc spectral radiance (Hattenburg, 1967) and the spectral slit width of the monochromator.

With the 0.6 mm slit width (2.4 nm spectral width at 650 nm) normally used, the mean effective wavelength differed from the wavelength setting of the monochromator when using a narrow slit and spectral line by only 0.01 nm with an uncertainty of about 0.05 nm. It was quite adequate, therefore, when realizing the low range of the IPTS, to use the wavelength corresponding to the peak of a narrow spectral line determined with a narrow slit. The maximum uncertainty due to mean effective wavelength in the calibration is estimated to be about 0.03°.

The ratio of spectral radiance $\mathscr{R}$ in Eq. (4.11) was obtained with the beam conjoiner shown in Fig. 4.6 (Erminy, 1963). By adjusting each source in the conjoiner to independently match the gold-point blackbody, all three sources together would represent an $\mathscr{R}$ of 3. By adjusting two sources to be equal to each other and together equal to three times the gold point, each would represent an $\mathscr{R} = 3/2$. In this manner, many different integral and fractional values of $\mathscr{R}$ were obtained without any assumption of linearity of the photomultiplier or electronic system.

Spectral radiance ratios relative to the gold point of $\frac{3}{4}$, 1, $\frac{3}{2}$, 2, 3, 4, 6, 8, 12, and 14 were realized on the vacuum pyrometer lamp as a function of the lamp current. Using these $\mathscr{R}$'s and the correct mean effective wavelength, Eq. (4.11) was used to determine the corresponding blackbody temperatures. A fourth-order equation of lamp current, as a function of temperature, was then fitted by a least squares criterion to the resulting ten temperatures and corresponding currents. The standard deviation of the fit was about 0.05°.

Three low-range calibrations have been performed to date, the second after 260 hr of pyrometer lamp burning time and the third after 710 hr. The first and second calibrations differed by a constant of 0.83°, with a standard

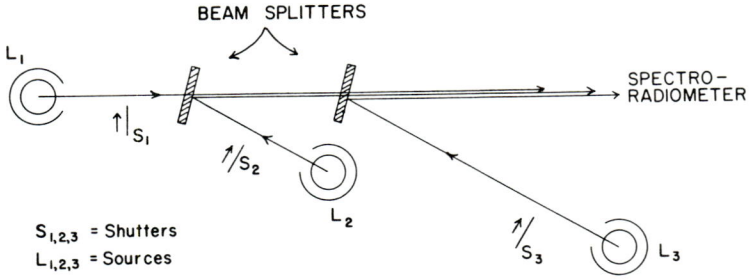

Fig. 4.6. Beam conjoiner.

deviation of the differences of the unfitted data of 0.061°; and the second and third differed by 0.91° with a standard deviation of 0.04°. In the 710 hr in which this lamp has been used, its calibration has changed by the same amount (in temperature) at all points in the low range. This makes it possible to check the entire low range with only one ratio determination, e.g., the gold point itself.

4.5.2. High-Range Calibration

The upper ranges of the pyrometer were calibrated by determining transmittances for the individual absorbing glasses using the low-range calibration. This was accomplished by matching a vacuum lamp at a brightness temperature of about 1320°C, observed through an absorbing glass, with a vacuum lamp observed directly, so that

(4.14) $$\tau(\lambda) L_{b\lambda}(T_H) = L_{b\lambda}(T_L)$$

Using Wien's equation, which has an error of only about 0.0001 % for the maximum value of λT used in the low range, one obtains

(4.15) $$\tau(\lambda) = \exp -(C_2/\lambda)(1/T_L - 1/T_H) = \exp -C_2 A/\lambda$$

where A is the so-called "A" value of the absorbing glass.

In pyrometry, absorbing glasses are usually selected which have an exponential transmittance, i.e.,

(4.16) $$\tau(\lambda) = \exp -(k/\lambda)$$

In this case, the A value is independent of the mean effective wavelength and a constant with respect to the temperature of the source that the pyrometer is observing. Thus, an A value determined in the low range may be used at higher temperatures. Also from Eq. (4.15), it can be seen that two or three absorbing glasses can be used in series; and, neglecting reflection, the resulting A value is the sum of the individual A values.

Using the above principles and correcting for reflectances, which were experimentally determined (~ 0.16 % correction in transmittance for two absorbing glasses), the A values were obtained for groups of one, two, and three absorbing glasses; and the higher-range calibrations, consisting of these A values and the low-range calibration, were thus obtained.

One additional point, the glass usually used in pyrometers for absorbing glasses is Corning Pyrometer Brown. This glass was found to have a temperature coefficient, transmittance decreasing with an increase in glass temperature, making it difficult to use with the precision desired. Another glass was found, however, Jena (Schott) NG-3, which also had an exponential transmittance but had a coefficient with the opposite sign. The two glasses were combined in a composite filter, with their thicknesses selected so that the temperature coefficient was negligible.

The pyrometer portion of the spectroradiometer has been calibrated at 2500°K with an estimated standard deviation uncertainty of about 0.6° relative to the IPTS, or the same uncertainty reported (loc. cit.) for the calibration of the NBS photoelectric pyrometer.

4.5.3. Correction to the Thermodynamic Scale

In order to obtain absolute values of spectral radiance, it is necessary to use the Thermodynamic Kelvin Temperature Scale (TKTS) rather than the IPTS. Table I gives the estimated differences between these scales for temperatures in our range of interest. In applying these differences as corrections to the IPTS, an additional uncertainty of the difference, about 0.9° at 2500°K, is incurred. The 0.9 and 0.6° uncertainties are independent, so that the total uncertainty on the TKTS is the square root of the sum of the squares of these individual standard deviation uncertainties or about 1.1°.

As a result of an extensive program in gas thermometry now in progress at NBS, the TKTS correction uncertainty is expected to be reduced by a factor of at least five in a few years. Thus, though the best currently available accuracy of a temperature measurement on the TKTS at 2500°K is about $4\frac{1}{2}$ times poorer than our criterion, it is only about $2\frac{1}{2}$ times poorer on the IPTS and there are high expectations that, in the near future, the correction to the TKTS will introduce an insignificant additional uncertainty.

TABLE I. Estimated differences between the Thermodynamic Kelvin Temperature Scale and the International Practical Temperature Scale.[a]

T(TKTS)	T(TKTS) − T(IPTS)
2100°K	2.9°
2200	3.1
2300	3.3
2400	3.5
2500	3.8
2600	4.0
2700	4.3
2800	4.6
2900	4.8
3000	5.1

[a] Terrien and Preston-Thomas, 1967.

4.6. Comparison of Blackbody and Tungsten-Strip Lamp

Many of the features required for using the spectroradiometer as an accurate pyrometer, at 654.7 nm, are obviously equally useful at other wavelengths where the high-temperature blackbody and a gas-filled tungsten-strip lamp are to be compared. These include rapid precise positioning of the sources on the slit mask, low scattering of radiation originating outside of the target area, and accurate wavelength determination. The insignificant spectral scattering, resulting from the use of a double rather than a single monochromator, is even more important at other wavelengths, particularly at the extremes of the wavelength range of interest, where either the photomultiplier response or the radiance is very low and scattered radiation constitutes a larger percentage of the spectroradiometer output.

The major uncertainty in the comparison of the blackbody and lamp is the lamp itself. The best gas-filled lamps observed drift at a rate of about 0.002 % per hour at a 2500°K brightness temperature in the red. However, some lamps have been observed to drift by a factor of 100 greater. Thus, only the best lamps available are stable enough not to limit the accuracy of the calibration. Even then their burning time should be kept to a minimum.

The gas-filled strip lamps normally used in these calibrations are manufactured by the General Electric Company (sole U.S. supplier) and are designated ultraviolet spectrum lamps. The tungsten-strip filament is about 3 mm wide and 30–50 mm long and has a small notch at about its midpoint. There is a quartz window permitting use down to about 200 nm. For the temperatures required for the ultraviolet spectral radiance calibrations, a direct current and voltage of about 40 A and 7–12 V is typical.

Before calibration, the optimum orientation for the lamp is determined. The radiation from the target area is very noisy and even oscillatory unless the lamp filament is oriented close to vertical. This is believed to be caused by convection of the argon gas cyclically cooling the filament. When the most favorable orientation is determined, an arrow is etched on the rear surface of the lamp; a line through the center of the notch and tip of the arrow serves as a reference axis. This axis is usually made horizontal and is also approximately the optic axis of the spectroradiometer, thereby insuring not only a reproducible orientation but also a reproducible direction of sighting on the lamp.

The reproducibility of the blackbody and tungsten-strip lamp comparison, using the lamp and blackbody as described, has a standard deviation ranging from about 0.05 % at 650 nm to 0.15 % at 300 nm and 0.3 % at 210 nm.

If the strip-lamp radiation and the response of the spectroradiometer are both polarized, an error will be produced in the comparison.

The spectroradiometer has a polarization $[(R_{max} - R_{min})/(R_{max} + R_{min})]$ of about 0.20, with a maximum response in a direction parallel to the slit at 210 nm; 0.0 at 500 nm; and about 0.15 at 850 nm, with a maximum in a direction perpendicular to the slit. At intermediate wavelengths, the value of the polarization varies approximately linearly with wavelength. The absence of polarization at 500 nm is brought about by the compensator shown in Fig. 4.4. It consists of two plates of calcium fluoride placed between the exit slit and the detector and so oriented that the response at 500 nm is independent of the polarization of the incident radiation.

In order to minimize the error due to the polarization, only lamps with a polarization of 1 % or less, as determined at 500 nm, are used. Under these conditions, the effect of polarization is small and can be corrected with sufficient accuracy so that the uncertainty in the comparison due to this factor is, at most, about 0.1 %.

4.7. Summary of Uncertainties

Table II summarizes the estimated uncertainties, in terms of per cent of spectral radiance, with which the best available tungsten-strip lamps have been calibrated. The various sources of uncertainty are listed separately. The uncertainties are given in terms of one standard deviation and combined in quadrature to give a total standard deviation uncertainty.

The reduction in the uncertainty of the correction to the TKTS by about five, expected in a few years, will result in lowering the total uncertainties by about 30 %. The probability of further improvement in the accuracy of spectral radiance determinations, in the wavelength range 210–850 nm, is

TABLE II. Estimated uncertainties, in per cent of spectral radiance, of calibrating a gas-filled tungsten-strip lamp.

Source of uncertainty	Wavelength (nm)			
	200	300	650	850
Blackbody quality (for $\varepsilon = 0.999$)	0.2	0.1	0	0.1
IPTS temperature	1.0	0.5	0.2	0.2
Correction to TKTS	1.0	0.7	0.3	0.2
Transfer from blackbody to lamp	0.3	0.2	0.1	0.1
Total standard deviation uncertainty	1.5	0.9	0.4	0.3

high if the stability of vacuum pyrometer lamps or some other conveniently usable reference source is improved. Otherwise, the probability is small. At any rate, without a corresponding improvement in the stability of the sources to be calibrated (i.e., the gas-filled tungsten-strip lamps) any increase in calibration accuracy is of limited value.

REFERENCES

—— (1963). New values for the physical constants—recommended by NAS–NRC. *Nat. Bur. Std. U.S. Tech. News. Bull.* **47**, 175–177.

Clementi, E. (1960). Transition probabilities for low-lying electronic states in C_2. *Astrophys. J.* **132**, 898–904.

deVos, J. C. (1954). Evaluation of the quality of a blackbody. *Physica*, **20**, 669–689.

de Vos, J. C. (1954). A new determination of the emissivity of tungsten ribbon. *Physica* **20**, 690–714.

Drowart, J., Burns, R. P., DeMaria, G., and Inghram, M. G. (1959). Mass spectrometric study of carbon vapor. *J. Chem. Phys.* **31**, 1131–1132.

Erminy, D. E. (1963). Scheme for obtaining integral and fractional multiples of a given radiance. *J. Opt. Soc. Am.* **53**, 1448–1449.

Hattenburg, A. T. (1967). Spectral radiance of a low-current graphite arc, *J. Appl. Opt.* **6**, 95–100.

Herzberg, G. (1944). "Atomic Spectra and Atomic Structure." 257 pp. Dover, New York.

Herzberg, G. (1950). "Molecular Spectra and Molecular Structure." 658 pp. Van Nostrand, Princeton, New Jersey.

Kostkowski, H. J. (1966) A new radiometric equation and its application. *J. Appl. Opt.* **5**, 1959.

Kostkowski, H. J., Erminy, D. E., and Hattenburg, A. T. (1965). Accuracy of spectral radiance calibrations increased. *Nat. Bur. Std. U.S. Tech. News Bull.* **49**, 14.

Kostkowski, H. J., Erminy, D. E., and Hattenburg, A. T. Higher accuracy spectral radiance measurements. *J. Res. Nat. Bur. Std.* (to be published).

Kostkowski, H. J., and Lee R. D. (1962). Theory and methods of optical pyrometry. *In* "Temperature, Its Measurement and Control in Science and Industry," Vol. 3, Part I (F. G. Brickwedde, ed.), pp. 449–481. Reinhold, New York.

Kreis, N. H., and Broida, H. P. (1956). Spectrum of the C_3 molecule between 3600 Å and 4200 Å. *Can. J. Phys.* **34**, 1471–1479.

Larrabee, R. D. (1959). The spectral emissivity and optical properties of tungsten. *J. Opt. Soc. Am.* **49**, 619–625.

Lee, R. D. (1966). The NBS photoelectric pyrometer and its use in realizing the International Practical Temperature Scale above 1063°C. *Metrologia* **2**, 150–162.

Stimson, H. F. (1961). International Practical Temperature Scale of 1948; Text revision of 1960. *J. Res. Nat. Bur. Std.* **65A**, 139–145.

Terrien, J., and Preston-Thomas, H. (1967). Progress in the definition and in the measurement of temperature. *Metrologia* **3**, 29–31.

— 5 —
NEW INFRARED DETECTORS

E. H. Putley

5.1. Introduction

A few words about the type of infrared applications with which we are concerned at the Royal Radar Establishment will help understanding my approach to infrared detectors. For many years now, one of our principal interests has been the use of thermal radiation for military purposes, both for the detection and location of hot or warm objects and the use of the emission from objects near room temperature for thermal mapping (thermography). For these applications, the region of the spectrum from the visible out to about 15 μm is the most useful. Atmospheric absorption does not permit us to employ the whole of this spectrum, but concentrates our interest on certain windows, mainly near 2 μm, near 4–5 μm, and at 8–14 μm. Longer wavelengths are, of course, of interest in spectroscopy, but most terrestrial applications are prevented by the high atmospheric absorption until the millimeter region is reached. One possible exception to this is the partial window at 337 μm.

Our interests in the infrared are not confined to military applications. In addition to an interest in solid-state spectroscopy, we have also studied the application of far infrared techniques to plasma diagnostics in controlled thermonuclear research (Kimmitt et al., 1965). This work took our interests into the submillimeter and millimeter regions of the spectrum. Thus, our various interests have taken us across the whole IR spectrum from the visible to millimeter waves.

One common requirement that these applications have shared has been the need to develop infrared detectors with suitable characteristics. In most cases we have needed fast, sensitive detectors, so that one of our main activities has been the development of photoconductive detectors, which are now available for the whole of the IR spectrum.

It is convenient to classify infrared detectors as thermal detectors or as photoconductive ones. In thermal detectors, the absorbed radiation raises the temperature of the detector, so changing some temperature-dependent property of it. In photoconductive detectors, the absorbed radiation induces electronic transitions which lead directly to a change in electrical conductivity. Workers in precision thermal radiometry will be very familiar with several types of thermal detector, but they may not be so familiar with the photoconductive ones. It is hoped that this contribution will cover the development of the photoconductive detector as well as mentioning some of the more recent developments of thermal detectors.

When we are choosing a detector for a particular application, the main factors we must consider are:

(1) spectral response,
(2) sensitivity,
(3) speed of response, and
(4) convenience in use.

Let us first quickly compare photoconductive and thermal detectors on these four counts.

(1) Thermal detectors have a much broader spectral response than photoconductive ones, although the spectral response of the latter is usually fairly broad on the short wavelength side of a threshold characteristic of the particular photoconductor.

(2) The sensitivity of the best photoconductive detectors (for $\lambda < 100~\mu$m, at least) is better than that of the best thermal detectors. As we shall see when we discuss the performance of an ideal detector, this is a consequence of the restricted spectral response of the photoconductive detector.

(3) The response time of thermal detectors is usually longer than that of photoconductive detectors, typical values being 1 sec–1 msec for the thermal detectors and 10–0.1 μsec for photoconductive detectors, although particular detectors of either type may lie outside these limits.

(4) In some applications, ruggedness and ease of use are more important than sensitivity. It is probably true that the most sensitive thermal detectors are more delicate than the correspondingly sensitive photoconductive ones, but one disadvantage of photoconductive detectors is that they often require cooling below room temperature, and for wavelengths longer than 10 μm, cooling to 4°K is usually required. On the other hand, far infrared thermal detectors can be obtained which operate at room temperature. It is worth remembering, however, that when the highest sensitivity in the 100 μm region is required, thermal detectors also require cooling. (Putley and Martin, 1967).

Some of the points mentioned in this summary will be brought out in the next section in which the factors limiting the sensitivity of a perfect (ideal) detector will be discussed. Sections 5.3 and 5.4 will discuss photoconductive detectors for wavelengths shorter than and longer than 10 μm, respectively. Section 5.5 will describe new thermal detectors, while Section 5.6 will discuss briefly the requirements for laser detectors.

5.2. The Performance of an Ideal Detector

In this section we will consider the sensitivity attainable by a perfect infrared detector. If we had a perfectly noise-free detector, then the best sensitivity that could be attained would be determined by the fluctuations in the background thermal radiation incident upon the detector. It is, of course, this fluctuation which also determines the sensitivity of a perfect radio receiver, but unlike the radio receiver, the result we obtain for the infrared detector depends upon some of the characteristics of the detector.

Fluctuations of thermal radiation have been discussed by several workers, one of the first of these being Lewis (1947). He considered the Planck distribution function. If the number of photons per unit volume in the frequency interval ν to $\nu + d\nu$ of an enclosure in thermal equilibrium at a temperature $T°K$ is q_ν, then

(5.1) $$q_\nu = (8\pi\nu^2\, d\nu/c^3)(\exp h\nu/kT - 1)^{-1}$$

and q_ν will have a mean square fluctuation which can be shown to be

(5.2) $$(\Delta q_\nu)^2 = q_\nu(\exp h\nu/kT)(\exp h\nu/kT - 1)^{-1}$$

Suppose our infrared detector occupies a small area A in the wall of the enclosure (Fig. 5.1). Then, in unit time the number of photons in the frequency interval between ν and $\nu + d\nu$, arriving at the area A from all directions within a complete hemisphere, will be

(5.3) $$Q_\nu = \tfrac{1}{4} q_\nu cA$$

If the detector were perfect, all of these would be absorbed usefully. However, let us assume that only a fraction η are absorbed. The number absorbed per unit time will be

(5.4) $$J_\nu = \eta Q_\nu = \tfrac{1}{4}\eta q_\nu cA$$

The fluctuation in q_ν will produce a corresponding fluctuation in J_ν

(5.5) $$\overline{(\Delta J_\nu)^2} = \tfrac{1}{4}\eta cA(\Delta q_\nu)^2$$
$$= J_\nu(\exp x)(\exp x - 1)^{-1}$$

where $x = h\nu/kT$.

For many purposes, we will not be interested in the total mean square fluctuation but rather the frequency spectrum of it. The relation between the total fluctuation and its spectrum for this case is the same as that for the

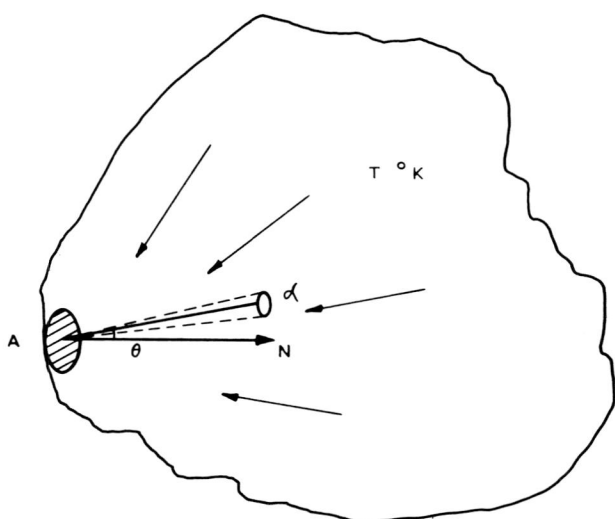

Fig. 5.1. Blackbody enclosure.

shot noise fluctuation of an electron stream, so that the component of the mean square fluctuation within a frequency interval Δf, centered about a modulation frequency f, will be

(5.6) $$\overline{(\Delta J_\nu(f))^2} = 2J_\nu \Delta f (\exp x)(\exp x - 1)^{-1}$$

To obtain the complete contribution to this component of the noise spectrum, we must integrate Eq. (5.6) over the complete spectral range to which the detector responds. The simplest case is that of an ideal thermal detector which responds to the whole spectrum. The complete component of the mean square fluctuation is then found by integrating over the whole spectral range from $\nu = 0$ to $\nu = \infty$. The quantity we have calculated so far is the fluctuation in the photon arrival rate. For many purposes, we are more interested in the power flowing into the detector. To obtain the corresponding mean square power fluctuation, the expression written down for the mean square photon fluctuation must be multiplied by $(h\nu)^2$.

Thus, the frequency spectrum of the mean square fluctuation of the total power absorbed by a thermal detector will be

(5.7) $$\overline{(\Delta W(f))^2} = \frac{4\pi\eta A \, \Delta f}{c^2} \int_0^\infty h^2 \nu^4 \, d\nu \, \frac{\exp x}{(\exp x - 1)^2}$$
$$= 4\pi\eta A \, \Delta f \frac{(kT)^5}{c^2 h^3} \int_0^\infty \frac{x^4 \exp x \, dx}{(\exp x - 1)^2}$$

The integral in Eq. (5.7) can be integrated:

$$\int_0^\infty \frac{x^4 \exp x \, dx}{(\exp x - 1)^2} = \frac{4\pi^4}{15}$$

so that

$$\overline{(\Delta W(f))^2} = \frac{16(\pi kT)^5}{15c^2 h^3} \eta A \, \Delta f$$

(5.8) $$= 8\sigma k T^5 \eta A \, \Delta f$$

where σ is Stefan's constant.

The quantity calculated is the mean square fluctuation in the power absorbed by the detector. If it is in thermal equilibrium with its surroundings it must, by the principle of detailed balancing, radiate an equal amount of power which will also have the same mean square fluctuation. Thus, the total mean square fluctuation is

(5.9) $$\overline{(\Delta W(f)(\text{absorption and emission}))^2} = 16\sigma k T^5 \eta A \, \Delta f$$

From this expression, we can calculate the minimum power which a perfect thermal detector can detect. This we will define as the power P, incident upon the detector, required for the detector to absorb an amount of power from the signal source to equal the root mean square power fluctuation. If P is the incident power, the absorbed power will be ηP, so that

(5.10) $$\eta P = |16\sigma k T^5 \eta A \, \Delta f|^{1/2}$$

or

$$P = \text{Noise Equivalent Power (NEP)}$$

(5.11)
$$= |16\sigma k T^5 A \, \Delta f / \eta|^{1/2}$$
$$= 3.56 \times 10^{-17} T^{5/2} (A \, \Delta f / \eta)^{1/2} \quad \text{W}$$

Thus, for a black detector ($\eta = 1$) at 300°K, of area 1 cm², and with $\Delta f = 1$ Hz

$$\text{NEP} = 5.5 \times 10^{-11} \quad \text{W}$$

To make a corresponding calculation for a photoconductive detector, we have to make some assumption about its spectral response. The simplest one is to assume that a fraction η (independent of frequency) of all photons of frequency $\geqslant \nu_c$ are absorbed while no photons of frequency $< \nu_c$ are absorbed. Since, in most cases, each absorbed photon will ionize one electron, irrespective of the photon's energy, the significant quantity to calculate is the fluctuation in the rate of absorption of photons from the background radiation. If, as with the simple thermal detector, we assume the photoconductive detector responds to radiation from over a complete hemispherical field-of-view, then the required frequency spectrum of the mean square rate of arrival of background photons is obtained by integrating Eq. (5.6) between the limits ν_c and ∞.

$$\overline{(\Delta J(f))^2} = \frac{4\pi\eta A \, \Delta f}{c^2} \int_{\nu_c}^{\infty} \nu^2 \, d\nu \, \frac{\exp x}{(\exp x - 1)^2}$$

$$= 4\pi\eta A \, \Delta f \frac{(kT)^3}{c^2 h^3} \int_{x_c}^{\infty} \frac{x^2 \exp x \, dx}{(\exp x - 1)^2}$$

(5.12)
$$= 4\pi\eta A \, \Delta f \frac{(kT)^3}{c^2 h^3} [J_2(\infty) - J_2(x_c)]$$

where

$$J_n(x) = \int_0^x \frac{x^n \exp x \, dx}{(\exp x - 1)^2}$$

The integral $J_2(x)$ cannot be evaluated in terms of simple functions, but it has been tabulated by Rogers and Powell (1958). The expression (5.12) gives the fluctuation in the rate of arrival and again there will be equal fluctuation associated with the departure (i.e., recombination of the photoelectrons (van Vliet, 1967; Petritz, 1959). If now we wish to calculate the noise equivalent power P at a frequency $\nu_s \geqslant \nu_c$, we note that if this power is incident upon the detector, the rate at which photons are absorbed will be $\eta P/h\nu_s$. We define the noise equivalent power by equating this number to the root mean square fluctuation in the concentration of photoelectrons. The factor $2^{1/2}$ is introduced to include fluctuations in the rate of recombination as well as generation. Thus

$$\eta P/h\nu_s = 2^{1/2}(4\pi\eta A \, \Delta f)^{1/2}\{(kT)^{3/2}/ch^{3/2}\}[J_2(\infty) - J_2(x_c)]^{1/2}$$

or

(5.13) $$P = \nu_s (8\pi A \, \Delta f)^{1/2} \{(kT)^{3/2}/c(\eta h)^{1/2}\}[J_2(\infty) - J_2(x_c)]^{1/2}$$

Since $P \propto \nu_s$, P will be smallest (for a given detector) when $\nu_s = \nu_c$. Thus, the sensitivity of a photoconductive detector will be highest at a wavelength just shorter than its threshold wavelength.

Putting in numerical values, and expressing the result in terms of wavelength, gives

(5.14) $$P = 1.0 \times 10^{-17}(T^{3/2}/\lambda_c)\left[J_2(\infty) - J_2\left(\frac{1.44}{\lambda T}\right)\right]^{1/2}\left(\frac{A \, \Delta f}{\eta}\right)^{1/2}$$

where $J_2(\infty) = 3.2899$.

To consider a numerical example, put $A = 1$ cm^2, $\Delta f = 1$ Hz, $\eta = 1$, $\lambda = 10$ μm, and $T = 300°$K. We obtain $1.44/\lambda T = 4.8$ and (from Roger and Powell's table) $J_2(4.8) = 3.00286$, so that for a photoconductive detector at 10 μm exposed to background radiation over a hemispherical field-of-view,

$$\text{NEP} = 2.8 \times 10^{-11} \text{ W}$$

The behavior of P as λ_c is varied is shown in Fig. 5.2. In the figure, $1/P$ ($= D$ the detectivity) is plotted against λ. This shows that the performance of a photoconductive detector is worst for wavelengths near the peak of the blackbody distribution for the background radiation (which is also shown in the figure). The performance improves rapidly as the wavelength is reduced, since the amount of background radiation received falls off. It also improves as the wavelength increases. The amount of background radiation received is now tending to a constant, but the amount of energy represented by an equivalent number of long wavelength photons is diminishing.

The expressions discussed so far apply strictly to a detector at the same temperature as the background. When considering a cooled detector, the area A can be taken to be that of an aperture admitting radiation to the cooled enclosure containing the detector. The calculation of the fluctuation in the background radiation falling on this aperture is still correct, except that the aperture may limit the field-of-view of the actual detecting element so that not all the radiation from the complete hemisphere will reach the detector. It is easy to modify Eqs. (5.7) and (5.12) to take this into account. If the detector receives radiation from a small element of the background, subtending a solid angle α at the aperture A in a direction making an angle θ with the normal to A, then we can write

$$\frac{\text{radiation from } \alpha}{\text{radiation from complete hemisphere}} = \frac{\alpha \cos \theta}{\pi}$$

Thus Eqs. (5.7) and (5.12) can be corrected for the restricted field-of-view of the detector by multiplying them by the factor $\int (\alpha \cos \theta)/\pi$, where the integral is evaluated over the field-of-view of the detector.

When a cooled detector is used, the effective spectral response may be limited by cooled filters. These will not contribute to the background radiation (unlike a filter at the same temperature as the background) so that this case can be dealt with by inserting suitable limits of integration in (5.7) or (5.12).

The quantity calculated so far is the fluctuation in the radiation received by the detector. When dealing with a thermal detector at the same temperature as the background, it is easy to see that there must be an equal fluctuation in its emission. The calculation for a photoconductive detector is more difficult, but Alkemade (1959) has shown that, in this case, also, there will be

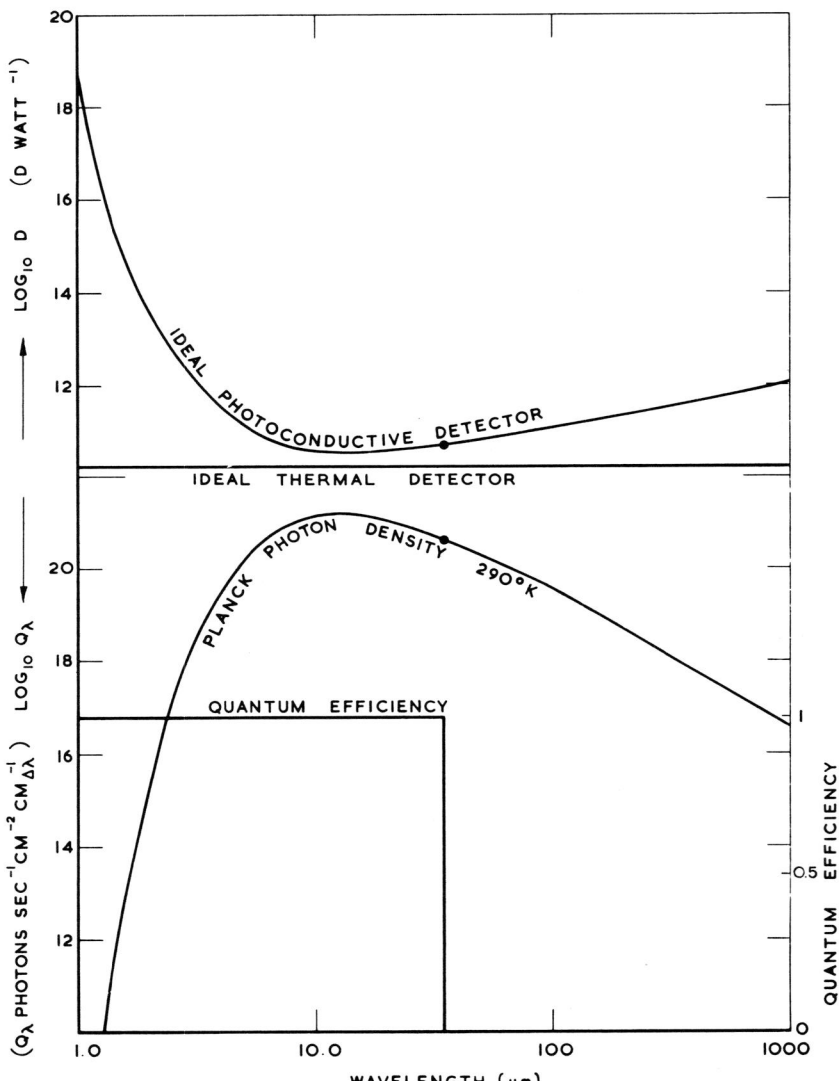

Fig. 5.2. Performance of an ideal photoconductive detector (Putley, 1966).

a fluctuation in the recombination of the photoelectrons which will be equal to that of the incident background radiation. I do not know of a formal calculation for a cooled thermal detector which brings out this point. Since in operation the detector will be in a steady state, the background energy received by it must be disposed of by some process; and it seems reasonable to associate an equal fluctuation with this. Therefore, in writing down expressions for the noise equivalent power of cooled detectors, the factor of 2 will be retained.

Taking these factors into account, the expression obtained for the noise equivalent power of a cooled photoconductive detector is

$$(5.15) \quad P = 1.0 \times 10^{-17} (T^{3/2}/\lambda_c) \left[\int \frac{\alpha \cos \theta}{\pi} \right]^{1/2} \left[\{J_2(x_1) - J_2(x_2)\} \left(\frac{A \, \Delta f}{\eta} \right) \right]^{1/2}$$

The corresponding expression for a thermal detector is

$$(5.16) \quad P = 6.94 \times 10^{-18} \, T^{5/2} \left[\int \frac{\alpha \cos \theta}{\pi} \right]^{1/2} \left[\{J_4(x_1) - J_4(x_2)\} \left(\frac{A \, \Delta f}{\eta} \right) \right]^{1/2}$$

In these expressions, T is the temperature of the background radiation not that of the detector, and uniform absorption of incident radiation is assumed between wavelengths λ_1 and λ_2 ($x = 1.44/\lambda T$).

Both at very short and at very long wavelengths these expressions can be simplified. At very short wavelengths, i.e., when $x = 1.44/\lambda T \gg 1$, the photon fluctuations can be treated using classical statistics, when the integrals simplify to expressions which can be evaluated easily. Consider the photoconductive case. Equation (5.5) now reduces to

$$\overline{(\Delta J_\nu(f))^2} = 2 J_\nu \Delta f$$

$$(5.17) \quad J_\nu = 2\pi \left(\frac{\eta A}{C^2} \right) \left(\frac{kT}{h} \right)^3 \int_{x_c}^{\infty} \frac{x^2 \, dx}{e^x - 1}$$

If $x \gg 1$ this becomes

$$(5.18) \quad J_\nu = 2\pi \left(\frac{\eta A}{C^2} \right) \left(\frac{kT}{h} \right)^3 \int_{x_c}^{\infty} x^2 e^{-x} \, dx$$

integrating to give

$$(5.19) \quad J_\nu = 2\pi \left(\frac{\eta A}{C^2} \right) \left(\frac{kT}{h} \right)^3 [x_c^2 + 2x_c + 2] e^{-x_c}$$

and hence

$$(5.20) \quad \overline{(\Delta J_\nu(f))^2} = 4\pi \left(\frac{\eta A}{C^2} \right) \left(\frac{kT}{h} \right)^3 [x_c^2 + 2x_c + 2] e^{-x_c} \, \Delta f$$

Taking into account the recombination fluctuation, the corresponding expression for the noise equivalent power is (cf. Eq. (5.13))

$$P = \nu_c (8\pi A \, \Delta f)^{1/2} [(kT)^{3/2}/c(\eta h)^{1/2}][x_c^2 + 2x_c + 2]^{1/2} \exp(-x_c/2)$$

(5.21) $\quad = 1.0 \times 10^{-17} (T^{3/2}/\lambda_c) \left(\dfrac{A \, \Delta f}{\eta}\right)^{1/2} [x_c^2 + 2x_c + 2]^{1/2} \exp(-x_c/2)$

This approximation is sufficiently accurate for use out to 10 μm. At long wavelengths ($x < 1$) the Rayleigh–Jeans approximation can be used for the blackbody distribution. Under this approximation, the expressions for the thermal and photon detectors reduce to the same expression. While the ones derived here are valid at all wavelengths, when applying them in the submillimeter region ($\lambda > 100$ μm) care must be taken to assign the correct value to the field-of-view factor. The problem is now very similar to that of a radio antenna (McLean and Putley, 1965).

For a fuller account of the problems discussed in this section, the books by Smith *et al.* (1968) and Kruse *et al.* (1962) should be consulted.

5.3. Intrinsic Photoconductive Detectors

Photoconductivity in selenium was observed as long ago as 1876. In the years between the World Wars both selenium and copper oxide photocells were widely used as visible light sensors. About this time, the thallous sulfide cell was introduced. This was the first to be developed as an infrared detector. It operated out to 1.1 μm and remained in use until the development of PbS photocells in the Second World War. These are still amongst the most useful detectors for wavelengths ⩽3 μm. PbTe and PbSe detectors were developed shortly afterwards but, in more recent years, these have been largely replaced by InSb detectors. All these detectors make use of intrinsic photoconductive effects. At the present time, efforts are being made to develop intrinsic photodetectors for use out to 10 μm. At longer wavelengths than this, the extrinsic photodetectors (to be described in the next section) are likely to remain in use.

When a semiconductor is irradiated with photons of energy greater than that separating the valence and the conduction bands, electrons will be excited from the valence to the conduction band (Fig. 5.3). If the semiconductor is maintained at a temperature low enough to ensure that the rate of thermal excitation across the gap is sufficiently small, the optical excitation will produce a significant change in conductivity.

In addition to causing an increase in conductivity, the optical absorption can be used to produce other electrical effects. The most important of these occur when a p-n junction is illuminated (Fig. 5.4).

fluctuations, there will be a fluctuation in the density of free carriers in the semiconductor (van Vliet, 1967). The electron concentration in semiconductors is usually sufficiently small for classical statistics to be valid. If the average number of carriers present is N_d, its rms fluctuation will be $N_\mathrm{d}^{1/2}$. van Vliet shows that if a current I is passing through the sample, the fluctuation in N_d will produce a corresponding fluctuation in I, which will have a spectrum

(5.27) $$\overline{(\Delta I)^2} = \frac{4I^2\tau}{N_\mathrm{d}(1+\omega^2\tau^2)}$$

$$= \frac{4I^2}{G(1+\omega^2\tau^2)}$$

where τ is the lifetime of the carriers and G is the thermal generation rate. If the specimen has a resistance R and the amplifier bandwidth is Δf, the corresponding noise voltage will be

(5.28) $$\overline{(\Delta V_\mathrm{gr})^2} = 4I^2R^2(\Delta f/G)$$

(assuming $\omega\tau \ll 1$).

In Section 5.2 we showed that for $\lambda < 10\ \mu\mathrm{m}$ the photon fluctuation could also be treated by classical statistics, and hence if the optical generation rate is ηQ, there will be a noise voltage associated with this,

(5.29) $$\overline{(\Delta V_\mathrm{op})^2} = 4I^2R^2(\Delta f/\eta Q)$$

When the thermal and optical generation rates are comparable, the noise voltage is given by an expression similar to (5.28) or (5.29) containing the total generation rate $G + \eta Q$. In addition to sources of nonideal noise which may or may not be present, there must also be Johnson noise present. The Johnson noise voltage will be

$$\overline{(\Delta V_\mathrm{J})^2} = 4kTR\,\Delta f$$

Combining Eqs. (5.27)–(5.29), we can write

(5.30) $$\frac{\overline{(\Delta V_\mathrm{J})^2}}{\overline{(\Delta V_{\mathrm{gr}+\mathrm{op}})^2}} = \frac{4kTR\,\Delta f(G+\eta Q)}{4I^2R^2\,\Delta f} = \frac{kT(G+\eta Q)}{I^2R}$$

This shows that to obtain the best results we want to make G small compared with ηQ, to make T small, and to make I and R are large as possible. Reducing T reduces the thermal excitation rate, and this is the principal reason why the performance improves on cooling. At a given temperature, the resistance of a semiconductor is a maximum when it contains a small excess of p-type impurities. This occurs because the mobility of holes is normally less than that of electrons. For this reason, slightly p-type material is found to make the best detectors.

With InSb at room temperature it is not possible to make the value of (5.30) small and Johnson noise is the dominant mechanism, but at 77°K, G becomes very small. The radiation fluctuation becomes the main source of noise and we approach very closely to the performance of an ideal detector. The performance obtained with typical InSb detectors is shown in Fig. 5.7.

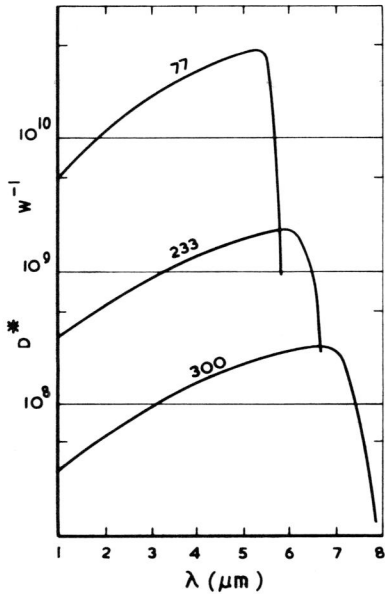

FIG. 5.7. Spectral detectivity of InSb detectors (Morten and King, 1965).

Figure 5.8 illustrates some of the recombination processes. These have been studied in some detail in InSb, both experimentally (Zitter *et al.*, 1959; Laff and Fan, 1961) and theoretically (Beattie and Landsberg, 1958). It is found that near room temperature the Auger process is the dominant one. This is responsible for a lifetime of about 10^{-8} sec at room temperature, which at first increases on cooling. Below 200°K the lifetime begins to fall. This is caused by the presence of deep impurity levels acting as recombination centers. At temperatures below 170°K, other impurity states take part by trapping holes, causing the hole lifetime to become longer than the electron lifetime. It is possible to obtain values for each lifetime by measuring both the photoconductive and photoelectromagnetic effects, since the effective lifetimes found from these measurements are given by different relations between hole and electron lifetimes (Blakemore, 1962). The low-temperature results of

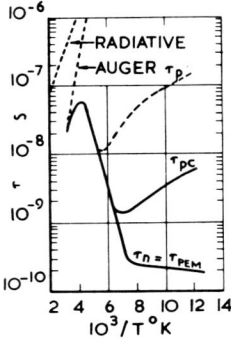

FIG. 5.8. Recombination processes in InSb (Zitter et al., 1959).

Fig. 5.8 were obtained by Zitter et al., (1959) for a sample containing an extrinsic hole concentration of 6×10^{15}. The photoconductive lifetime shown there is in fact somewhat shorter than that found in the most sensitive detectors at 77°K, which usually have response times of 1–10 μsec. Room temperature photoconductive detectors have response times less than 10^{-7} sec (as shown in the figure).

Indium arsenide is another III-V compound which is very similar to InSb, but it has a larger energy gap so that $\lambda_c \sim 3$–4 μm. The spectral detectivities of InAs detectors are illustrated in Fig. 5.9. These are p-n junction detectors. At 77°K their performance approaches the ideal detector and is comparable with the PbS detector. At room temperature, the InAs detector

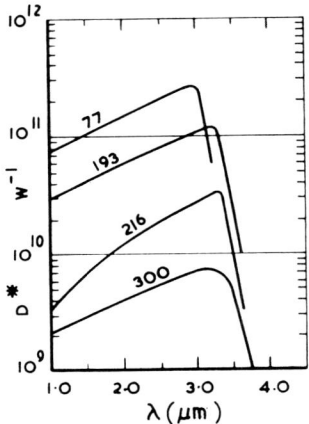

FIG. 5.9. Spectral detectivity of InAs detectors (Putley, 1966).

TABLE I.

Type	Operating temp. (°K)	Useful wavelength range (μm)	Wavelength of max. detectivity	Detectivity D^* (or D) at optimum wavelength
Si–PV	295	<1.2	0.9	2.5×10^{12}
Tl$_2$S–PC layer	295	<1.2	0.9	2.2×10^{12}
Ge–PV	295	<1.8	1.6	4×10^{11}
PbS–PC layer	295	<3	2.5	8×10^{10}
PbS–PC layer	193	<3.4	2.6	4×10^{11}
PbSe–PC layer	295	<4.5	3.5	10^9
PbSe–PC layer	77	<7	5	10^{10}
PbTe–PC layer	77	<5.2	4	2.7×10^9
InAs–PV	295	<3.7	3.4	7×10^9
InAs–PV	196	<3.5	3.2	7×10^{10}
InAs–PV	77	<3.4	3.0	3×10^{11}
InSb–PC	295	<7.8	6	3×10^8
InSb–PC	77	<5.5	5	5×10^{10}
InSb–PV	77	<5.5	5.4	7×10^{10}
InSb–PEM	295	<7.8	6.2	3×10^8
Ge–Au–IPI	60	<9	5	3×10^{10}
Ge–Hg–IPI	30	<14	12	3×10^{10}
Ge–Cu–IPI	15	<30	25	4×10^{10}
Ge–Zn–IPI	4	<40	35	2×10^{10}
Ge–Ga–IPI	4	<125	104	3×10^{11}
Ge–B–IPI	4	<140	108	2×10^{11}
InSb[a]	1.8	>200	1000	2×10^{12} (D)
InSb[b]	1.8	>200	1000	2×10^{11} (D)
InSb[c]	4	determined by magnetic field	150 μm (12 kG) 26 μm (73 kG)	1.2×10^{11} (D) 2×10^{10} (D)

is still inferior to the older PbS detector. At this temperature both are limited by Johnson noise. PbS will then give the better performance because it has a longer lifetime.

Promising 10 μm intrinsic detectors are now being developed using (Hg-Cd)Te (Lawson *et. al.*, 1959; Kruse, 1965; Vérié and Ayar, 1967) and (Pb-Sn)Te junction devices (Melngailis and Calawa, 1966). Bartlett *et. al.* (1969) have recently described a background limited 77°K (Hg-Cd)Te detector comparable in performance to a 30°K Hg doped Ge detector.

Some relevant properties of photoconductive detectors are listed in Table I.

Infrared detectors.

Response time (μsec)	Resistance	Remarks: availability
1000	450 kΩ	Manufactured by Texas Instruments (U.S.A).
500	5 MΩ	Not available commercially
		Available from several manufacturers
300	500 kΩ	⎫ Available Mullard (U.K.), Santa Barbara Research
3500		⎭ Center (U.S.A.)
2		
40		⎫ Available S.B.R.C. (U.S.A.)
25	30 MΩ	⎭ Available from S.A.T. (France)
1	20 Ω	⎫
1	50 Ω	⎬ Available from several manufacturers including Texas
2	≤10 MΩ	⎭ Instruments, S.B.R.C., and Philco (U.S.A.)
0.05	5 Ω	⎫
5–10	2000 Ω	⎬ Available from several manufacturers including
<1		⎭ Mullard and Philco (U.S.A.)
0.2	20 Ω	
<1	1 MΩ	⎫ Available from several manufacturers including
<1	>1 MΩ	⎬ Mullard (U.K.), S.A.T. (France), Texas Instru-
0.1	≤3000 Ω	⎭ ments, S.B.R.C., Philco and Raytheon (U.S.A.)
~1	~1 MΩ	⎫
0.04	~10 kΩ	⎬ Not available commercially
0.1–0.01	~10 kΩ	⎭
100		
0.2	~5 kΩ	⎫ Available from Mullard (U.K.)
<1	~10⁴–10⁵ Ω	⎭

PV, photovoltaic; PC, photoconductive; PEM, photoelectromagnetic effect; IPI impurity photoionization. [a] Hot electron, no magnetic field. [b] Hot electron in weak magnetic field. [c] Magnetoptical resonance.

5.4. Extrinsic Photoconductive Detectors

Although a continuous range of semiconductors exists with small energy gaps down to zero, problems of preparing these materials to the necessary purity have prevented the development of practical intrinsic photoconductive detectors for wavelengths longer than about 7 μm—as mentioned in the last section, 10 μm intrinsic detectors are still in the experimental stage. The use of impurity photoconductivity was suggested some years ago (Burstein et. al., 1951) as a means of detecting far infrared radiation. As is well known, the

TABLE I.

Type	Operating temp. (°K)	Useful wavelength range (μm)	Detectivity D^* or D (W^{-1})
Semiconductor pin thermocouple	295	Vis.→30	3×10^9
Blackbody thermocouple	295	Vis.→45	
Golay cell	295	Vis.→several mm	3×10^9 (D)
Thermistor bolometer	295	Vis.→40	$1.6 \times 10^8\ \tau^{1/2}$
Pt foil bolometer	295	Vis.→25	1.4×10^8
Superconducting Sn bolometer	3.7	>10	3×10^{11} (D)
Carbon bolometer	≤2.1	>10	3×10^{10} (D)
Ge bolometer	≤2.1	>10	3×10^{11} (D)
TGS pyroelectric detector	295	>2	2×10^8 (D)
BaTiO$_3$ dielectric bolometer	295	Vis.→20	1.1×10^8

behavior of semiconductors at low temperatures is determined by the residual impurities. Localized energy states occupying positions between the valence and conduction bands are associated with the impurity atoms (Fig. 5.10); and at very low temperatures the impurity electrons or holes will condense onto these levels, and the behavior of the material will approach that of an insulator. If now the material is irradiated with photons of energy less than

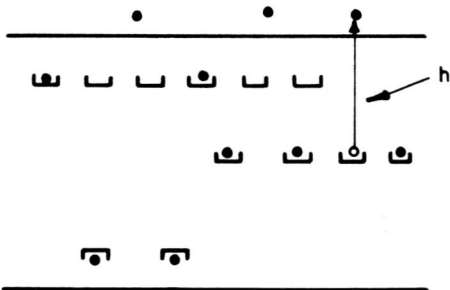

FIG. 5.10. Extrinsic photoconductivity.

Infrared detectors (cont.).

Response time (msec)	Resistance	Remarks: availability
30	10–100 Ω	Mounted in vacuum. Spectral response determined by window. Also available mounted in air. Manufactured by Hilger and Watts (U.K.) and by Reeder (U.S.A.)
170		Not available commercially
15		Manufactured by Unicam (U.K.) and by Eppley (U.S.A.)
1–10	2.5 MΩ	Manufactured by Barnes Engineering (U.S.A.)
16	40 Ω	Manufactured by Baird-Atomic (U.S.A.)
1250	100 Ω	Not available commercially
10	100 kΩ	Not available commercially
10	100 kΩ	Manufactured by Texas Instruments (U.S.A.)
1	>10 MΩ	Becoming commercially available
1	125 kΩ	Becoming commercially available

that of the band gap but greater than the impurity ionization energy, intrinsic ionization cannot occur but the impurity states can be ionized. This effect has been observed in several semiconductors, but the only one in which it has been widely exploited is Ge.[2] The properties of a considerable number of impurities in Ge have been studied in detail. The shallowest have ionization energies corresponding to $\lambda_c > 100$ μm while several come between 10 and 100 μm, so that extrinsic Ge detectors are very convenient for this region of the spectrum.

Figure 5.11 shows the ionization energies of most of the impurities in Ge. The behavior of group V or group III atoms is the simplest. The outer shell of these atoms contains one electron more or less, respectively, than that of the Ge atoms. This extra electron or hole together with the corresponding unit of nuclear charge behaves like a hydrogen atom embedded in the Ge lattice. Its ionization energy can be obtained from that of a hydrogen atom in

[2] Note added in proof: Some development of extrinsic Si and Ga As detectors is now taking place.

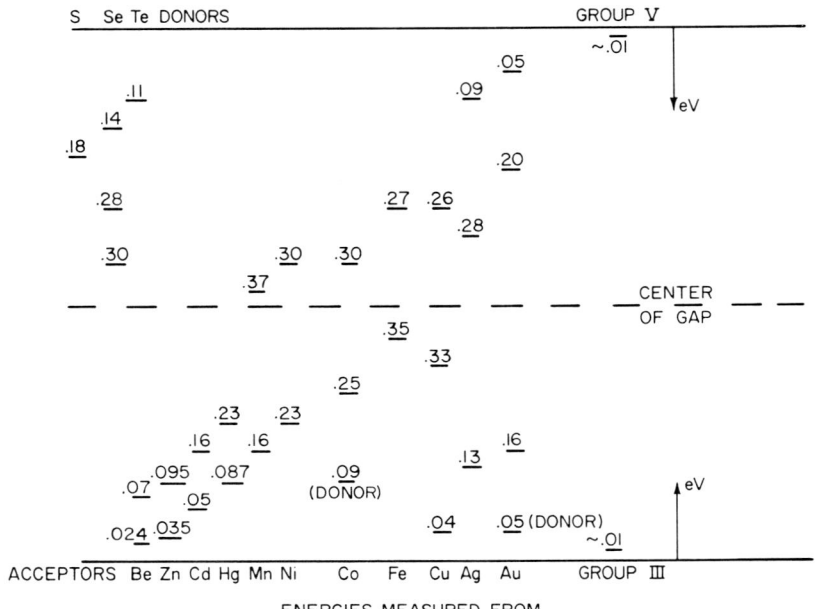

Fig. 5.11. Impurity levels in Ge.

free space, by scaling to allow for the dielectric constant of the medium and the appropriate effective mass for an electron in the Ge lattice. Thus, according to the Bohr theory, the ionization energy will be

$$\varepsilon = 13.6 \left(\frac{m^*}{m}\right) \frac{1}{K^2} \quad \text{eV} \tag{5.31}$$

For Ge, $K \sim 16 (m^*/m) \sim 0.2$, giving $\varepsilon \sim 0.01$ eV, corresponding to $\lambda_c \sim 120$ μm.

The behavior of other impurity atoms is more complex. These can exhibit two or more states of ionization. These different states can be studied by adding suitable amounts of compensating impurities; photoconductive effects are found with all states.

The impurities most commonly used in detectors are B and Ga ($\lambda_c > 100$ μm), Zn ($\lambda_c \sim 40$ μm), Cu ($\lambda_c \sim 30$ μm), Hg ($\lambda_c \sim 14$ μm), and Au ($\lambda_c \sim 9$ μm). For the longer wavelength detectors, cooling with liquid He is essential, but Hg-doped detectors operate satisfactorily at 35°K and Au-doped ones with liquid nitrogen. Figure 5.12 shows a compact metal cryostat suitable for a helium-cooled detector. Figure 5.13 shows the spectral detecti-

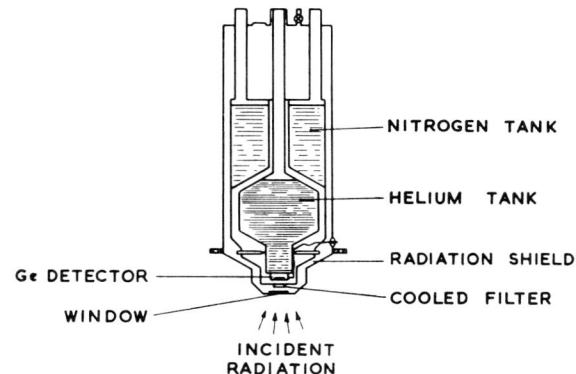

FIG. 5.12. Cryostat for Ge detectors (courtesy Mullard Ltd., London).

vities for several doped-Ge detectors, which at their peak wavelengths approach closely the ideal detector. The fact that measured detectivities can be higher than that calculated for an ideal detector, with a hemispherical field-of-view, does not mean that some of these detectors are better than perfect. It is simply a consequence of a restricted field-of-view and of the use of cooled filters.

There are several ways in which the extrinsic detectors differ from the intrinsic ones. The maximum concentrations of impurities which can be introduced are very small (usually less than 1 ppm). This is partly determined by the solubility of the impurities and partly by the need to avoid interactions between nearest neighbors. Hence, the impurity absorption coefficients are much smaller than the intrinsic ones, values of 10–100 cm^{-1} being typical. This means that a much larger piece of material is required. Because only small concentrations of active impurities can be used, the purity required for the host material is more critical than that for intrinsic detectors. It is for this reason that only Ge has so far been widely used as a host material. Another consequence of the small impurity concentration is that the cooling requirements are more critical than for intrinsic detectors (Long, 1967).

The most important recombination process is the direct capture of the free carrier by the impurity center, the carrier's energy being dissipated via lattice vibrations. The recombination time depends upon the concentrations and the capture cross sections of the impurities. Values for the capture cross sections can be found from hot electron measurements as well as from studies of photoconductivity. There is now good agreement between the values measured in different ways and derived theoretically. Typical values for the response times of doped-Ge detectors are 10–100 nsec.

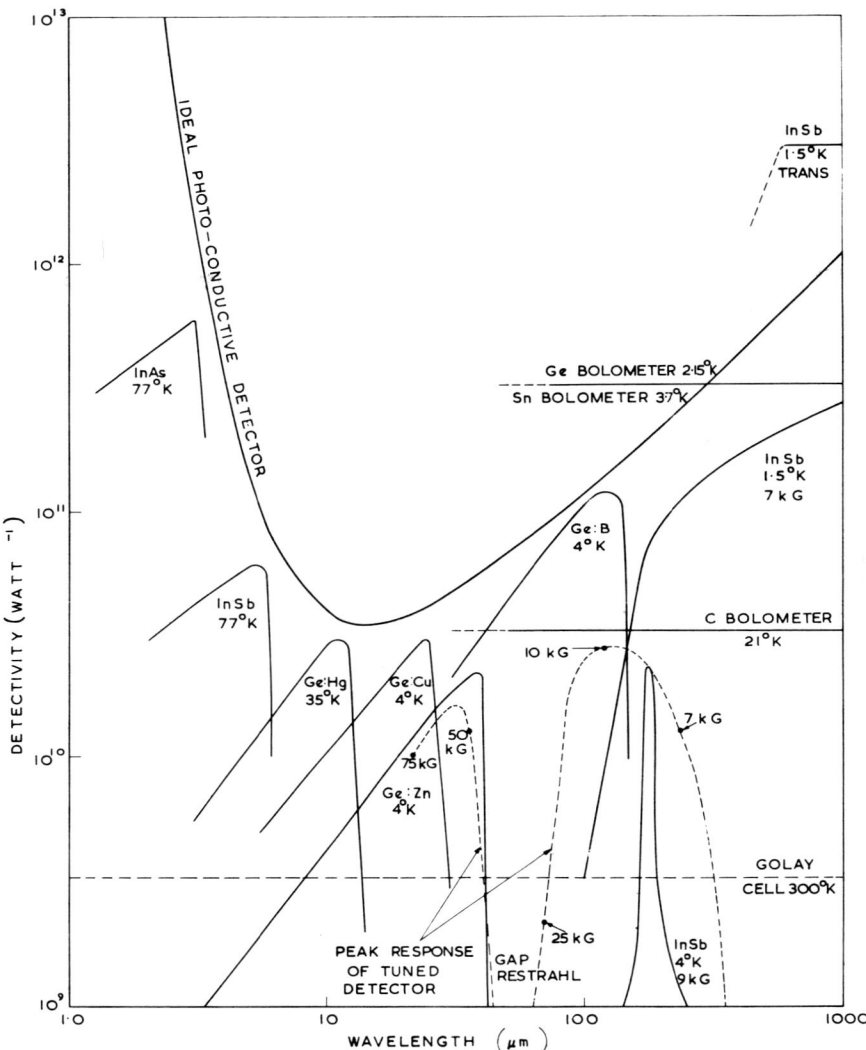

Fig. 5.13. Performance of far infrared detectors (Putley, 1966).

It is not possible to use extrinsic Ge detectors beyond about 120 μm since this corresponds to the ionization energy of the shallowest impurities. Another type of impurity photoconductivity can be used at longer wavelengths and, indeed, beyond 1 mm. In this third process, the incident radiation is absorbed by the free carriers in the conduction band of a high mobility semiconductor such as n-type InSb. One would not, at first sight, expect this process to lead to a change in conductivity. Free carrier absorption is, of course, a well-known property of semiconductors, but at most temperatures the energy absorbed by the electrons will be rapidly shared with the lattice, thus warming up the whole crystal. In fact, Novak (1959) has used this effect to make an infrared bolometer. When the temperature is lowered to 4°K or less, the lattice-free electron interactions (electrons are not frozen out even in the purest available n-type InSb) become very weak, so that when the electrons are irradiated their mean energy can be raised significantly above the thermal equilibrium value before a steady state is set up. Since the electron mobility depends upon the mean energy, this process leads to a change in mobility and hence conductivity. This type of behavior is usually studied by applying a relatively large electric field to the semiconductor, when the change in conductivity appears as a deviation from Ohm's law. In InSb at 4°K fields of the order of 1 V cm^{-1} are large enough.

The most convenient way to show how this process can be used to detect far infrared radiation is to write down an expression for the high-frequency conductivity

(5.32) $$\sigma(\omega) = \sigma_0(1 + \omega^2 \tau_e^2)^{-1}$$

Here σ_0 is the zero frequency conductivity and τ_e is the electronic relaxation time, and

(5.33) $$\sigma_0 = ne^2 \tau_e / m^*$$

where n is the electron concentration and m^* the effective mass. At very high frequencies, the conduction component of the current will become small compared with the displacement current.

The optical absorption coefficient is then

(5.34) $$\alpha = 4\pi\sigma/c(K)^{1/2}$$

where K is the dielectric constant. This expression shows that when $\omega\tau_e < 1$, α will be a constant, but when $\omega\tau_e > 1$, α will vary as ω^{-2} or as λ^2. Thus, free carrier absorption will become very small at short wavelengths when $\omega \gg 1/\tau_e$. Typical values for the parameters of a fairly pure sample of InSb at 4°K are: $n \sim 5 \times 10^{13}$ cm^{-3}, $\mu = 10^5$ cm^2 V^{-1} sec^{-1}, and $m^* \sim 1.4 \times 10^{-2}$ gm. These values give $\tau_e = 8.5 \times 10^{-13}$ sec, so that $\omega\tau_e = 1$ for $\lambda = 1.6$ mm. Hence, at wavelengths as short as 1 mm we are still in the low frequency regime. At 1 mm the absorption coefficient is $\alpha = 22$ cm^{-1}, but at 100 μm it has fallen to 0.30 cm^{-1}.

The electric field dependence of the conductivity can be described by the equation

(5.35) $$\sigma = \sigma_0(1 + \beta E^2)$$

and analysis of the mechanism for detecting radiation (Putley, 1965) shows that the responsivity and the response time are given by the expressions

(5.36) $$R = \beta V/v\sigma$$

and

(5.37) $$\tau = 3/2(k/e)\beta \, dT/d\mu$$

where V is the applied voltage and v the volume of the element. Typical values found for R and τ are 500 V W^{-1} and 0.2 μsec, respectively.

At this point, it must be mentioned that the performance of the device is strongly influenced by the presence of a magnetic field. The argument developed so far has ignored this. When a magnetic field is applied, the resistance increases rapidly and this leads to an increase in R, improving the performance, but the response time does not change very much. The reason for the rapid increase in R is that magnetic inductions of a few dT's split up the conduction band into Landau levels and also split off a shallow impurity state into which the conduction electrons descend. Thus, when large enough fields are applied, the number of free electrons is so much reduced that the performance falls off. The best results are obtained with fields of about 7 dT. When larger fields are applied, another photoconductive effect appears. This occurs over a narrow band near the cyclotron resonance frequency

(5.38) $$\omega_e = eB/m^*$$

Electrons are excited from the ground state of the impurity into a higher state and from this they fall into the lowest group of Landau levels, thus increasing the conductivity. The position of the peak can be varied by adjusting the magnetic field. Figure 5.13 shows typical behavior of both the broadband and the narrowband modes of operation.

The detectivity of the InSb submillimeter detector is limited by the noise in the first stage of its amplifier, but it is still considerably more sensitive than the best uncooled thermal detectors and it is considerably faster than any thermal detector. Figure 5.14 shows a simple cryostat for this detector.

Fuller details of both the Ge and InSb extrinsic detectors have been given in recent review articles (Putley, 1964, 1965). Data on these detectors are included in Table I.

5.5. New Types of Thermal Detectors

5.5.1. Low-Temperature Bolometers

There are several advantages in operating a bolometer at very low temperature. Equation (5.11) shows that the sensitivity of an ideal thermal detector

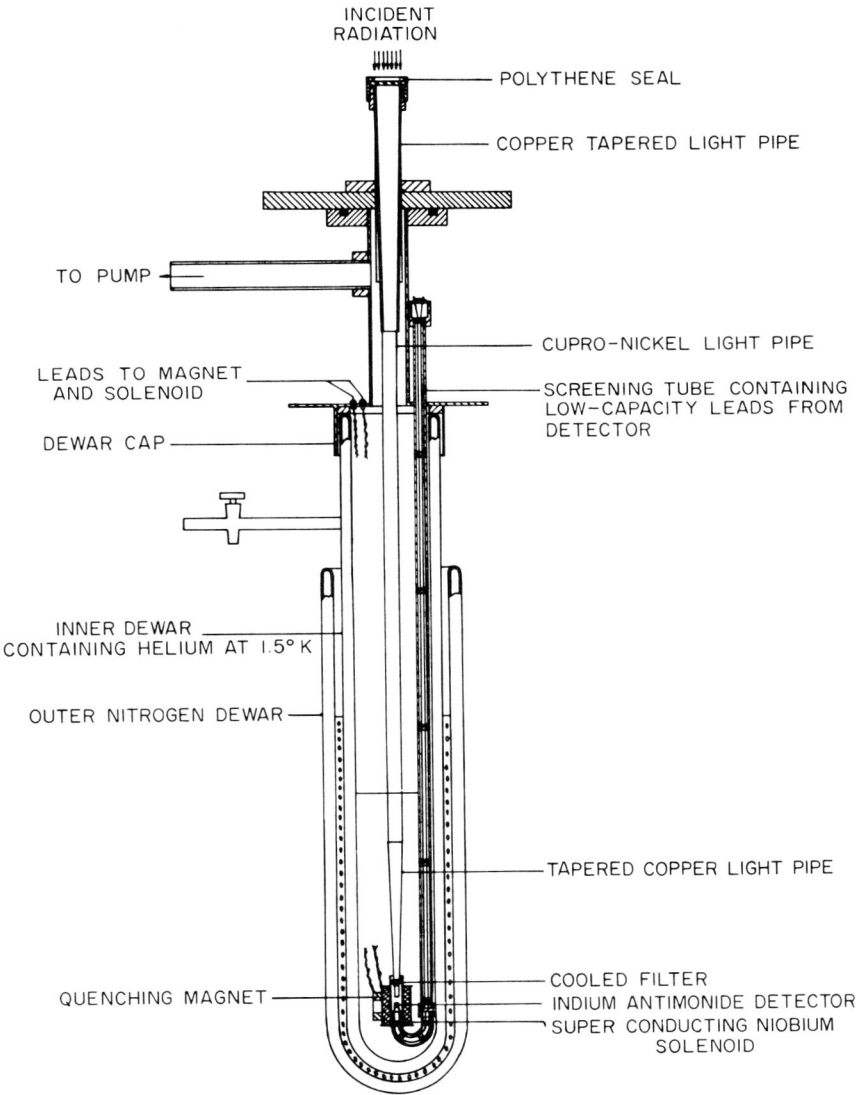

FIG. 5.14. Cryostat for InSb submillimeter detectors.

will be increased by lowering the temperature. Since normally some part of the experiment will take place at an elevated temperature, the best sensitivity that can be attained will not be found by putting the detector's temperature in Eq. (5.11) but rather by integrating Eq. (5.16) over the waveband of interest, assuming that the detector is shielded from radiation outside this band by suitable cooled filters.

It will only be possible to take advantage of this reduction in background radiation and detect very small signals if the responsivity of the detector is sufficiently large that noise (from the amplifier or other external sources) does not become troublesome. It is therefore fortunate that lowering the temperature improves the characteristics of bolometer materials.

The responsivity R and time constant τ of a bolometer are given by (Smith et al., 1968; Jones, 1953):

$$R = \alpha V(G - \alpha Q) \tag{5.39}$$

$$\tau = C(G - \alpha Q) \tag{5.40}$$

where α is the resistance temperature coefficient, V the applied voltage, and Q the electrical power dissipated in the element. C is the thermal capacity of the element and G the thermal conductance coupling it to its surroundings. These equations are valid provided $\omega \tau < 1$. They show that α should be large and G small. If G is made small then C must also be kept small, otherwise τ will become excessively long. In the helium temperature range, not only are thermal capacities much reduced but the largest values of α are found, for instance at superconducting transitions and also in suitably doped semiconductors.

The superconducting bolometer was the first type of cooled bolometer to be developed. Originally niobium nitride was used (Andrews et al., 1946; Fuson, 1948) but more recently tin has been employed (Martin and Bloor, 1961). In this type of bolometer a layer of tin a few μ thick is evaporated onto a mica substrate. Lead contacts are evaporated over the ends of the tin film. In operation, the bolometer must be maintained near the superconducting transition temperature (approximately 3.7°K for tin). At this temperature, lead is superconducting so that the resistance of the connection to the element will be zero. Thermal contact to the heat sink is made by nylon threads used to support the element. Optimum results are obtained when the resistance r of the film lies between 1 and 10 Ω, when dr/dT is found to lie between 100 and 1000 Ω (°K)$^{-1}$, making $\alpha \sim 100$ (°K)$^{-1}$. One of the problems with this type of detector is to get the radiation absorbed efficiently, since at long wavelengths metals are very good reflectors and "blacking" is very difficult. A second difficulty is that of designing a sufficiently good amplifier. The best so far produced use a step-up transformer cooled to the detector's temperature, followed by a low-noise amplifier at room temperature. Even with the best available amplifiers, the performance is limited by amplifier noise rather than by radiation fluctuations. Despite this limitation, Martin and Bloor (1961) achieved a noise-equivalent-power of about 3×10^{-12} with $\tau = 1.25$ sec. With more recent developments in amplifier design, this figure can now be improved somewhat or, alternatively, a faster and more robust element can be used.

The main disadvantages of the superconducting bolometer of the type described are that the element is fragile, its time constant is very long, and very accurate temperature control (to within 10^{-5} °K) is required. For these reasons, the superconducting bolometer has not been very widely used, and it is now possible to obtain a comparable performance using helium-cooled semiconducting bolometers which are both simpler to construct and more robust in operation.

The first type of cooled semiconducting bolometer to be introduced was constructed from the carbon composition material used in the manufacture of certain types of commercial resistor, as produced by Allan and Bradley and other manufacturers. This material has a high temperature coefficient below 20°K and is often used as a low-temperature resistance thermometer. A bolometer can be constructed by mounting a plate cut from an Allan–Bradley resistor of thickness about 0.3 mm onto a heat sink. Best results are obtained by operating below the helium λ-point (2.17°K). Not only does the responsivity increase on cooling, but also operation below the λ-point eliminates noise from the cryostat of partly mechanical, partly thermal, origin, produced by bubbling in the refrigerant and by temperature gradients. When the liquid helium is reduced to its superfluid state, these effects are very much reduced. The detector of the type described has a resistance of about 100 kΩ, and with a time constant of 1 msec, a NEP of 3×10^{-11} W can be achieved. Its advantages are that it is robust and can be constructed from easily obtainable materials.

More recently, even better semiconducting bolometers have been produced. In Ge containing a concentration of about 10^{16}–10^{17} cm^{-3} of shallow impurities, the photoconductive effects described in Section 5.4 are small because the impurity centers are interacting with each other. The localized impurity states are broadening into a conducting band. However, as with the InSb sub-mm detector, the absorption of the free carriers becomes large in the sub-mm region, but at shorter wavelengths down to the visible appreciable absorption is obtained which can be enhanced by "blacking" if necessary. Unlike the InSb detector, the electrons in this case are tightly coupled to the lattice so that radiation absorbed by them heats the whole crystal. To obtain the optimum performance, the correct concentration of compensating impurities is essential. The choice of correctly doped material is therefore somewhat critical. Nevertheless, very satisfactory bolometers have been described by Low (1961) and also Richards (1964). Low used Ga as the dominant impurity and has made considerable use of the device in infrared astronomy both in the 10 and 1000 μm bands. Richards used In-doping and has employed his detector with various far infrared interferometers. These detectors have a noise-equivalent-power below 10^{-11} W, thus having better performance than the carbon bolometer. Operation below the He λ-point (2.17°K) is again desirable. Response time of 1–10 msec and element resistance of about 100 kΩ are typical. This detector is robust and easy to use. It has a better performance than the C bolometer but has the disadvantage that correctly doped-Ge is more difficult to obtain.

Somewhat similar performance has also been obtained with doped-GaAs (Wheeler and Hill, 1966). In this case, the difficulty of obtaining correctly doped-GaAs is greater than that for Ge.

In all these bolometers, the amplifier noise is the main factor limiting the sensitivity, although in the most favorable cases it should be of the same order as the radiation noise.

A final method of detection which should be mentioned is use of the InSb detector (described in Section 5.4) with a cooled transformer and narrowband amplifier as for the superconductor bolometer. Kinch and Rollin (1963) have pointed out that this arrangement has a sensitivity, at 1000 μm, superior to that of the cooled bolometers. Since, in this arrangement, the response time of the system is limited by the amplifier to about 1 msec, the speed of the InSb detector is sacrificed, but for many applications this will be justified. The disadvantage of this mode of operation of the InSb detector is that the performance will fall below 200–300 μm, so that at shorter wavelengths its performance will fall below that of the best Ge bolometer. Richards (private communication) has shown that, at the longer wavelengths, the InSb detector has a better NEP than the Ge bolometer. Its performance has a greater wavelength dependence, which is a disadvantage for applications where uniform performance over a wide spectral range is desirable.

5.5.2. The Pyroelectric Detector

Certain materials possess a unique axis along which a permanent electric dipole exists. Typical examples are tourmaline, barium titanate, triglycine sulfate, lithium sulfate, and lithium niobate. The presence of this dipole implies that surfaces cut normal to the axis should be electrically charged. However, the presence of this charge cannot usually be observed because it becomes neutralized by stray charge attracted to and trapped at the surfaces. In some of these materials, the presence of the moment can be detected by applying a large external electric field which can reverse the direction of the moment. The stray surface charges cannot redistribute themselves quickly, so that the change in surface charge produced by the reversal of the moment can be measured. Materials with this property are known as "ferroelectric." At room temperature, barium titanate and triglycine sulfate behave in this way but the other materials do not. If, however, the temperature is changed slightly the lattice spacings will change. This will cause a small change in the internal dipole moment. Again, because the stray surface charges cannot redistribute themselves quickly, the change in surface charge density can be measured. Thus, although the dipole moment cannot be measured directly, its temperature coefficient can (Cady, 1946). This latter quantity is called the pyroelectric coefficient P. Although all ferroelectric materials show the pyroelectric effect, pyroelectric materials are not necessarily ferroelectric.

The values of the pyroelectric coefficients at room temperature for the materials mentioned are as follows:

	Tourmaline	$BaTiO_3$	TGS	Li_2SO_4	$LiNbO_3$
$P(C\ cm^{-2}\ °C^{-1})$	4×10^{-10}	2×10^{-8}	3×10^{-8}	8×10^{-9}	4×10^{-9}

Since a modern electrometer can detect a charge of 5×10^{-16} C, it follows that a change in temperature of less than 10^{-6} °C is sufficient to produce an observable effect. If this change is brought about by the absorption of infrared radiation, a type of thermal detector is obtained. This possibility was suggested as long ago as 1938 (Yeou Ta) and was used by Chynoweth to study the behavior of ferroelectric materials. A detailed analysis of the mechanism of this type of infrared detector was first made by Cooper (1962a,b). Cooper's calculations indicated that it should be possible to construct detectors of performance approaching the ideal. Unfortunately, the results obtained so far have fallen somewhat short of this expectation. The difficulty is that in addition to a large pyroelectric coefficient, it is necessary to have a small dielectric constant, a small dielectric loss (or high resistivity), and good infrared absorption. Since some of these properties depend upon the perfection of the sample used, it is likely that improved methods of material preparation will lead to results approaching Cooper's calculations. Even with presently available material, some promising results have been reported. Thus Stanford (1965) has obtained an NEP of 1.5×10^{-9} W using a TGS detector. Hadni et al. (1965) have also obtained promising results with a TGS detector and have shown that it is very suitable for use in the 10–1000 μm region. Similar results have also been obtained by Beerman (1967). Ludlow et al. (1967) have constructed detectors using Li_2SO_4, which have an effective response time of <1 μsec and which are therefore suitable for detecting pulses from far infrared lasers. The NEP of these detectors is about 10^{-9} W.

That thermal detectors can have such a short effective time constant will at first seem surprising to everyone familiar with thermal detectors. In fact, the true thermal time constant of the pyroelectric detector is of the same order as that for other types of thermal detector. However, the dominant source of noise in the pyroelectric detectors so far produced appears to be Johnson noise. The frequency spectrum of this is similar to that of the signal at frequencies above the thermal cutoff frequency. Hence, over an extended range, the signal-to-noise ratio is independent of frequency. Finally, amplifier noise will become important, and it is this consideration which determines the maximum useful operating frequency.

At present, the pyroelectric detector has certain specialized applications. It works effectively at long wavelengths without cooling and is fast enough to detect laser pulses. If its noise-equivalent-power can be improved somewhat, it could well become a serious rival to the Golay cell.

5.6. Laser Detectors

It is not proposed to give a detailed account of the techniques for detecting laser radiation, but to compare the problems of detecting laser and thermal radiation.

The sensitivity of the best thermal detectors approaches closely the limit set by background radiation fluctuations. Nevertheless, this limit ($\sim 10^{-11}$ near 10 μm wavelength) is poor compared with that of an ideal radio receiver, for which the noise-equivalent-power $\sim kTB$, so that for $B = 1$ Hz, NEP $\sim 4 \times 10^{-21}$ W at 300°K. The reason for this big difference is that in a simple superheterodyne receiver, the effective radio frequency bandwidth is equal to that of the i.f. amplifier, but an infrared detector is equivalent to a video receiver using a square law detector in which the input rf bandwidth preceding the detector is much wider than the bandwidth of the output circuit. In this case, the noise-equivalent-power for a perfect receiver is $\sim kT(B_{\text{input}}B_{\text{output}})^{1/2}$. In fact, when Eq. (5.15) is calculated for a detector whose response is limited to the submillimeter region, it reduces to an expression of this form. Therefore, if we can reduce the effective infrared bandwidth to which the detector responds, the performance should approach that of an ideal radio receiver. This statement needs several qualifications. The first is to consider how far the simple expression for a radio receiver, NEP $\sim kTB$ is valid. This has been discussed in detail by several authors, including McLean and Putley (1965). It turns out that this result will not be seriously in error until the wavelength becomes so short that the energy of an individual signal photon becomes $\sim kT$. As the wavelength is reduced beyond this point, the fluctuation of the background radiation from the narrowband B of the optical spectrum becomes small compared with the fluctuation in the rate of arrival of the signal photons. This crossover occurs at about 10 μm for room temperature background.

The next question is how the narrow optical bandwidth can be achieved. If we use a narrowband transmission filter at the same temperature as the background, this will itself provide background radiation so that it will be ineffective in reducing the noise. When we are using a cooled detector, we can usefully employ a cooled filter and this is a useful technique where we want to use a relatively broad spectral band. To produce bandwidths of a few megahertz or less (easily achieved in a radio receiver) is not practicable by this technique.

There are some detection systems based on narrowband effects, such as the quantum counter and the analogous nonlinear optical frequency converter. Both these systems convert an infrared signal into a visible one which is then detected by a photomultiplier. The only system which at present can approach the ideal limit is the simple heterodyne one in which a laser is used to provide the coherent local oscillator radiation. Receivers of this type have been operated both in the visible (Ross, 1966) and at 10 μm (Teich et al., 1966; Buczek and Picus, 1967) which closely approach the performance of the perfect narrowband receiver.

On reading some papers on optical heterodyne receivers, one sometimes gets the impression that very specialized detectors are required. In fact, any of the standard broadband detectors can be used, as Gebbie et al. (1967) have shown. Of course, it is true that one may wish to use a heterodyne system for an application with a very high information handling rate so that a bandwidth at the detector of 1 GHz or more is stipulated. Then, a specially designed detector will be required.

The final point to be considered is the application which this type of receiver has in radiometric and related measurements. As is well known, devices such as the Dicke microwave radiometer provide a very sensitive means of detecting small changes in temperature. Would there be any advantage in applying this technique at 10 μm where, using the CO_2 laser as the local oscillator, receivers of high sensitivity have been built? If we compare the performance we could obtain with a 10 μm narrowband receiver, with that obtainable with a broadband detector at 10 μm (which we can do using the expressions developed in Section 5.2), we find that, in fact, the best performance is achieved with the broadband detector. For a heterodyne radiometer to be of use, it must operate at a wavelength where the photon energy is small compared with kT where T is the background temperature. Thus, with room temperature background, a heterodyne radiometer might be useful at 100 μm wavelength, but not at 10 μm or shorter wavelengths. Although for radiometry, near infrared heterodyne receivers are not useful, they could have other applications. For example, one is often uncertain as to the optical characteristics, reflectivity, and emissivity of the surface being studied. Use of a laser and heterodyne detector could perhaps lead to a simple method of making these measurements; systems for measuring distance, by interferometeric methods, and velocities, by Doppler shifts, could also be important in certain studies.

References

Alkemade, C. T. J. (1959). On the excess photon noise in single beam measurements with photo-emission and photoconductive cells. *Physica* **25**, 1145–1158.
Andrews, D. R., Milton, R. M., and De Sorbo, W. J. (1946) A fast superconducting bolometer. *J. Opt. Soc. Am.* **36**, 518–524.
Bartlett, B. E., Charlton, D. E., Dunn, W. E., Ellen, P. C., Jenner, M. D., and Jervis, M. H. (1969). Background limited photoconductive HgCdTe detectors for use in the 8–14 micron atmospheric window. *Infrared Phys.* **9**, 35–36.
Beattie, A. R., and Landsberg, P. T. (1958). Auger effect in semiconductors. *Proc. Roy. Soc. (London)* **A249**. 16–29.
Beerman, H. P. (1967). Pyroelectric infrared radiation detector. *Am. Ceram. Soc. Bull.* **46**, 737–740.
Blakemore, J. (1962). " Semiconductor Statistics," 381 pp. Pergamon Press, New York.
Buczek, C. J., and Picus, G. S. (1967). Heterodyne performance of Hg-doped Ge. *Appl. Phys. Letters* **11**, 125–126.
Burstein, E., Oberly, J. J., Davisson, J. W., and Henvis, B. W. (1951). The optical properties of donor and acceptor impurities in silicon. *Phys. Rev.* **82**, 764.
Cady, W. G. (1946). " Piezoelectricity," 806 pp. McGraw-Hill, New York.
Cooper, J. (1962a). Minimum detectable power of a pyroelectric thermal receiver. *Rev. Sci. Instrum.* **33**, 92–95.
Cooper, J. (1962b). A fast-response pyroelectric thermal detector. *J. Sci. Instrum.* **39**, 467–471.
Fuson, N. (1948). The infra-red sensitivity of superconducting bolometers. *J. Opt. Soc. Am.* **38**, 845–853.
Gebbie, H. A., Stone, N. W. B., Putley, E. H., and Shaw, N. (1967). Heterodyne detection of sub-millimetre radiation. *Nature* **214**, 165–166.
Hadni, A., Henninger, Y., Thomas, R., Vergnat, P., and Wyncke, B. (1965). Sur les propriétés pyroelectriques de quelques materiaux et leur application à la détection de l'infrarouge. *J. Phys.* **26**, 345–360.
Hilsum, C., and Simpson, O. (1959). The design of single-crystal infrared photocells, *Proc. Inst. Elect. Engrs.* **106B**, 398–401.
Jones, R. C. (1953). The general theory of bolometer performance. *J. Opt. Soc. Am.* **43**, 1–14.
Kimmitt, M. F., Prior, A. C., and Roberts, V. (1965). Far-infrared techniques. *In* " Plasma Diagnostic Techniques " (R. H. Huddlestone and S. L. Leonard, eds.), Chap. 9, pp. 399–430. Academic Press, New York.
Kinch, M. A., and Rollin, B. V. (1963). Detection of millimetre and sub-millimetre wave radiation by free carrier absorption in a semiconductor. *Brit. J. Appl. Phys.* **14** 672–676.
Kruse, P. W. (1965). Photon effects in $Hg_{1-x} Cd_x Te$. *Appl. Opt.* **4**, 687–692.
Kruse, P. W., McGlauchlin, L. D., and McQuistan, R. B. (1962). " Elements of Infrared Technology," 448 pp. Wiley, New York.

Laff, R. A., and Fan, H. Y. (1961). Carrier lifetime in indium antimonide. *Phys. Rev.* **121**, 53–62.
Lawson, W. D., Nielsen, S., Putley, E. H., and Young, A. S. (1959). Preparation and properties of HgTe and mixed crystals of HgTe–CdTe. *J. Phys. Chem. Solids*, **9**, 325–329.
Lewis, W. B. (1947). Fluctuations in streams of thermal radiation. *Proc. Phys. Soc. (London)* **59**, 34–40.
Long, D. (1967). Generation-recombination noise limited detectivities of impurity and intrinsic photoconductive 8–14 μ infrared detectors. *Infrared Phys.* **7**, 121–128.
Low, F. J. (1961). Low-temperature germanium bolometer. *J. Opt. Soc. Am.* **51**, 1300–1304.
Ludlow, J. H., Mitchell, W. H., Putley, E. H., and Shaw, N. (1967). Infrared radiation detection by the pyroelectric effect. *J. Sci. Instrum.* **44**, 694–696.
Martin, D. H., and Bloor, D. (1961). The application of superconductivity to the detection of radiant energy. *Cryogenics* **1**, 159–165.
McLean, T. P., and Putley, E. H. (1965). The performance of ideal receivers of optical, infrared and radio frequency radiation. *Roy. Radar Establishment J.* **52**, 5–34.
Melngailis, L., and Calawa, A. R. (1966). Photovoltaic effect in Pb_xSn_{1-x} Te diodes. *Appl. Phys. Letters* **9**, 304–306.
Morten, F. D., and King, R. E. J. (1965). Photoconductive indium antimonide detectors. *Appl. Opt.* **4**, 659–663.
Novak, R. (1959). A fast InSb bolometer. *Fifth Conf. Intern. Comm. Optics*, Stockholm.
Petritz, R. L. (1959). Fundamentals of infrared detectors. *Proc. I.R.E.* **47**, 1458–1467.
Putley, E. H. (1964). Far infrared photoconductivity. *Phys. Stat. Sol.* **6**, 571–614.
Putley, E. H. (1965). Indium antimonide submillimetre photoconductive detectors. *Appl. Opt.* **4**, 649–657.
Putley, E. H. (1966). Solid state devices for infrared detection. *J. Sci. Instrum.* **43**, 857–868.
Putley, E. H., and Martin, D. H. (1967). "Detectors Spectroscopic Techniques" (D. H. Martin, ed.), Chap. 4, pp. 113–151. North-Holland, Amsterdam.
Richards, P. L., (1964). High-resolution Fourier transform spectroscopy in the far infrared *J. Opt. Soc. Am.* **54**, 1474–1484.
Rogers, W. M., and Powell, R. L. (1958). "Tables of Transport Integrals." *Nat. Bur. Std. Circ.* **595**, July 3. 46 pp.
Ross, M. (1966). "Laser Receivers," 405 pp. Wiley, New York.
Smith, R. A., Jones, F. E., and Chasmar, R. P. (1968). "The Detection and Measurement of Infrared Radiation," 2nd. ed., 503 pp. Oxford Univ. Press, London.
Stanford, A. L. (1965). Detection of electromagnetic radiation using the pyroelectric effect. *Solid-State Electron.* **8**, 747–755.
Teich, M. C., Keyes, R. J., and Kingston, R. H. (1966). Optimum heterodyne detection at 10.6 μm in photoconductive Ge : Cu. *Appl. Phys. Letters* **9**, 357–360.
van Vliet, K. M. (1967). Noise limitations in solid state photodetectors. *Appl. Opt.* **6**, 1145–1169.
Vérié, C., and Ayas, J. (1967). Cd_xHg_{1-x} Te infrared photovoltaic detectors. *Appl. Phys. Letters* **10**, 241–243.
Wheeler, R. G., and Hill, J. C. (1966) Spectroscopy in the 5 to 400 wavenumber region with the Grubb Parsons Interferometric spectrometer. *J. Opt. Soc. Am.* **56**, 657–665.
Yeou, Ta. (1938). Effets des radiation sur les cristaux pyroelectrique (Possibilité de leur utilisation comme detecteurs de radiations infrarouges). *Compt. Rend. Acad. Sci.* **207**, 1042–1044.
Zitter, R. N., Strauss, A. J., and Attard, A. E. (1959). Recombination processes in p-type indium antimonide. *Phys. Rev.* **115**, 266–273.

— 6 —
BLACKBODIES AS ABSOLUTE RADIATION STANDARDS

R. E. Bedford

6.1. Introduction

All matter, by reason of its temperature, continuously emits and absorbs electromagnetic energy in the form of *thermal radiation*, so-called because it originates from the thermal excitations of the constituent atoms and molecules. That there is an intimate relationship between these reciprocal processes of emission and absorption is expressed by Kirchhoff's law,[1] which states that the ratio of the *spectral radiant emittance* $H(\lambda)$ (radiant power emitted per unit area per unit wavelength interval at wavelength λ) to the *spectral absorptivity* $\alpha(\lambda)$ (fraction of the total incident radiant power per unit wavelength interval at wavelength λ which is absorbed) is the same for all thermal radiators at the same temperature:

(6.1) $\qquad H(\lambda)/\alpha(\lambda) = \text{constant} \qquad$ (for a given temperature)

Of all possible thermal radiators, we shall direct our attention first to an ideal case—that of a body (called a *blackbody*) which completely absorbs all radiation incident upon it. It follows from Eq. (6.1) that for a blackbody (since $\alpha(\lambda) = 1$)

(6.2) $\qquad H_b(\lambda) = \text{constant} \qquad$ (for a given temperature)

or, stated otherwise, the spectral radiant emittance of a blackbody ($H_b(\lambda)$) at a given wavelength is determined only by its temperature. The specific

---

[1] There is by no means unanimous opinion on the implications of Kirchhoff's law or on the conditions under which it is strictly valid. For example, most derivations of the law require the system to be in thermodynamic equilibrium, although the law is in reality a more general relationship applying also for some nonequilibrium conditions. A detailed discussion of the various interpretations of Kirchhoff's law is given by Agassi (1967).

equation relating $H_b(\lambda)$ to temperature [T, K (see notes added in proof at end of chapter)] and wavelength (λ) is Planck's law[2]:

(6.3) $$H_b(\lambda)\, d\lambda = c_1 \lambda^{-5}\, d\lambda/[\exp(c_2/\lambda T) - 1]$$

where c_1 and c_2 are constants, called the first and second radiation constants respectively (see also the discussion at the end of this section). If Planck's law is integrated over all wavelengths to give the total radiant emittance of a blackbody (H_b), the Stefan–Boltzmann law results:

(6.4) $$H_b = \sigma T^4$$

where σ is the Stefan–Boltzmann constant, and $\sigma = \pi^4 c_1/15 c_2^4$ The constants c_1, c_2, and σ are related to the fundamental atomic constants by the relations

(6.5) $$c_1 = 2\pi h c^2$$

(6.6) $$c_2 = hc/k$$

(6.7) $$\sigma = 2\pi^5 k^4/15 h^3 c^2$$

where h is Planck's constant, k is the Boltzmann constant and c is the velocity of electromagnetic radiation in vacuum.

A blackbody such as we have been considering does not exist; it may, however, be approximated arbitrarily closely by an isothermal cavity bounded by opaque walls with a small opening for the escape of radiation. In such a cavity, temperature equilibrium of the surface (a necessary condition for blackbody radiation) is attained through multiple reflections. It follows from Eqs. (6.1) and (6.2) that the spectral radiant emittance of any real body cannot be greater than that of a blackbody at the same temperature, that is

(6.8) $$H(\lambda) \leqslant H_b(\lambda).$$

For a real body, let us define the *spectral emissivity*[3] $\varepsilon(\lambda)$ to be

(6.9) $$\varepsilon(\lambda) = H(\lambda)/H_b(\lambda)$$

---

[2] We should point out here that Planck's law may equally well be formulated in terms of *spectral radiance* $N(\lambda)$ (radiant power emitted in a given direction per unit solid angle per orthogonally projected unit area per unit wavelength interval at wavelength λ) or *spectral radiant energy density* $u(\lambda)$ (radiant energy per unit volume per unit wavelength interval at wavelength λ). In either case, the functional form of Eq. (6.3.) is unchanged; only definition of the constant c_1 is altered. Further, Eq. (6.3) is valid for either unpolarized or polarized radiation, so long only as c_1 is properly defined. In Eq. (6.5) c_1 is defined for unpolarized radiation; for completely plane polarized radiation c_1 is smaller by a factor of 2.

[3] Current literature sometimes distinguishes between *emissivity* and *emittance*, the former being a property of a *material* and the latter a property of a *real surface*. The two are numerically equal if the real specimen of the material is opaque and has an optically smooth surface free from contamination, defects, etc. Confusion may arise, however, between the use of *emittance* in this context, and its use to denote a flux density. Here we will always use emittance in the latter sense, and denote the ratio of Eq. (6.9) by emissivity.

where $\varepsilon(\lambda) \leqslant 1$. Then, for a given temperature it follows from Eqs. (6.1), (6.2), and (6.9) that

(6.10) $$\varepsilon(\lambda) = \alpha(\lambda)$$

If Kirchhoff's law[4] be formulated in terms of *total radiant emittance* (H), then by similar reasoning we obtain

$$H/\alpha = \text{constant} = H_b$$

and

(6.11) $$\varepsilon = \alpha \quad \text{(at a given temperature)}$$

where ε and α are *total emissivity* and *total absorptivity*, respectively.

The relations (6.3), (6.4), (6.10), and (6.11) form the basis of absolute blackbody radiometry. Let us consider some of their implications. From Eqs. (6.3) and (6.4) the total and spectral radiant emittances of a blackbody are determined solely by its temperature. From thermodynamic reasoning, one can also show that blackbody radiation is isotropic (i.e., it is unpolarized and diffusely emitted). None of these qualities are even approximately true for radiation from many real substances. For example, radiation from a polished metal surface is strongly polarized, partially or wholly specularly reflected, and is highly dependent on the surface structure. The spectral distribution of the radiant emittance is generally irregular; the total radiant emittance may be obtained from Eqs. (6.3) and (6.9) as

(6.12) $$H = \int_0^\infty \varepsilon(\lambda) H_b(\lambda) \, d\lambda.$$

For the important special case of $\varepsilon(\lambda)$ independent of wavelength, $\varepsilon(\lambda) \equiv \varepsilon$, and from Eqs. (6.4) and (6.12), we have

(6.13) $$H = \varepsilon \sigma T^4$$

(where ε may be dependent on T). Such radiators are called *gray bodies* and include most "cavity radiators." If $\varepsilon(\lambda)$ varies with λ, Eq. (6.13) is only an approximation to Eq. (6.12). Many nonconductors such as oxides, paints, etc. behave approximately as gray bodies over rather broad spectral regions.

If we define the *spectral reflectivity* $\rho(\lambda)$ and the *spectral transmissivity* $\tau(\lambda)$ as the fractions of the total incident radiant power per unit wavelength interval at wavelength λ, which are reflected and transmitted respectively, then

(6.14) $$\alpha(\lambda) + \rho(\lambda) + \tau(\lambda) = 1$$

---

[4] Kirchhoff originally formulated the law in this way; the generalization represented by Eq. (6.1) is now usually also ascribed to him.

For an opaque body, $\tau(\lambda) = 0$; and from Eqs. (6.10) and (6.14) we obtain

(6.15) $$\varepsilon(\lambda) = 1 - \rho(\lambda)$$

Similarly, we may write

(6.16) $$\varepsilon = 1 - \rho$$

where ρ is the *total reflectivity*. Values of emissivities are frequently obtained from experimental measurements of reflectivities by means of Eqs. (6.15) and (6.16).

For most opaque bodies, radiation is absorbed by and emitted from a thin surface layer. Consequently, in applications of the preceding equations, one must ascertain that it is the surface temperature of the body, and not the interior temperature, which is being used. The two may differ substantially.

It is important to consider the limitations of Eq. (6.1) when the radiators are not in thermal equilibrium. In most experimental measurements of emissivities and reflectivities, this is in fact the case; Eqs. (6.10)–(6.16) must then be used with caution with selective radiators. For example, Ashby and Shocken (1965) write Eq. (6.11) in the form

$$\varepsilon = \eta\alpha$$

where η is a function dependent on the temperatures of the radiator (T_s) and the incident radiation (T_r). For the extreme case of a highly polished metal and $T_s/T_r \sim 10$, they compute $\eta \sim 0.9$; if $T_s \ll T_r$, η may be as high as 3.

Thus far, we have implicitly assumed the radiation transfer to be occurring in vacuum. In reality, the spectral radiant emittance is also dependent on the index of refraction (n) of the medium in which the radiating body is immersed. As a consequence, if Planck's law be formulated in terms of frequency (ν) rather than wavelength, the spectral radiant emittance varies with n^2. If further, n is independent of frequency, the transformation of $H_b(\nu)$ to $H_b(\lambda)$ leads to a factor n^2 in the numerator on the right-hand side of Eq. (6.3), and a factor n in the denominator of the exponent. Similarly, by integration, the total radiant emittance becomes $n^2\sigma T^4$. For those cases, when n is wavelength dependent, the dependence of radiant emittance upon index of refraction is not so simply expressed. Seldom in radiometry is either the source or detector immersed in other than vacuum, air, or a simple gas, so that in practice the error incurred by neglecting the index of refraction is usually negligible (for air, $\sim 0.06\%$).

The radiation laws (Eqs. (6.3) and (6.4)) may also be modified to include a relativistic correction arising from the velocities of the electrons producing the radiation. This correction is negligible for our purposes, and only becomes significant for wavelengths smaller than 0.1 Å and temperatures greater than 10^8 K.

6.2. Principles of Absolute Total Blackbody Radiometry

Absolute blackbody radiometry may be defined as the measurement in absolute magnitude of the total or spectral radiant flux or radiant flux density from a blackbody. Such measurements would appear to be impossible since an ideal blackbody does not exist. However, as we have pointed out, the theoretical blackbody may be closely approximated experimentally. Henceforth in this discussion, we shall understand by the term blackbody a good experimental approximation to a theoretical blackbody.

A blackbody is an ideal source for absolute radiation measurements; the total energy radiated, and its spectral distribution, are determined as simple functions of the temperature; the spatial distribution is given by Lambert's law. From an initially impossible measurement, we have seemingly arrived at a rather simple one. Wherein lies the fallacy?

A typical radiation measurement consists of alternately comparing the radiant flux density from a blackbody, at a high temperature, with that from a blackbody at a lower temperature, using some type of detector as the comparison device. It is generally not possible to obtain a meaningful result with only a single source, because the detector inevitably records as well the radiation from the surround, the magnitude of which is difficult to estimate; further, we must always bear in mind that it is the *net radiant exchange* between the detector and the source (including all the surround) which is being measured. The use of two sources implies some mechanism to provide alternate viewing by the detector; commonly, the cooler source forms part of a movable shutter, allowing the detector and hotter source to remain stationary. Let us consider then the *total* radiant heat transfer in such a system, subject to the following assumptions:

(a) The detector is a gray body whose emissivity is independent of temperature,
(b) the surround is a gray body at constant temperature,
(c) the measurements are made in vacuum or in a nonabsorbing medium, and
(d) an absolute measurement of the *total radiation* is desired, with an uncertainty approaching 0.1 %.

When the detector views the hotter blackbody, and when interreflections between the detector and the blackbody are insignificant, the net radiant flux density W_1 absorbed at the detector surface is given by

(6.17) $$W_1 = \sigma \varepsilon_0 \varepsilon_1 F_{01}(T_1^4 - T_{01}^4) + R$$

where

ε_0 is the emissivity of the detector,
ε_1 is the emissivity of the blackbody,
F_{01} is a nondimensional geometrical factor,
T_1 is the temperature of the blackbody,
T_{01} is the temperature of the detector surface when exposed to radiation from the blackbody, and
R is the net radiant flux density due to the surround.

When the hot blackbody is replaced by the cooler one (which will be called the shutter), we may write similarly

(6.18) $$W_2 = \sigma \varepsilon_0 \varepsilon_2 F_{02}(T_2^4 - T_{02}^4) + R$$

where

ε_2 is the emissivity of the shutter,
T_2 is the temperature of the shutter, and
T_{02} is the temperature of the detector surface when exposed to radiation from the shutter.

It requires some care to ensure that the quantity R is the same in both Eqs. (6.17) and (6.18). It depends upon, for example, proper location of the shutter. Many radiation measurements contain systematic errors because R has inadvertently changed between the two measurements.

One can usually ensure that $F_{01} \equiv F_{02}$. Then, subtraction of Eq. (6.18) from Eq. (6.17) yields

(6.19) $$W_1 - W_2 = \sigma \varepsilon_0 F_{01}[\varepsilon_1(T_1^4 - T_{01}^4) - \varepsilon_2(T_2^4 - T_{02}^4)]$$

Equation (6.19) gives the absolute net radiant flux density at the detector surface due to the blackbodies at known temperatures, thus allowing an absolute calibration of the detector. The equation is frequently written in the simpler form

(6.19a) $$W_1 - W_2 = \sigma \varepsilon_0 \varepsilon_1 F_{01}(T_1^4 - T_2^4).$$

If T_1 is sufficiently high, this approximation is valid; however, for lower temperature sources and a desired accuracy of 0.1%, the transition from Eq. (6.19) to Eq. (6.19a) must be made with caution. Ideally, $T_2 = T_{02}$, and the second term on the right side in Eq. (6.19) vanishes. If $T_2 \neq T_{02}$, the magnitude of this second term can usually be made small enough for its contribution to be negligible. This is not necessarily true, however, if ε_2 is large; for example, a few tenths of a degree difference between T_{01} and T_{02} may cause Eq. (6.19a) to be in error by 0.1% if $T_1 = 200°C$. Again, it is frequently true that $\varepsilon_1 \sim \varepsilon_2 \sim 1$. However, if a plane black shutter is used, ε_2 may be as low as 0.85, and the effect of this on Eq. (6.19a) must be examined.

calculations in each experiment of type (a) neglected a number of important (and seemingly obvious) corrections such as absorption by the surrounding medium, reflection from the receiving surface, thermal impedance of the black coating on the receiving surface, improper use of the shutter, emissivity of the cavity, and elimination of assorted stray radiation (sometimes, these corrections were later made by the original author, sometimes by others—the work of Coblentz is about the only exception); (iii) the measurements of type (b) give a lower average for σ_0 and are subject to fewer corrections.

In view of the general accuracy and "after-thought" corrections, the only experiment of the above group giving σ_0 significantly different from σ is that of Coblentz, who found $\sigma_0 = (5.722 \pm 0.012) \times 10^{-12}$ W cm^{-2} deg^{-4}. Because his range of σ_0 covers about 2 %, his estimate of uncertainty may be low; Gerlach (1912, 1913a, b, and c, and 1916a, b, and c) estimated it at closer to ± 0.06. Also, Coblentz operated his blackbody around 1080°C and measured the temperature with thermocouples calibrated on a scale essentially the same as the IPTS. Consequently, the correction of Section 6.3(c) would lower his result about 0.5 % to $\sigma_0 = 5.69$.[7]

In summary, the measurements of σ_0 prior to 1920 give a probable value of $\sigma_0 = 5.7 \pm 0.1$. We might also observe that in 1920 σ_c, computed from Eq. (6.7), had the value 5.72. Hence, workers at that time had no reason to suspect that $(\sigma_0 - \sigma_c)$ was significantly different from zero.

The later determinations of σ_0 are characterized not only by higher apparent accuracy, but also by higher values for σ_0 (ranging from 5.723 to 5.795). Two of these used the electrical compensation method. Kussman (1924) used a furnace in the temperature range 850–1250°C and observed $\sigma_0 = 5.650$ as the mean of 178 observations lying between the limits 5.50 and 5.69. He corrected this value by 0.2 to 0.8 % for absorption by water vapor (the partial pressure of which was recorded), and by 2.5 to 3.2 % for reflection from his receiver surface, finally obtaining $\sigma_0 = 5.795$. He estimated the effect of several systematic errors (the sum of which was 2.3 %) and claimed an overall accuracy of about 1 %. We note that his correction for reflection is about double that of Coblentz for rather similar surfaces.

Müller (1933) used a furnace at the gold point and a black receiver. His work seems to have been particularly carefully done and has the advantage of *no* correction for either reflection or absorption for some of the measurements. As the mean of 37 determinations between 5.67 and 5.81, he obtained $\sigma_0 = 5.771$.

Hoare (1928) performed a novel variation of the electrical compensation method. He used a Callendar radio-balance to measure the radiation from a

---

[7] In discussions of the numerical value of σ we will frequently neglect writing both the units and the factor 10^{-12}.

furnace at 800 to 1000°C, and obtained $\sigma_0 = 5.735$ as the mean of 38 observations lying between 5.681 and 5.760. This experiment is important because the compensation method differs in principle from the others, and because the receiver was truly black. In a second paper (1932a), he reported another set of measurements, this time using a blackbody at 100°C. The mean of 50 observations in the range 5.689–5.804 was $\sigma_0 = 5.737$ [see also Hoare (1932b)].

The first three results (but not Hoare's second value) should all be corrected by about 0.5 % for the difference between the IPTS and the thermodynamic temperature scale. In view of this correction, the close agreement between Hoare's two values must be considered fortuitous.

Hoffmann (1923) measured σ_0 by the emissivity substitution method at temperatures of 100 to 200°C, obtaining $\sigma_0 = 5.764 \pm 0.052$ as the mean of 24 observations in the range 5.52–5.80.

Wachsmuth (1921) used a substitution method with blackened concentric spheres, the inner one being maintained hotter (200 to 400°C) than the outer one with electrical power. Instead of using different surface emissivities, however, he varied the heat losses by changing the pressure and the type of gas between the spheres, obtaining finally $\sigma_0 = 5.73$.

Mendenhall's (1929) experiment is another variation of the substitution method. He contained a blackbody with a lid inside a blackened, temperature-controlled enclosure, and measured the power difference necessary to keep the blackbody temperature constant (near 100°C) when the lid was opened and closed. His result, $\sigma_0 = 5.79$, is subject to several "residual errors difficult to evaluate operating to reduce the observed value of σ." We also note that he used both gold and platinum lids, the former leading to $\sigma_0 = 5.79$ and the latter to $\sigma_0 = 5.63$. Errors due to lack of knowledge of the emissivity of the lids seem quite probable.

The only recent work[8] bearing on σ_0 is that of Eppley and Karoli (1957) who estimated, from absolute radiation measurements, $\sigma_0 = 5.77$. Corrections for reflection, absorption, and temperature scales are also involved here.

How to suitably weight these numerous experiments to obtain a "best" value of σ_0 is perplexing. Müller finds $\sigma_0 = 5.77$, Hoare 5.742 ± 0.027, Birge 5.735 ± 0.011, Wensel 5.69 ± 0.13, and Bearden and Thomsen 5.74 ± 0.03.

The chief uncertainty mentioned in most of the reports is associated with the reflectivity of the receiving surface. Certainly, such low reflectivities are not easily determined accurately. To what extent the spectral selectivity of these surfaces affects the results is seldom considered. Again, although some mention it, none of the authors using the electrical compensation method

---

[8] E. J. Gillham (private communication, 1965) has obtained $\sigma_0 \sim 5.74$ as the result of some preliminary experiments using an absolute radiometer and a low-temperature blackbody. (See notes added in proof at the end of this chapter.)

applied a correction to his observed results for the nonsymmetric method of heating the receiving surface. The radiant power is applied to the blackened face of the receiver, but the electrical power is usually applied directly to the substrate. That the thermal impedance of this black coating is not negligible has been discussed occasionally in the literature, most recently by Blevin and Brown (1966); they report the results of actual measurements of the thermal resistivities of several black coatings.

Another source of systematic error is stray radiation. From our experience, this may contribute significantly unless extreme precautions are taken; stray radiation in these experiments would lead to σ_0 being observed too high. None of the authors estimate the magnitude of this effect. Temperature scale corrections also reduce σ_0.

If we take as the average experimental value $\sigma_0 = 5.731$ and apply a correction of 0.5 % for the temperature scale, we obtain $\sigma_0 = 5.702$, which differs from the computed value (σ_c) by 0.6 %. The accuracy of σ_0 is not better than this.

6.5. Emissivities of Blackbody Cavities

Total radiation measurements to within 0.1 % require knowledge of the emissivities ε_0, ε_1 of Eq. (6.19a) to within 0.1 %. This is particularly difficult in the case of the emissivity of the detector, ε_0, if the detector is a plane black surface (for example, many thermopiles and bolometers). Even more important, the detector blackening (paint, metal black, etc.) is usually spectrally selective. For truly absolute radiation measurements, knowledge of ε_0 may be the limiting factor (as, for example, in the previously discussed determinations of σ); it is then advisable to make the detector also in the form of a blackbody cavity. On the other hand, many absolute measurements employ the detector only as a transfer or comparison device; ε_0 may then be included in the responsivity of the detector, its precise value is not required, and only the spectral selectivity is critical.

A number of methods have been described for calculating the apparent emissivity ε_1 of a blackbody cavity as a function of its surface emissivity ε, its geometry, and the nature of the internal reflections. Assuming diffuse reflection and a nonabsorbing medium in the cavity, Buckley (1927, 1928, 1934) developed an expression for the emissivity of a cylinder in the form of an integral equation, of which he obtained an approximate solution by substituting an approximate kernel in the integral. The method is not readily generalized to other configurations.

Introducing the further assumption that radiant flux entering the cavity is uniformly distributed after two reflections, Gouffé (1945) developed a closed expression[9] for the emissivity:

(6.21) $$\varepsilon_1 = [1 + (1-\varepsilon)(a/A - f)]/[\varepsilon(1-a/A) + a/A]$$

where a is the area of the opening, A is the total internal surface area (including the opening), and f is the fraction of the radiant flux leaving the cavity after a single reflection. For a spherical cavity, Eq. (6.21) is exact; however, because of its simplicity, it is widely used for any cavity for which $a \ll A$ and a and A are calculable.

De Vos (1954) devised a much more elaborate method in which he accounts for the nature of the internal reflections (and to some extent for any temperature gradients). He defines the following quantities:

r_s^{bc} is the partial (or directional) reflectivity of a surface element ds; that is, the fraction of the radiation reflected by ds in the direction c *per unit solid angle* when the radiation is incident from direction b;

$d\Omega_s^b$ is the solid angle subtended by surface element db at surface element ds;

ρ_s^b is the hemispherical reflectivity of a surface element ds, i.e., the fraction of the radiation incident on ds from db which is reflected into a hemisphere;

ε_s^b is the emissivity of ds in the direction b;

θ_s^b is the angle between the normal to ds and a line joining ds to db.

It follows from these definitions that

$$\rho_s^b = \int_{2\pi} r_s^{bn} \, d\Omega_s^n$$

Now consider a cavity of arbitrary shape having an opening da in the wall, and consider a small surface element ds on the interior wall. De Vos arrived at an expression for the apparent emissivity of ds, as seen through da, by treating in turn (a) radiation emitted by ds and reaching da directly, (b) radiation emitted from another surface element dn and reaching da by reflection from ds, and (c) radiation emitted from a surface element dm and reaching da after reflections from first dn, and then ds, and so on. He obtained, finally, for the apparent emissivity

(6.21a)

$$\varepsilon_a = 1 - r_s^{aa} \, d\Omega_s^a - \int r_n^{sa} \, r_s^{an} \, d\Omega_n^a \, d\Omega_s^n - \int r_m^{na} \, r_n^{sm} \, r_s^{an} \, d\Omega_m^a \, d\Omega_n^m \, d\Omega_s^n - \cdots$$

[9] Gouffé's original expression contained a factor a/A_0 in place of f, in which A was defined as the surface area of a sphere of diameter equal to the depth of the blackbody in the direction normal to the opening. Although the expression given by Gouffé for a/A_0 is the same as that for f in Eq. (6.21), it cannot be derived from his definition.

6. BLACKBODIES AS ABSOLUTE RADIATION STANDARDS

In deriving Eq. (6.21a), de Vos assumed the validity and made multiple use of the Helmholtz reciprocity relation

$$r_s^{na} \cos \theta_s^n = r_s^{an} \cos \theta_s^a$$

He also assumed that the partial reflectivity r_s^{na} is constant within the solid angle $d\Omega_s^a$. The validity of both these assumptions should be examined. The latter is certainly not true for some of the extreme types of reflection used by de Vos, unless the solid angle $d\Omega_s^a$ is small.

Equation (6.21a) should, in principle, give a good approximation to the apparent emissivity of a cavity and has the advantage that the type of reflection from the walls is explicitly included. Also, by suitably defining $d\Omega_s^a$, one can compute the apparent emissivity of that portion of the cavity actually viewed by the detecting system. Further, by including the wavelength dependence of the partial reflectivities, one can determine the spectral dependence of ε_a, although the calculation is tedious if the area viewed is of substantial size.

De Vos applied Eq. (6.21a) to a sphere, a cylinder, and an open-ended cylinder viewed through a hole in the side wall, using diffuse reflection and three types of partially specular reflection. He computed only the second-order correction [i.e., he neglected all correction terms after the second on the right-hand side of Eq. (6.21a)], although his results appear to indicate that, in most cases, at least one more term should have been included. His accuracy was further impaired because he used only approximate expressions for the solid angles $d\Omega_s^a$.

Edwards (1956) used the same reflectivities and applied the theory to a cone. Also, he claimed an arithmetic error in de Vos' calculations, presenting corrected values for the sphere and cylinder. These corrections considerably lowered de Vos' values for the cases of specular reflection and, if valid, imply that several terms in Eq. (6.21a) may be necessary. The difficulty of the calculation is thereby greatly increased.

Campanaro and Ricolfi (1966a) also reviewed de Vos' calculations for a sphere and claim to have improved them by eliminating two of his approximations and by including the third-order correction term. In spite of the fact that successive terms in Eq. (6.21a) are inherently positive, these authors obtained values for the apparent emissivity of a sphere even higher than de Vos' original values. This could suggest arithmetic errors in de Vos' values. Campanaro and Ricolfi (1966b) also discuss the effect of temperature gradients along the cavity wall on the effective emissivity, and in a later paper (1967) consider cylindrical and conical cavities.

Williams (1961) questions the validity of some of these calculations.

Perhaps the chief conclusion which can be drawn from these discrepant calculations is that application of Eq. (6.21a) is, in practice, formidable. However, a good deal of the confusion has recently been resolved by Quinn (1967). He has applied essentially the de Vos method to a closed-end cylinder and computed the emissivity of the bottom of the cylinder. He avoided several of the approximations of the previous authors; in particular, he used an exact expression for the solid angle $d\Omega_s{}^a$ and allowed r_s^{na} to vary within $d\Omega_s{}^a$. For the cylinder with diffusely emitting walls, Quinn used only the first- and second-order correction terms and obtained exactly the same result as did Sparrow *et al.* (1962) by the type of calculation outlined below. He concludes that de Vos' calculations are incorrect both because of the approximations included and because of arithmetic errors. Further, Quinn's results indicate that, at least for cylinders with length/radius > 6, correction terms in Eq. (6.21a) beyond the second are negligible.

Recently, Sparrow and his co-workers (1960, 1962, and 1963) have paid a good deal of attention to the problem. Their method is similar to Buckley's; they derive essentially the same integral equation, but solve it by a different method. The emittance of any point on the cavity wall is considered to include both emitted and reflected radiant flux. Assuming diffuse reflection, they obtain an integral equation for the apparent emissivity and solve it numerically with the aid of a computer. Let the total radiant flux per unit area leaving an area element dA_x at position x on the cavity wall be $B(x)$. $B(x)$ is sometimes called the *radiant exitance* of dA_x. Then, for a typical location $x = x_0$, we may write

$$(6.22) \qquad B(x_0) = \varepsilon \sigma T^4 + \rho G(x_0)$$

where $\rho = 1 - \varepsilon$ and $G(x_0)$ is the radiant flux per unit area incident on dA_{x_0} due to reflections from all other parts of the cavity. ($G(x_0)$ is the *irradiance* of dA_{x_0}.) The total radiant flux leaving dA_x at x is $B(x)\,dA_x$, and of this an amount $B(x)\,dA_x\,d^2F_{x,x_0}$ reaches dA_{x_0}, where d^2F_{x,x_0} is a geometric factor giving the fraction of the radiant flux leaving dA_x which reaches dA_{x_0}. For diffuse radiation, it may be shown that

$$(6.23) \qquad dA_x\,d^2F_{x,x_0} = dA_{x_0}\,d^2F_{x_0,x}$$

It follows that

$$(6.24) \qquad G(x_0) = \int B(x)\,d^2F_{x_0,x}$$

Then, Eq. (6.22) may be written

$$(6.25) \qquad B(x_0) = \varepsilon \sigma T^4 + (1-\varepsilon)\int B(x)\,d^2F_{x_0,x}$$

The apparent emissivity $\varepsilon_a(x_0)$ of dA_{x_0} is defined by

(6.26) $$\varepsilon_a(x_0) = B(x_0)/\sigma T^4$$

which, using Eq. (6.25), becomes

(6.27) $$\varepsilon_a(x_0) = \varepsilon + (1-\varepsilon) \int \varepsilon_a(x)\, d^2 F_{x_0,x}$$

(If the cavity under consideration is composed of multiply connected surfaces (e.g., a closed-end cylinder), a set of simultaneous equations is obtained; each equation of the set is of the form of Eq. (6.27) and may contain several terms on the right-hand side.)

An initial assumed distribution for $\varepsilon_a(x)$ is used in the integral on the right-hand side of Eq. (6.27), and the integral evaluated numerically. $\varepsilon_a(x_0)$ is obtained by iteration, the first solution being inserted into the integral to generate a second, etc. It is found, in practice, that convergence is quite rapid, even if the initial choice for $\varepsilon_a(x)$ is a poor one. The calculation of the quantities $d^2 F_{x_0,x}$ for given geometrical configurations is tedious but not difficult. The emissivity of the cavity as a whole, ε_1, may be obtained by suitable integration of $\varepsilon_a(x_0)$ over the whole cavity surface, allowance being made for the fraction of the radiant flux reaching the cavity opening from dA_{x_0}.

In some cases of extreme configurations (e.g., a cone with small angle or a cylinder with large ratio of depth to radius), inaccuracies may appear in the numerical integration of $\varepsilon_a(x_0)$ for certain values of x_0 at which the kernel of Eq. (6.27) is singular, unless a large number of points are used. Sometimes, even this may be precluded by limited computer storage facilities. We have recently (Bedford, 1967) developed an alternative method of solution of this equation, which largely eliminates these difficulties and is applicable to most commonly used cavity shapes. In brief, we consider $\varepsilon_a(x)$ to be constant over a small range of the co-ordinate x; this allows $\varepsilon_a(x)$ to be taken outside the integral, which in turn may be evaluated analytically over this small range of x. We obtain $\varepsilon_a(x_0)$ as the sum of many such terms and iteration proceeds as before.

One advantage of Sparrow's method is that for any particular source-detector geometry, the emissivity of that portion of the cavity actually viewed by the detector may be obtained by suitable integration. Such a quantity is not as readily obtained with the other methods.

Lin and Sparrow (1965) have generalized their method for the case of diffusely emitting but specularly reflecting walls and obtain, in general, higher emissivities for the cavities. They (Sparrow and Lin, 1965) also consider various intermediate types of reflection containing both specular and diffuse components.

Where results for similar configurations have been computed, all of these methods agree quite closely and indicate that ε_1 is not very sensitive to changes in ε when ε itself is high. Cavity emissivities may probably be computed to the desired accuracy of 0.1 % and may be made to approach closely to unity.

Experimental verification of these formulas is difficult. Michaud (1948) used cylinders of varying depths and verified Eq. (6.21) to within about 2 %. Vollmer (1957) used shallow cylinders and verified Buckley's result to within a few percent. Bauer (1961) compared de Vos' and Gouffé's results with experimental measurements made on cylinders, obtaining reasonable agreement. He measured the flux radiated normally when the cylinder was diffusely illuminated by means of a lamp in an integrating sphere at the opening, as well as the hemispherically emitted flux (using an integrating hemisphere at the cylinder opening) when the cavity was normally illuminated. O'Brien and Heckert (1965) measured, in a direction normal to the opening of a V-groove cavity, the luminous radiant flux from various positions along the cavity wall. To obtain diffuse illumination of the cavity, they mounted it near the wall of an 8-ft diameter integrating sphere. Their experimental values are from 2 to 4 % higher than the predictions of Sparrow *et al*. The most extensive test seems to be that of Kelly and Moore (1965) who also used shallow cylinders and measured reflectances in the visible region. For a depth to radius ratio varying from 0.5 to 2.0 and diffusely reflecting walls, their results agreed with the theories of Gouffé and Sparrow *et al*. Buckley's results predicted ε_1 up to 7 % too high. For specularly reflecting walls the deviations were greater.

6.6. Configuration Factors

The final quantity to be considered in Eq. (6.19a) is the geometrical factor F_{01}, which involves the areas of the source and detector and the solid angle subtended by the source at the detector. Suppose we first consider the general case of radiant heat exchange between two small surface elements dA_1 and dA_2, with emissivities ε_1 and ε_2 respectively, temperatures T_1 and T_2 respectively, diffusely radiating and arbitrarily situated. Let dA_1 be distant l from dA_2, and let the line joining dA_1 and dA_2 make an angle θ_1 with the normal to dA_1 and an angle θ_2 with the normal to dA_2. Then, we may write for the total radiant flux ($dP_{1,2}$), emitted by dA_1 and absorbed by dA_2,

(6.28) $$dP_{1,2} = \varepsilon_2 I_{\theta_1} dA_1 d\omega_2$$

where $d\omega_2$ is the solid angle subtended by dA_2 at dA_1, and I_{θ_1} is the radiant intensity per unit area in direction θ_1. For diffuse radiation,

(6.29) $$I_{\theta_1} = I_n \cos \theta_1$$

where I_n is the normal radiant intensity per unit area, given by

(6.30) $$I_n = \varepsilon_1 \sigma T_1^4 / \pi$$

We also have

(6.31) $$d\omega_2 = dA_2 \cos \theta_2 / l^2$$

Using Eqs. (6.29)–(6.31), Eq. (6.28) may be rewritten

(6.32) $$dP_{1,2} = \varepsilon_1 \varepsilon_2 \sigma T_1^4 \cos \theta_1 \cos \theta_2 \, dA_1 \, dA_2 / \pi l^2$$

In a similar way, we could have obtained for the radiant flux, emitted from dA_2 and absorbed by dA_1,

(6.33) $$dP_{2,1} = \varepsilon_1 \varepsilon_2 \sigma T_2^4 \cos \theta_1 \cos \theta_2 \, dA_1 \, dA_2 / \pi l^2$$

Subtraction of Eq. (6.33) from Eq. (6.32) gives the net absorbed radiant flux at dA_2.

For two finite areas A_1 and A_2, the net absorbed radiant flux at A_2 is obtained by integration over dA_1 and dA_2, to give

(6.34) $$P_{1,2} - P_{2,1} = \varepsilon_1 \varepsilon_2 \sigma (T_1^4 - T_2^4) A_1 f_{1,2}$$

where

(6.35) $$f_{1,2} = (1/A_1) \int_{A_1} \int_{A_2} (\cos \theta_1 \cos \theta_2 \, dA_1 \, dA_2 / \pi l^2)$$

Equations (6.34) and (6.35) are strictly valid only when multiple reflections between the areas A_1 and A_2 are not significant, and when the emittance is constant over the surface. If, for the emittance $\varepsilon \sigma T^4$, we substitute the radiant exitance (defined by Eq. (6.22)), then we obtain a more general result. Equations (6.34) and (6.35) can also be generalized to include the case of an absorbing medium between A_1 and A_2. For example, if the absorption follows the Beer–Lambert law (which is *not* the case for the earth's atmosphere), a term $e^{-\gamma l}$ appears under the integral, where γ is the absorption coefficient. f_{12} is termed the configuration factor and equals the fraction of the radiant flux emitted by A_1 which reaches A_2. The quantity F of Eq. (6.19a) is related to f by

(6.36) $$F_{01} = A_1 f_{10} / A_0.$$

If the dimensions of the source and detector are small relative to their distance apart, F_{01} may be approximated by $A_1 / \pi l^2$, the inverse square law. However, if this simplification is not sufficiently accurate, F_{01} may be evaluated explicitly. For example, for two coaxial disks of radii $0.1l$, the inverse square law differs by 2 % from the true value of F_{01}. In such a system, an error of 0.1 % may be incurred by a slight tilt of one disk relative to the other, or by a small displacement of one disk normal to their common axis.

Several approximate methods for the numerical calculation of configuration factors have been devised, including a number of graphical methods and an electrical analog "network method." More recently, Toups (1965, a and b) has prepared computer programs (Confac I and Confac II) which will evaluate the configuration factor for radiant exchange between any general polygon and another general polygon or polyhedron. Since many curved surfaces may be approximated sufficiently closely by substituting a series of plane segments these programs will handle a large number of cases.

6.7. Low-Temperature Blackbody as an Absolute Standard of Total Radiation

We will now discuss some of the features involved in the design of a blackbody to serve as an absolute standard of total radiation using, as an illustration, the blackbody standard of the National Research Council of Canada (NRC) (Bedford, 1960) and taking as a feasible goal the "magic number" of 0.1 % uncertainty. Generally speaking, no single blackbody can adequately cover the whole spectral range—one designed to operate at, say, 1000°C does not usually function satisfactorily at 200°C, if for no other reason than the radiant power becoming too small. The first step then is to match the blackbody to its specific task. Is infrared radiation to be measured predominately, or visible radiation, etc.?

Thus, the design problems for a low-temperature blackbody (which we shall arbitrarily define as one operating between 0 and 200°C) differ, in many respects, from those for a higher temperature blackbody. The first consideration is size. To obtain a reasonable level of radiant flux densities at the detector, the cavity and its aperture must be relatively large. The blackbody is usually constructed in the shape of a cone or cylinder, or some modification thereof; the practical difficulties in the construction of a sphere, for example, usually far outweigh any small theoretical gain in its performance. How to maintain and measure a uniform temperature over this large cavity is another major task.

The NRC low-temperature, total radiation standard, designed to operate between 40 and 150°C, is shown on the right-hand side of Fig. 6.1. It was felt that, for a large blackbody in this temperature range, isothermal conditions could best be maintained with a stirred liquid bath, rather than with, say, an electric heater wound directly on the outer surface of the cavity. Efficient stirring in the bath dictated a conical rather than a cylindrical cavity. To obtain the desired length to diameter ratio in a reasonable size, we used a cylindro-conical cavity—a cone 56 cm long of half-angle 9°, with an 18-cm cylindrical extension at its base and the end partially closed by a plate containing a beveled aperture 12.7 cm in diameter. The blackbody is immersed

6. BLACKBODIES AS ABSOLUTE RADIATION STANDARDS 185

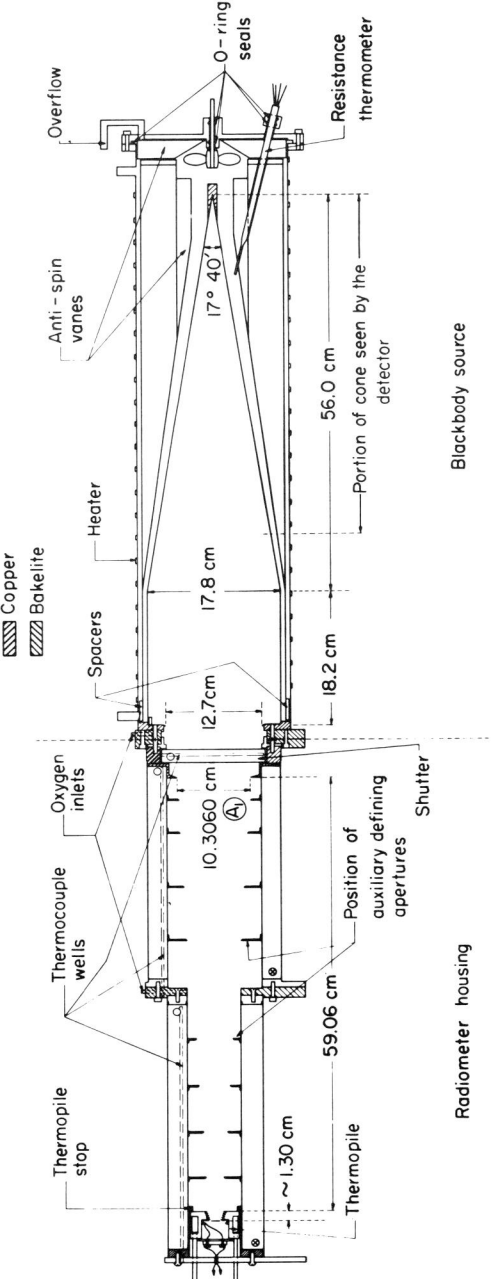

Fig. 6.1. Blackbody source and radiometer housing [reproduced by permission of the National Research Council of Canada from Bedford (1960)].

in stirred oil in a brass tank; to obtain efficient heat transfer to the inner cavity surface, the cavity wall is constructed of thin (0.10 cm) copper sheet. The oil is circulated over the outer surface of the blackbody from apex to base by a propeller, appropriate channeling of the oil flow being obtained with suitable antispin vanes mounted on an inner can, as shown in the figure. Heat is supplied to the oil by a chromel ribbon, uniformly wound on the outer surface of the brass tank and powered by a 2-kW variable autotransformer. The temperature of the oil is measured with a platinum capsule resistance thermometer. The inner surface of the cavity is blackened with commercial flat black paint.

Before the blackbody is used as an absolute standard, a number of checks are necessary.

(a) *Emissivity:* Given a cavity of the shape shown in Fig. 6.1, we estimated its departure from an ideal blackbody on the basis of the several methods outlined in Section 6.5. The reflectivity of the paint on the walls was measured to be 0.05 in the visible region; on the basis of published data on similar paints, we estimate its emissivity (ε) in the infrared to be 0.85 to 0.90. From this and the dimensions, Gouffé's analysis (Eq. (6.21)) gives, for the blackbody emissivity, $\varepsilon_1 = 0.998$. Using Edwards' (1956) extensions to a cone of de Vos' (1954) theory for partially specular reflection, we estimate for our blackbody $\varepsilon_1 \sim 0.999$. Since our original publication (Bedford, 1960), Sparrow and Jonsson's (1963) results for a cone have become available. Applying their data to our blackbody (with suitable extrapolation), we estimate $\varepsilon_1 \sim 0.996$–0.999. On the basis of these calculations, we take $\varepsilon_1 = 0.998 \pm 0.001$. An exact calculation of ε_1, for our geometry, assuming diffuse reflection, is presently in progress. Some idea of the blackness of the cavity may be obtained from Fig. 6.2, which is a view of the cavity through a portion of the radiometer housing.

(b) *Temperature gradients:* Probably the chief cause of nonblackbody conditions in a source such as ours is lack of temperature uniformity over the whole cavity. To measure the gradients, we soldered differential copper-constantan thermocouples at suitable locations on the blackbody surface (both inner and outer). A number of interesting features were observed which it may be useful to record here.

1. The brass tank must be kept completely full of oil for temperature gradients in the liquid to be smaller than 0.01 °C. Even a small amount of air in the tank adversely affects the stirring. This caused some complications (due to oil expansion and contraction), resulting from the desire to raise and lower the blackbody temperature. A system whereby the oil expands into an exterior reservoir and automatically siphons itself back has proved satisfactory for keeping the tank full.

Fig. 6.2. View of the blackbody cavity through a portion of the radiometer housing. The baffle defining the black cavity is coated with commercial flat black paint.

2. To keep temperature gradients in the oil smaller than 0.01 °C, a minimum propeller speed is necessary, which however was not very high in our case.

3. With the blackbody in a horizontal position, the temperature of the blackbody surface near the open end was as much as 1 °C lower than that of the oil. This cooling was caused by convection losses from the cavity. When the blackbody is operated vertically (base down), the conical surface is isothermal to within 0.01 °C, while the extreme open end is not more than

0.1°C lower. In any blackbody with a large aperture it would appear that convection losses represent an extremely important problem, yet the topic is seldom discussed in the literature.

6.8. Experimental Difficulties in the Transfer of a Total Radiation Scale to Secondary Devices

Although the blackbody itself defines the scale of total radiation, we must of course have some means of measuring its radiant flux. Thus, in the design of an absolute radiation standard, the source and detector really form an integrated unit; the "standard detector" then becomes a secondary standard to which the radiation scale is transferred. The calibration of other types of detectors is most easily carried out by comparing them with the standard detectors; direct calibrations against the blackbody are usually awkward (or impossible). Another set of secondary standards may be generated by calibrating, in turn, with the standard detectors a set of tungsten-filament lamps (for example), operated under specific electrical and geometrical conditions. Which secondary standards are better, depends on their long-term stability, their particular application, etc.

At NRC the standard detectors are a set of Eppley circular-type, 8- and 16-junction, bismuth-silver thermopiles. Each thermopile is enclosed in its own low-reflection type case, as shown in Figs. 6.3 and 6.4. This, in turn is mounted in the radiometer housing, as shown on the left-hand side of Fig. 6.1. The water-cooled housing contains the shutter at the extreme right, the circular aperture A_1 which defines the portion of the blackbody seen by the detector, a set of eight baffles for limiting the transmission of reflected radiation, and the detector whose position is reproducibly determined by a mechanical stop. The thermopile and radiometer housing are maintained at a uniform temperature (near room temperature) by circulating water from a temperature-controlled reservoir through the shutter, the first section of the housing, the thermopile case, and the second section of the housing, in series in this order. The temperatures of these various components (as well as the temperature differences between them) are measured with copper-constantan thermocouples inserted in German silver wells, as indicated.

With the apparatus shown we desire to measure absolute total radiation, hopefully to within 0.1 %. The difficulties encountered in approaching this goal are probably typical of those involved in any radiation measurements using low-temperature blackbodies. Hence, we shall enumerate several of these possible causes of systematic errors.

(1) The temperature of the blackbody (T_1) is readily measured with the platinum resistance thermometer to within 0.01 °C on the IPTS. The surface

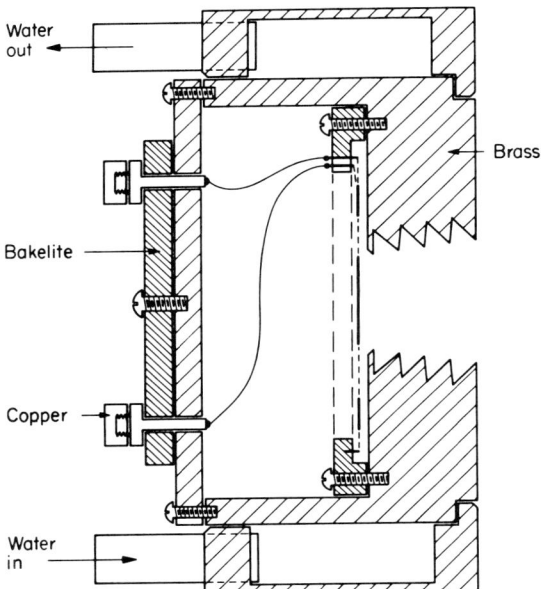

Fig. 6.3. Thermopile case [reproduced by permission of the National Research Council of Canada from Bedford (1960)].

temperature of the cone does not differ from the thermometer temperature by more than 0.01°C (discussed in Sec. 6.7(b)). In this temperature range, differences between the IPTS and the thermodynamic scale are not significant for our purpose.

(2) Transfer of heat by convection from the blackbody to the housing and thermopile must be carefully examined in any setup. Accurate measurements are impossible with our system used horizontally. When the shutter is opened the thermopile reaches a quasi-equilibrium, and then a slow steady increase in signal is superposed on considerable random drifting. At the same time, T_1 drops steadily, while various components of the housing heat appreciably. No really stable condition is ever reached. By the simple expedient of turning the system vertical with the source uppermost, we are able to reduce these effects to an insignificant level. Figure 6.5 is a photograph of the blackbody and thermopile housing mounted vertically, together with the instruments for recording the blackbody temperature and the thermocouples for measuring the housing temperatures.

(3) The thermopile signal is continuously recorded and, except for very low signals, is measured with a precision better than 0.1 % using standard

Fig. 6.4. Photograph of a thermopile case showing the serrated wall.

potentiometric procedures. If amplification is required, a photocell-galvanometer amplifier is used which, to attain the desired zero stability, is mounted on a spring-suspended vibration-free iron block. The equipment is shown in Fig. 6.6.

(4) Proper operation of our apparatus requires that the temperatures of the shutter, the housing, the various baffles, and the thermopile case do not change when the shutter is opened and closed, and that all these temperatures be closely the same. This was tested using differential copper-constantan thermocouples. Although the temperature of the whole unit changes several hundredths of a degree as the blackbody temperature changes from 40 to 150°C, no significant temperature differences are observed.

6. BLACKBODIES AS ABSOLUTE RADIATION STANDARDS 191

FIG. 6.5. Total radiation standard. On the right are shown the blackbody and the radiometer housing mounted in a vertical position. On the left is the readout equipment for the platinum resistance thermometer which measures the blackbody temperature.

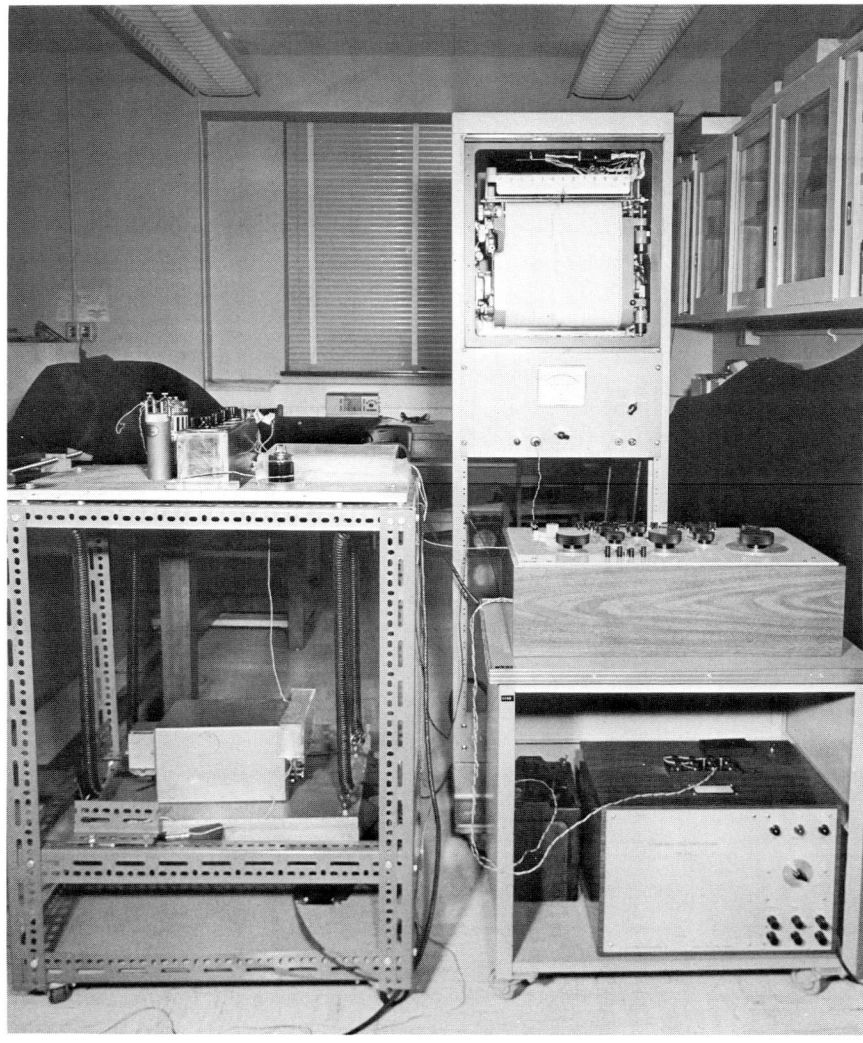

Fig. 6.6. Equipment used for recording the thermopile output. The photocell galvanometer amplifier is shown on its antivibration mount, while the 5-figure vernier potentiometer, thermostatted standard cell enclosure and recording potentiometer appear at the right.

(5) In our apparatus, the shutter is plane with an emissivity $\varepsilon_2 \sim 0.85$. In Eq. (6.19), then, we do not have $\varepsilon_1 = \varepsilon_2$. However, as we discussed earlier, T_2 is sufficiently close to T_{02} that we may neglect the second term in parentheses. We have also shown (Bedford, 1960) that T_{01} in the first term may be replaced by T_0, the temperature of the thermopile case. T_0 is measured to within 0.01°C with a copper-constantan thermocouple inserted in a German silver well in the thermopile case.

(6) From Eq. (6.19), the radiant flux measured by the thermopile is directly proportional to the emissivity ε_0 of the detecting surface. In our application, where the thermopile is used as a transfer device, we need not known ε_0 explicitly. It is simply included in the value of the responsivity. For utmost accuracy in the transfer, however, it is vitally important that ε_0 be independent of wavelength (graybody); if it is not, the responsivity obtained by calibration against the blackbody is valid only when the thermopile views a source with the same spectral energy distribution. We use both gold-black and carbon-black thermopiles. Over the temperature range of our blackbody, both are gray. Various tests with incandescent light sources indicate that the former are also gray to much shorter wavelengths. The latter are not, with differences of up to 2% or more in responsivity being obtained if the source emits predominately visible radiation (see Chapter 3, this volume).

(7) One of the most probable causes of systematic error, and one which is most difficult to evaluate quantitatively, is radiation reflected from the housing to the detector surface. Consider first the thermopile case itself. A small amount of reflection may be tolerated here providing its magnitude is the same for all types of sources; it is included in the responsivity. Reflection from the serrated wall of our thermopile (Fig. 6.3) is estimated to be somewhat less than 1%. Reflection from the housing and baffles (Fig. 6.1) must be eliminated. To evaluate the amount of reflection, measurements were made with the housing painted and unpainted, with several different kinds of paints and with various baffle positions, using different defining apertures at the positions indicated. We estimate that less than 0.1% of the thermopile signal is due to radiation reflected from the housing. Our measurements indicated two important sources of stray radiation:

(a) A large part of it arises from reflections from the face of the limiting aperture A_1 itself. (It is important that the baffles shield as much as possible of A_1 without reducing the aperture of the system; their alignment is critical.)

(b) Reflections from the edges of the baffles themselves can be surprisingly large. (When the edge thicknesses of the baffles averaged 0.02–0.05 cm, radiation reflected from them contributed about 0.3% of the observed signal; reduction of the edge thicknesses to about 0.003 cm made the reflection undetectable.) A photograph of this baffle system is presented in Fig. 6.7.

Fig. 6.7. View of the baffles in the radiometer housing. (The bright spot at the rear is on the face of the thermopile case.)

(8) An ideal blackbody is a Lambertian emitter. For small solid angles, we tested our blackbody by using four different limiting apertures (smaller of course than A_1) at the positions indicated in Fig. 6.1. Within the accuracy of the measurements, we obtained the same thermopile responsivity with each. When using these apertures, we have an example of the importance of the geometrical factor F_{01} (Eq. (6.36)). It was necessary to measure, accurately, the relative alignment of the apertures and the thermopile surface to compute F_{01} to within 0.1 %. In some cases, the effect of tilt and displacement of the aperture effectively offset the thermopile center as much as 0.7 cm from the normal axis of the aperture, giving F_{01} about 0.2 % different from the coaxial disk expression.

(9) Accurate total radiation measurements demand that no absorption occurs in the medium separating the blackbody and the detector. Water vapor and carbon dioxide, both present in varying amounts under normal atmospheric conditions, have strong absorption bands in the near infrared. Ideally, the radiation should be measured in vacuum. In our case, this is inconvenient; we therefore fill the blackbody and radiometer housing with a dry, nonabsorbing gas, usually oxygen. After a preliminary flushing, a

continuous slow trickle of gas is sufficient to prevent re-entry of air. We measure an absorption of 4 to 4.5 % in our apparatus when it contains air with a relative humidity of about 35 %.

(10) The responsivities of most thermal detectors are dependent on both the thermal conductivity and the pressure of the surrounding medium. We have calibrated our thermopiles in nitrogen, dry air, argon, and helium, as well as in oxygen, and have made relative measurements in these gases at pressures ranging from atmospheric to vacuum using a tungsten-filament lamp as source. The changes in responsivity with pressure and thermal conductivity are not inconsiderable, as shown in Fig. 6.8. The responsivity is 1.0 % higher in nitrogen than in oxygen.

(11) We have observed another effect associated with the removal of moist air from the apparatus. A stable thermopile responsivity is not reached until about 24 hours after introduction of the dry gas; during this period it systematically decreases about 1 %. We have not isolated the origin of the effect,

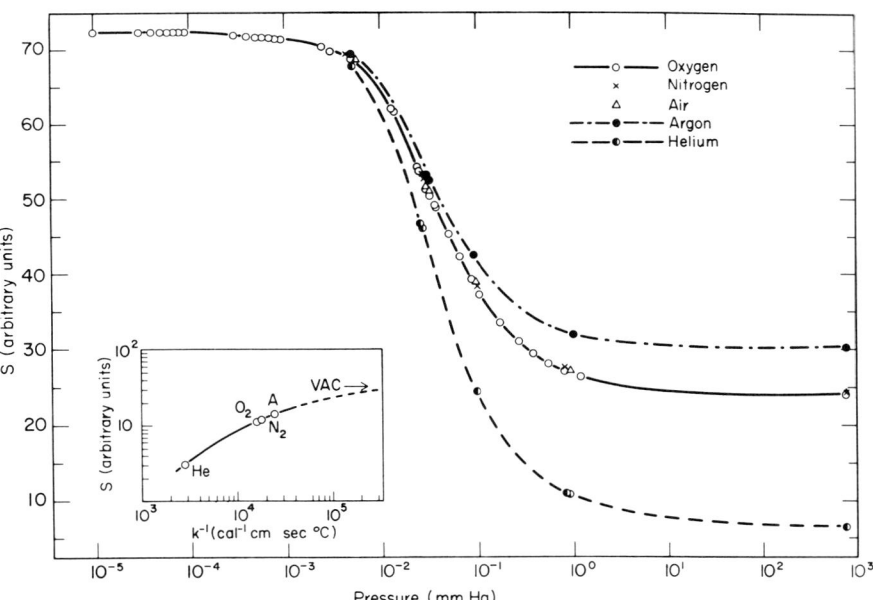

FIG. 6.8. Responsivity of a thermopile as a function of the pressure. The responsivity (S) is plotted linearly in arbitrary units and the pressure on a logarithmic scale in mm Hg. The solid line gives results for oxygen, the broken line for helium, and the broken-dotted line for argon. Some points for nitrogen (crosses) and dry air (triangles) are also given. The inset is a logarithmic plot of the responsivity against the reciprocal thermal conductivity of the gas, at atmospheric pressure [reproduced by permission of the National Research Council of Canada from Bedford (1960)].

although it seems to be associated with the removal of absorbed water vapor from the receiving surface.

(12) We have found that thermopiles of the type used here sometimes suffer a change (usually a decrease) in responsivity by as much as 1 % after an initial evacuation. Thereafter, they appear to remain stable over extended periods. This is illustrated by the data presented in Table III. Thermopile number 2 was evacuated subsequent to the 1960 calibration.

TABLE III. Responsivities of three thermopiles.

Thermopile	Type of Black	Responsivity ($\mu V/\mu W$ cm^{-2})		
		1960	1966	1967
1	carbon black	0.1231	0.1234	
2	carbon black	0.1163	0.1156	0.1157
3	gold black	0.0942	0.0943	

(13) Many thermal detectors are used with windows. The chief advantage of this is to produce generally a more stable output; the detector is much less sensitive to pressure changes, convection currents, etc. Also, it allows for easy evacuation or special gas filling of the detector. However, for absolute measurements of total radiation we feel that lack of adequate knowledge of the window transmission, and the difficulty of correcting for it if it were known, introduces too large an uncertainty. We prefer to use no window and tolerate the atmospheric disturbances.

(14) The responsivity of a thermopile is also dependent on its temperature—one cannot calibrate the instrument at one temperature and use it at another without knowing its "temperature coefficient." The responsivities of our thermopiles decrease linearly with temperature (over a considerable range near room temperature) at 0.2 to 0.4 %/°C. Examples are shown in Fig. 6.9.

(15) In some radiation measuring systems, diffraction by the various apertures may introduce systematic errors. In our apparatus the effect is negligible.

6.9. Accuracy of Total Irradiance Standards

In the preceding sections, we have discussed, in detail, one method of establishing an absolute scale of total irradiance using a low-temperature blackbody as the primary source. Such scales are maintained in other national standards laboratories, based either on blackbodies operating at considerably higher temperatures (e.g., near the freezing temperature of gold) or on absolute

6. BLACKBODIES AS ABSOLUTE RADIATION STANDARDS

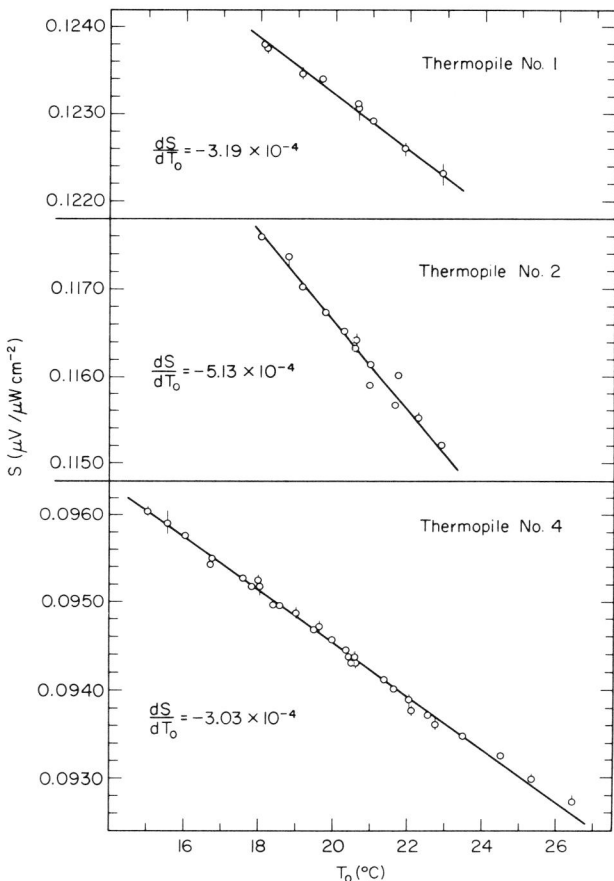

FIG. 6.9. Responsivities (S) of three thermopiles as functions of the temperature (T_0) of the cold junctions [reproduced by permission of the National Research Council of Canada from Bedford (1960)].

detectors as the primary standards. (Examples of the latter are described elsewhere in this volume.) It is obviously desirable that these national scales be mutually consistent, and agree with other related references such as the International Pyrheliometric Scale (IPS)—see Chapter 1.

About three years ago, an international comparison of irradiance scales was organized by the London National Physical Laboratory (NPL) on behalf of the Comité Consultatif de Photométrie of the Bureau International des Poids et Mesures, with eight national laboratories participating. Each laboratory received three gas-filled tungsten-filament lamps, whose irradiance was measured under specific geometrical and electrical conditions. All of these lamps received similar prior and subsequent calibrations at the NPL. Although some problems with lamp stability occurred, a fairly reliable estimate of the divergence of the national scales was obtained. (See notes added in proof.)

The results of this comparison indicate that six of the eight scales agree to within $\pm 0.5\%$, and that the overall agreement of the eight scales is within $\pm 1.2\%$. The former figure is more realistic; of the two scales diverging most from the mean, one is not based on a primary standard and the other is not reliably represented (the three lamps on which its value is based were found to be unstable).

This disagreement ($\pm 0.5\%$) among the national scales probably gives a realistic estimate of the absolute accuracy of present-day total radiation measurements. It is of more than passing interest that this uncertainty is somewhat smaller than the difference between the experimental and computed values of the Stefan–Boltzmann constant. Further measurements to resolve this latter discrepancy would be in order.

6.10. Applications

Blackbodies find wide application in a variety of radiation measurements, both as sources and as detectors. All of these applications depend on one or more of the blackbody's calculable radiant emittance, spectral energy distribution and emissivity.

Notes Added in Proof

1. On January 1, 1969, the IPTS (more explicitly, the IPTS-48) described herein was replaced by the IPTS-68 [Comité International des Poids et Mesures (1969)]. The numerical values of temperature on this new scale approximate as closely as possible those on the TTS described herein. For example, in Table II, column 5, the differences TTS − IPTS are, to within sufficient accuracy, the same as IPTS-68 − IPTS-48. Whenever the term IPTS appears in the text it is to be understood to mean IPTS-48. Radiometric measurements based upon the new IPTS-68 will not require the correction discussed in Section 6.3 (c). A corresponding pertinent change in nomenclature was introduced by the IPTS-68: The unit of thermodynamic temperature is now the *kelvin* (symbol K) and no longer the degree kelvin (symbol °K).

2. Another recent determination of σ has been described by Kendall (1968), who obtained, as the mean of several measured values, $\sigma = 5.6866$. This is higher than the theoretical value by only 0.3%. Novel features of Kendall's method are: (a) both the source and detector are blackbodies; (b) the detector is always hotter than the source, and at a temperature in the range 20°C to 200°C; (c) the source is at 77 K; (d) the irradiance is determined by measuring the electric power required to maintain the detector, which is radiating to the colder source, exactly at the preset temperature of a guard; (e) the experiment was done in vacuum (5×10^{-6} torr).

3. For a description of this international comparison of radiometric scales see Betts and Gillham (1968) and Bedford and Ma (1968).

References

References preceded by an asterisk are not cited in the text; they are included because of their relevance to the subject.

Agassi, J. (1967). The Kirchhoff–Planck radiation law. *Science* **156**, 30–37.
Ashby, N., and Schocken, K. (1965). The theory of emissivity of metals. *In* "Symposium on Thermal Radiation of solids" (S. Katzoff, ed.), pp. 63–71. NASA SP-55, Washington, D.C.
Bauer, G. (1961). Reflexionmessungen an offenen Hohlräumen. *Optik* **18**, 603–622.
*Bauer, E., and Moulin, M. (1909). Sur la constante de la loi de Stefan. *Compt. rend. Acad.* **149**, 988–990.
*Bauer, E., and Moulin, M. (1910a). Sur la constante de la loi de Stefan et le rayonnement du platine. *Compt. rend. Acad. Sci.* **150**, 167–169.
*Bauer, E., and Moulin, M. (1910b). La constante de la loi de Stefan. *J. Phys. Radium* **9**, 468–490.
Bearden, J. A., and Thomsen, J. S. (1957). A survey of atomic constants. *Nuovo Cimento Suppl.* **5**, 267–360.
Bedford, R. E. (1960). A low temperature standard of total radiation. *Can. J. Phys.* **38**, 1256–1278.
Bedford, R. E. (1967). Unpublished.
Bedford, R. E., and Ma, C. K. (1968). International comparison of measurements of irradiance. *Metrologia* **4**, 111–116.
Betts, D. B., and Gillham, E. J. (1968). An international comparison of radiometric scales. *Metrologia* **4**, 101–111.
Birge, R. T. (1929). Probable values of the general physical constants. *Rev. Mod. Phys.* **1**, 1–73.
Blevin, W. R., and Brown, W. J. (1966). Black coatings for absolute radiometers. *Metrologia* **2**, 139–143.
Buckley, H. (1927). On the radiation from the inside of a circular cylinder. *Phil. Mag.* **4**, 753–762.
Buckley, H. (1928). On the radiation from the inside of a circular cylinder. Part II. *Phil. Mag.* **6**, 447–457.
Buckley, H. (1934). On the radiation from the inside of a circular cylinder. Part III. *Phil. Mag.* **17**, 576–581.
Campanaro, P., and Ricolfi, T. (1966a). Effective emissivity of a spherical cavity. *Appl. Opt.* **5**, 929–932.
Campanaro, P., and Ricolfi, T. (1966b). Radiant emission characteristics of a nonisothermal spherical cavity. *Appl. Opt.* **5**, 1271–1273.
Campanaro, P., and Ricolfi, T. (1967). New determination of the total normal emissivity of cylindrical and conical cavities. *J. Opt. Soc. Am.* **57**, 48–50.
Coblentz, W. W., and Emerson, W. B. (1916). Studies of instruments for measuring radiant energy in absolute value. *Bull. Bur. Std.* **12**, 503–552.
Coblentz, W. W. (1916). Present status of the determination of the constant of total radiation from a black body. *Bull. Bur. Std.* **12**, 553–579.
Coblentz, W. W. (1922). Present status of the constants and the verification of the laws of thermal radiation of a uniformly heated enclosure. *Sci. Papers Bur. Std.* **17**, 7–48.
Comité International des Poids et Mesures (1969). The international practical temperature scale of 1968. *Metrologia* **5**, 35–44.

Committee on Fundamental Constants of the NAS–NRC. (A. G. McNish, chairman). (1963). New values for the physical constants. *Nat. Bur. Std. U.S. Tech. News Bull.* **47** (No. 10), 175–177.

de Vos, J. C. (1954). Evaluation of the quality of a blackbody. *Physica* **20**, 669–689.

du Mond, J. W. M. (1958). Present status of precise information on the universal physical constants. Has the time arrived for their adoption to replace our present arbitrary conventional standards? *I.R.E. Trans. Instr.* **1–7**, 136–175.

Edwards, D. F. (1956). The emissivity of a conical blackbody. *Engineering Research Institute Report* 2144-105-T, 18 pp. Univ. of Michigan Press, Ypsilanti, Michigan.

Eppley, M., and Karoli, A. (1957). Absolute radiometry based on a change in electrical resistance. *J. Opt. Soc. Am.* **47**, 748–755.

*Féry, Ch. (1909). Détermination de la constante de la loi de Stefan. *Compt. Rend. Acad. Sci.* **148**, 915–918.

*Féry, Ch., and Drecq, M. (1911). Sur la constante de la loi du rayonnement. *J. Phys. Radium* **1**, 551–560.

*Féry, Ch., and Drecq, M. (1912a). Sur la constante du rayonnement. *Compt. Rend. Acad. Sci.* **152**, 590–592.

*Féry, Ch., and Drecq, M. (1912b). Sur le pouvoir diffusif du noir de platine et le coefficient de la loi de Stefan. *Compte. Rend. Acad. Sci.* **155**, 1239–1241.

*Féry, Ch., and Drecq, M. (1913). Sur une nouvelle methode pour mesurer le coefficient de la loi de Stefan. *J. Phys. Radium* **3**, 380–384.

Gerlach, W. (1912). Eine methode zur Bestimmung der Strahlung in absoluten Maas und die Konstante des Stefan–Boltzmannschen Strahlungsgesetzes. *Ann. Phys.* **38**, 1–29.

Gerlach, W. (1913a). Zur Kritik der Strahlungsmessungen. I. *Ann. Phys.* **40**, 701–710.

Gerlach, W. (1913b). Zur Kritik der Strahlungsmessungen. II. *Ann. Phys.* **41**, 99–114.

Gerlach, W. (1913c). Zur Kritik der Strahlungsmessungen. III. *Ann. Phys.* **42**, 1163–1166.

Gerlach, W. (1916a). Über die Absorption der schwarzen Strahlung im Wasserdampf und Kohlensäuregehalt der Luft. *Ann. Phys.* **50**, 233–244.

Gerlach, W. (1916)b. Über die Verwendung von Russ und Platinmohr als Schwärzungsmittel des Empfängers bei absoluten Strahlungsmessungen. *Ann. Phys.* **50**, 245–258.

Gerlach, W. (1916c). Die Konstante des Stefan–Boltzmannschen Strahlungsgestzes; neue absolute Messungen zwischen 20 und 450°C. *Ann. Phys.* **50**, 259–269.

Gillham, E. J. (1965). Private communication.

Gouffé, A. (1945). Corrections d'ouverture des corps-noirs artificiels compte tenu des diffusions multiples internes. *Rev. Optique* **24**, 1–7.

Hall, J. A. (1965). The radiation scale of temperature between 175°C and 1063°C *Metrologia* **1**, 140–158.

Heusinkveld, W. A. (1966). Determination of the differences between the thermodynamic and the practical temperature scale in the range 630 to 1063°C from radiation measurements. *Metrologia* **2**, 61–71.

Hoare, F. E. (1928). A determination of the Stefan–Boltzmann radiation constant using a Callendar radio balance. *Phil. Mag.* **6**, 828–839.

Hoare, F. E. (1932a). A determination of the Stefan–Boltzmann radiation constant using a Callendar radio balance. *Phil. Mag.* **13**, 380–392.

Hoare, F. E. (1932b). Note on the most probable value of the Stefan–Boltzmann radiation constant. *Phil. Mag.* **14**, 445–449.

Hoffmann, K. (1923). Bestimmung der Strahlungskonstanten nach der Methode von Westphal. *Z. Phys.* **14**, 301–315.

*Kahanowicz, M. (1917). The constant σ in the Stefan–Boltzmann law. *Nuovo Cimento* **13**, 142–167.
*Keene, H. B. (1913). A determination of the radiation constant. *Proc. Roy. Soc.* (*London*) **A88**, 49–60.
Kelly, F. J., and Moore, D. G. (1965). A test of analytical expressions for the thermal emissivity of shallow cylindrical cavities. *Appl. Opt.* **4**, 31–40.
Kendall, J. M. (1968). The JPL standard total-radiation absolute radiometer. Tech. Rept. 32-1263, 10 pp. Jet Propulsion Laboratory, California Institute of Technology, Pasadena, California.
*Kurlbaum, F. (1898). Über eine Methode zur Bestimmung der Strahlung in absoluten Maass und die Strahlung des schwarzen Körpers zwischen 0 und 100 Grad. *Ann. Phys.* **65**, 746–760.
*Kurlbaum, F., and Valentiner, S. (1913). Erwiderung an Hrn. Gerlach auf die Abhandlung "Zur Kritik der Strahlungsmessungern. II." *Ann. Phys.* **41** 1059–1063.
Kussman, A. (1924). Bestimmung der Konstanten σ des Stefan–Boltzmannschen Gesetzes. *Z. Phys.* **25**, 58–82.
Lin, S. H., and Sparrow, E. M. (1965). Radiant interchange among curved specularly reflecting surfaces—application to cylindrical and conical cavities. *J. Heat Transfer* **87**, 299–307.
Mendenhall, C. E. (1929). A determination of the Stefan–Boltzmann constant of radiation. *Phys. Rev.* **34**, 502–512.
Michaud, M. (1948). Facteur d'emission des cavités de formes géométriques simple. *Compt. Rend. Acad. Sci.* **226**, 999–1000.
*Millikan, R. A. (1916). A direct photoelectric determination of Planck's "h." *Phys. Rev.* **7**, 355–388.
Moser, H. (1965) Gasthermometrie bei höheren Temperaturen. *Metrologia* **1**, 68–73.
Müller, C. (1933). Bestimmung der Konstanten σ des Stefan–Boltzmannschen Strahlungsgesetzes. *Ann. Phys.* **82**, 1–36.
O'Brien, P. F., and Heckert, B. J. (1965). Effective emissivity of a blackbody cavity at nonuniform temperature. *Illum. Eng.* **60**, 187–195.
*Puccianti, L. (1912a). Radiation constant. *Nuovo Cimento* **4**, 31–48.
*Puccianti, L. (1912b). Radiation constant. *Nuovo Cimento* **4**, 322–330.
Quinn, T. J. (1967). The calculation of the emissivity of cylindrical cavities giving near blackbody radiation. *Brit. J. Appl. Phys.* **18**, 1105–1113.
*Shakespear, G. A. (1912). A new method of determining the radiation constant. *Proc. Roy. Soc.* (*London*) **A86**, 180–196.
Sparrow, E. M., and Albers, L. U. (1960). Apparent emissivity and heat transfer in a long cylindrical hole. *J. Heat Transfer* **82**, 253–254.
Sparrow, E. M., Albers, L. U., and Eckert, E. R. G. (1962). Thermal radiation characteristics of cylindrical enclosures. *J. Heat Transfer* **84**, 73–79.
Sparrow, E. M., and Jonsson, V. K. (1963). Radiant emission characteristics of diffuse conical cavities. *J. Opt. Soc. Am.* **53**, 816–821.
Sparrow, E. M., and Lin, S. L. (1965). Radiation heat transfer at a surface having both specular and diffuse reflectance components. *Intern. J. Heat Mass Transfer* **8**, 769–779.
*Todd, G. W. (1909). Thermal conductivity of air and other gases. *Proc. Roy. Soc.* (*London*) **A83**, 19–39. (See especially Sec. 3; A determination of the radiation constant.)
Toups, K. A. (1965a). "A General Computer Program for the Determination of Radiant-Interchange Configuration and Form Factors." Confac I. SID 65-1043-1, 169 pp. North American Aviation, Inc.

Toups, K. A. (1965b). "A General Computer Program for the Determination of Radiant-Interchange Configuration and Form Factors." Confac II. SID 65-1043-2, 331 pp. North American Aviation, Inc.

*Valentiner, S. (1910). Vergleichung der Temperaturmessung nach dem Stefan–Boltzmannschen Gesetz mit der Skale des Stickstoffthermometers bis 1600°. *Ann. Phys.* **31**, 275–311.

*Valentiner, S. (1912). Über die Konstante des Stefan–Boltzmannschen Gesetzes. *Ann. Phys.* **39**, 489–492.

Vollmer, J. (1957). Study of the effective thermal emittance of cylindrical cavities. *J. Opt. Soc. Am.* **47**, 926–932.

Wachsmuth, R. (1921). Neubestimmung der Konstanten des Stefan–Boltzmannschen Strahlungsgesetzes. *Verhandl. Deut. Phys. Ges.* **2**, (No. 3), 36–37.

Wensel, H. T. (1939). International temperature scale and some related physical constants. *J. Res. Nat. Bur. Std.* **22**, 375–395. (See especially Sec. V: Values of c_2 from σ.)

*Westphal, W. H. (1912). Constant of the Stefan law of radiation. *Verhandl. Deut. Phys. Ges.* **14**, 987–1012.

*Westphal, W. H. (1913). Constant of the Stefan radiation law. *Verhandl. Deut. Phys. Ges.* **15**, 897–902. (See also Millikan (1916), p. 379.)

Williams, C. S. (1961). Discussion of the theories of cavity-type sources of radiant energy. *J. Opt. Soc. Am.* **51**, 564–571.

for simple systems of two surfaces. In addition to the desirability of approximating point-source conditions, it is essential that the limiting rays from the detector passing through the source-defining aperture intercept only the interior isothermal surface of the blackbody radiating cavity.

When deviation from a point source is negligible, the net radiant flux density W at the detector may be given by

(7.5) $$W = (\sigma A_1/\pi D^2)(T^4 - T_0^4) \quad [\text{W cm}^{-2}]$$

where σ is the Stefan–Boltzmann radiation constant, A_1 the area of the limiting aperture which acts as the virtual source, D the distance between the limiting aperture and the receiving surface of the detector, T the temperature of the blackbody (°K), and T_0 the temperature of the shutter (°K).

Equation (7.5) holds if we assume that the detector is at the same temperature as the shutter, that the emissivity of the shutter is near unity, and that the temperature of the surrounding does not change when the shutter is opened and closed. For source temperatures above 1000°C (1273°K), correction for T_0 near 300°K is very small, while at low source temperatures, the T_0 correction is significant.

To obtain an absolute calibration (watts) the right-hand side of Eq. (7.5) should be multiplied by the area of the detector receiving surface.

(g) The correction for absorption by atmospheric gases in the source-detector path will depend on the source temperature, path length, and quantity of CO_2 and water vapor (both of which have strong absorption bands in the near infrared) present at the time observations are made. Corrections will range from 0.5% for source temperatures near 1000°C to near 5% for a 50°C source where the path length is approximately 50 cm and the relative humidity is about 40%. Most accurate total radiation measurements require that no absorption takes place in the medium between the source and detector. As vacuum conditions are impractical to obtain, purging with a nonabsorbing gas such as dry nitrogen is necessary.

From the above discussion of the sources of error, it is easily understood why an accuracy of 1% is difficult to achieve in experimental blackbody radiometry.

7.4. Current Status of Cavity Sources

In recent years, because of the increased need to calibrate detectors and systems, largely brought about through the expanding programs in the atmospheric and space sciences, a number of blackbody calibration sources, covering the temperature range of 77° to 3000°K, have been developed. As most of these sources are only approximations to absolute blackbodies, they probably should be referred to as graybodies or cavity-type radiators.

Liquid nitrogen-cooled (77°K) cavity radiators are employed to reasonably simulate the background temperatures of outer space. One, at the National Environmental Satellite Laboratory of ESSA in Suitland, Maryland (Hilleary et al., 1966), has a conical-shaped cavity with an aperture of about 100 cm^2; the interior surface is coated with Eppley–Parsons optical lacquer, and an effective emissivity of at least 0.998 is claimed (Fig. 7.12).

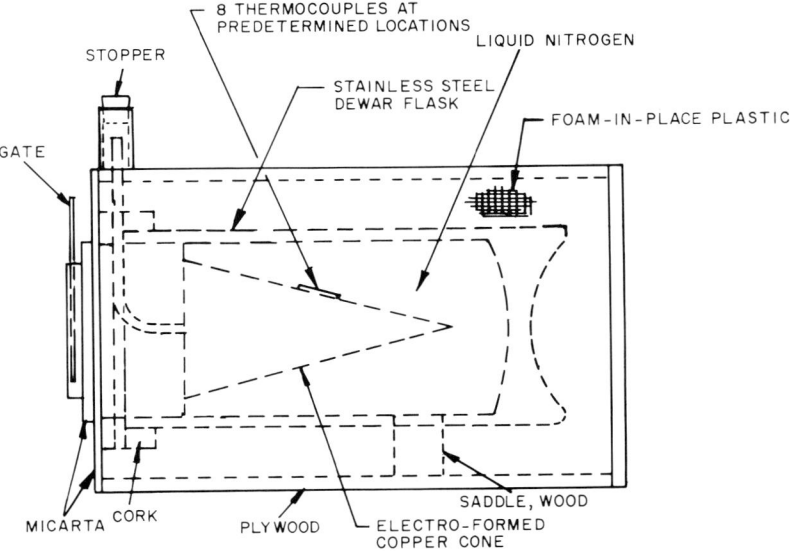

FIG. 7.12. Liquid nitrogen blackbody (courtesy of ESSA). A: thermocouples; B: housing; C: vermiculite insulation; D: refractory sphere; E: cavity sighting tube; F: heater winding.

7.4.1. Types Commercially Available

In addition to the Eppley 1000°C blackbody described above, the Barnes Engineering Company has marketed a series of cavity-type radiation reference sources covering the range 20°C above ambient to 1000°C. Most of these units consist of two separate components, viz., a blackbody simulator and a temperature controller. The simulator consists of an insulated, electrically heated alloy having a conical-shaped radiating cavity, and on a few models, a wheel containing several apertures up to 0.65 cm in diameter. One model is equipped with a collimator which provides an effective 10-cm aperture. A bead thermistor or platinum resistance thermometer senses the temperature which is read from a calibrated dial. Based on the stated emissivity and temperature accuracies, these radiation sources should duplicate absolute blackbody radiation within 5 to 10%.

Electro Optical Industries also manufacture a line of blackbody radiation sources several of which are available with cylindrical or spherical, as well as conical-shaped, cavities having a stated emissivity of 0.99 and openings from 0.1 to 7.5 cm in diameter. Some of the models are miniature in size and weigh as little as 1 lb. The temperature range extends from 50 to 1725°C. Separate controllers are employed for temperature regulation, while a few models are supplied with a thermocouple for measuring cavity temperature. These sources are similar in operation to those of Barnes Engineering and should be expected to have about the same absolute accuracy.

Of interest is a commercially available spherical blackbody manufactured by Land Pyrometers, England, and distributed in the U.S. through Atlantic Pyrometers (see Fig. 7.13). It appears to have the features necessary for an absolute blackbody source. The radiating cavity, spherical in shape, is lined with silicon carbide and heated with a Kanthal wire winding. Models having higher temperature limits of 1200 and 1600°C are available. Two Pt-Pt10%Rh thermocouples, with junctions extending into the cavity, measure the temperature of the source, which is observed through a cylindrical sighting tube and has a specified emissivity greater than 0.999. Employing calibrated thermocouples, this source can probably duplicate an absolute radiator to within 2%.

A high-temperature source capable of 3000°K operation has been designed by Rocketdyne, North American Aviation (Schumacher, 1964). Essentially, it consists of a heated graphite cylindrical tube about 12.5 mm in diameter with a 6.3-mm aperture and an emissivity of 0.98. Features of this source are rapid response time and ease in replacing graphite elements. Water cooling is required for the housing and a constant flow of argon is needed to prevent

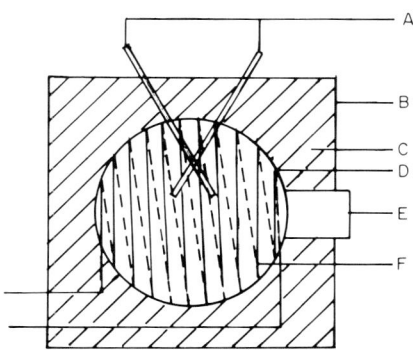

FIG. 7.13. Schematic of spherical blackbody radiator (courtesy of Atlantic Pyrometers).

erosion of the graphite tube. Flexibility in placing the blackbody unit is afforded by interconnecting hoses and electrical cables to the power control-cooling system console. A 5-cm aperture blackbody of similar design has also been built by this company.

A blackbody with a range of 1000 to 3000°C is the high-temperature model of the series of sources produced by Electro Optical Industries. Basically, it has the same design features and principal of operation as the Rocketdyne 3000°K source. An emissivity of 0.99 is claimed for the graphite conical cavity radiator. A removable calcium fluoride window (spectral transmittance range 0.2–11 μm) is provided to reduce argon consumption. Three models employing 1.2-, 2.5-, and 5-cm diameter cavities have maximum apertures of 0.5, 1.5, and 2.5 cm, respectively. With an upper operating limit of 3000°C (3273°K), this is the highest temperature source commercially available.

Recently, the Barnes Engineering Company announced a 3000°K radiation standard consisting of a graphite cylindrical cavity heated by rf induction and enveloped in an atmosphere of argon. The housing is water cooled. An emissivity of at least 0.99 is claimed for the source, with a usable aperture of 12.5 mm diameter. Warm-up time is 25 min from ambient to 3000°K.

It is difficult to estimate the absolute accuracy of the above high-temperature sources without knowledge of cavity temperature uniformity and calibration of the sensor.

7.4.2. Development of Large-Area, Low-Temperature Blackbodies

A low-temperature blackbody source designed to calibrate infrared detectors and systems, such as the Satellite Infrared Spectrometer of ESSA, has been developed by the Eppley Laboratory (Karoli *et al.*, 1967). Essentially, the design incorporates an effective source area of 65 cm^2 uniformly temperature controlled, over an operating range of -35 to $+50°$C, through the utilization of thermoelectric heat pumping. In order to provide the desired source area without constructing a large cone which would be difficult to temperature-control uniformly, a contiguous cavity array was selected. The hexagonal shape was chosen because it approximates a cylinder and because hexagonal structures are commercially available as "honeycomb."

The aluminum honeycomb employed as the blackbody source has a cavity depth of 2.5 cm, an opening of 0.63 cm, and a wall thickness of 0.005 cm. The wall end area of the cavity array is approximately 1.5% of the total source area. Blackened with 3M 101 C10 velvet coating, an emissivity of better than 0.997 has been calculated. The cavity array, mounted on a copper plate, is closely coupled to four matched, two-stage (cascade-type) thermoelectric modules, and is located within a temperature-regulated baffle system which reduces thermal gradients over the source area and along the cavity walls. To improve the temperature stability of the system, the module heat sinks are stabilized by circulating water from the mains. The realizable lower limit is dependent on thermal loading and sink (water) temperature. Nine copper-constantan thermocouples are mounted at five locations on the cavity baseplate and at the center of each of the four baffle units. Figure 7.14 shows the front and back views of the blackbody.

Heat pumping of the target and baffles is controlled, independently, by two interference-free, proportional regulators which provide linear thermal control in both the heating and cooling modes of operation. Cycling at the control point is $\pm 0.05°C$.

A similar unit, designed for operating in a vacuum, has been constructed. By eliminating convective effects, a temperature range of $-45°$ to $+60C$ was achieved.

Relatively large-area, compact blackbodies can be built combining the cavity array and thermoelectric heat pumping concepts. Through an increase in the number of modules, a source with 225 cm^2 usable area, with no loss in performance, has been assembled.

An intensive study of the temperature and emissivity characteristics of the source has been carried out (Hilleary et al., 1968). Of prime importance was the determination of the Δt correction to the base thermocouple measurements, in order to establish a more accurate value for the effective blackbody temperature. Temperature gradients along the cell wall of the cavity array were measured at five operational temperatures, with a maximum Δt value of 0.8°C from the baseplate monitoring thermocouples observed for the lower half of the cavity at the temperature limits. Uniformity of temperature across the cavity base was $\pm 0.25°C$ or better.

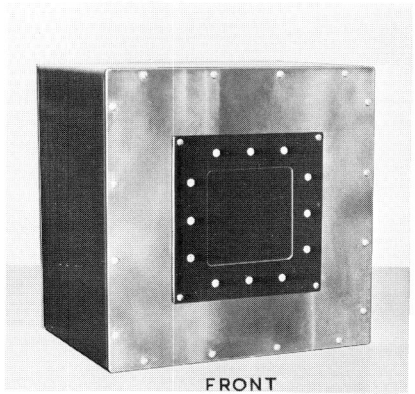 

Fig. 7.14. Photograph of a low-temperature (-35 to $+50°C$) blackbody with cavity array [after Karoli et al. (1967)].

Radiance distribution of the source, both with respect to area and viewing angle, was determined using a Golay infrared detector with an attachment to limit the field of view to 3°. Measurements at +40 and −25°C, sampling five small areas, showed a ±0.5% uniformity in radiance distribution. Viewing angle uniformity was of the same order of magnitude. This test was conducted by viewing the blackbody at angles of from 0 to 45° inclination, so that the detector viewed depths of approximately 1 to 0.13 of the cavity wall. As the accuracy of the measurement was better than 1.0%, and the data constant to within 0.5%, a temperature gradient exceeding 1.0°C would have produced a change greater than 1% in the measurement at the 45° angle of inclination.

A specially designed thermopile detector was also employed to measure the radiance properties of the source. Two important characteristics of the thermopile are a sensitivity constant to within ±0.2%, over an ambient temperature range of 5 to 45°C, and consistent to within 1%, when calibrated, directly, against the NBS reference and the Eppley–Karoli high-temperature blackbody, and, indirectly, against the IPS[2] reference. The output of the thermopile was compared to a computed irradiance at the detector receiver and found to be constant to within 1% (see Table I). The source was operated at selected points between +50 and −23°C, while the detector temperature ranged from +35 to +6°C. This test was further evidence that, within the limits of the experiment, no Δt correction larger than 1°C would be justified, and that the ±0.6°C value determined from temperature measurements was realistic.

Analysis of data obtained, using the SIRS spectrometer and two of the low-temperature blackbodies, showed that corrected and uncorrected spectral radiance calculations differed by only 1.2% when the source temperature separation was 74°C. Applying a Δt correction of ±0.6° would place the reliability of the blackbody radiance at about 0.5%.

7.4.3. Future Requirements

A series of blackbodies utilizing freezing metal are being developed by the National Bureau of Standards, Washington, the design of which is reported by Kostkowski et al. (Chapter 4). Presently, three such standards are in use employing tin, zinc, and gold with freezing points on the IPTS-68 of 231.97, 419.58, and 1064.4°C, respectively. Although each providing only a single calibration point, these sources feature excellent temperature definition and uniformity as well as high emissivity (0.999+). They have been adopted as primary standards by the U.S. Department of Defense for calibrating working standards of spectral radiance, from the visible region to 15 μm, with an uncertainty of about 1%.

[2] International Pyrheliometric Scale (Drummond, 1970).

TABLE I. Response of monitoring thermopile (No. 7137) when irradiated by the Eppley low-temperature blackbody (both source and detector at selected temperatures).

| Series | Approximate Temperatures (°C) | | Operating temperatures (°K) | | | A | B | A/B |
	Source	Detector	T_1 Source	T_2 Thermopile	T_3 Shutter	$\varepsilon_1 T_1^4 - \varepsilon_3 T_3^4$	EMF (μV) thermopile[a]	
1	+50	+21	322.7	294.5	295.0	32.7×10^8	101	0.324×10^8
2	+28	+6	301.2	279.0	297.7	21.2	64.5	0.328
3	+4	+27	277.3	300.3	299.9	−21.8	−67.5	0.323
4	−8	+24	265.2	297.3	296.7	−28.1	−87	0.323
5	−22	+25	250.9	298.7	298.0	−39.4	−121	0.325
6	−23	+35	250.2	308.7	307.8	−50.6	−156	0.324
Mean								$0.325 \pm 0.5\%$

[a] Thermopile response is essentially independent of ambient temperature.

A variable temperature blackbody, consisting basically of a graphite cavity radiator in a block of heated copper, has also been developed by NBS as a secondary standard. With a maximum operating temperature approaching 1000°C, it produced the same flux as the zinc freezing-point blackbody at 419.6°C.

Emphasis in absolute radiometry in the future will most likely be placed on the following:

(a) Development of blackbody standards operating above 3000°C and below 0°C;

(b) Improving the accuracy in transfering blackbody calibrations to working sources and detectors of both total and spectral radiance; and

(c) Experimental determination of the Stefan–Boltzmann constant of total radiation and verification of the Planck equation.

High-temperature standards will be difficult to construct and impractical to use because of the lack of suitable materials for the blackbody radiator as well as the excessive cost and size of power supplies required. It is reasonable to assume that an absolute detector will be the future standard for calibrating high-temperature sources.

Low-temperature standards present problems of a different nature. Larger areas are generally necessary to compensate for the loss in radiant flux. Working sources, combining liquid nitrogen cooling with electrical heating, operating between −195 and +50°C, have found considerable use in calibrating infrared detecting systems at the National Environmental Satellite Center. The radiator is a blackened plate of desired area with parallel or concentric " V " grooves to improve its emissivity.

A blackbody standard based on the same principle of operation and covering the same temperature range as above has recently been constructed by the Eppley Laboratory (Karoli *et al.*, 1968) and is shown schematically in Fig. 7.15. Employing cryogenic techniques, it consists, essentially, of a modular Dewar body surrounding a liquid nitrogen reservoir and a chamber to contain the blackbody radiator. Nitrogen is admitted from the main reservoir to the blackbody zone through a throttle valve, heat exchanger, and diffuser. Gaseous nitrogen then acts as an exchange medium flowing around, and in intimate contact with, the cavity-array radiator. Added to the controlled cooling is sufficient electrical heating to achieve the desired temperature. The design concept maintains near ideal thermal conditions, and since the blackbody radiator zone is isolated from the liquid nitrogen reservoir, high sample temperatures (up to 350°K) are achieved with little increase in refrigerant consumption.

Kostkowski has pointed out (Chapter 4) the probability of reducing the present uncertainty in spectral radiance calibrations in the near future. However, he cites the futility of this goal unless there is an improvement in the

7. EXPERIMENTAL BLACKBODY (ABSOLUTE) RADIOMETRY

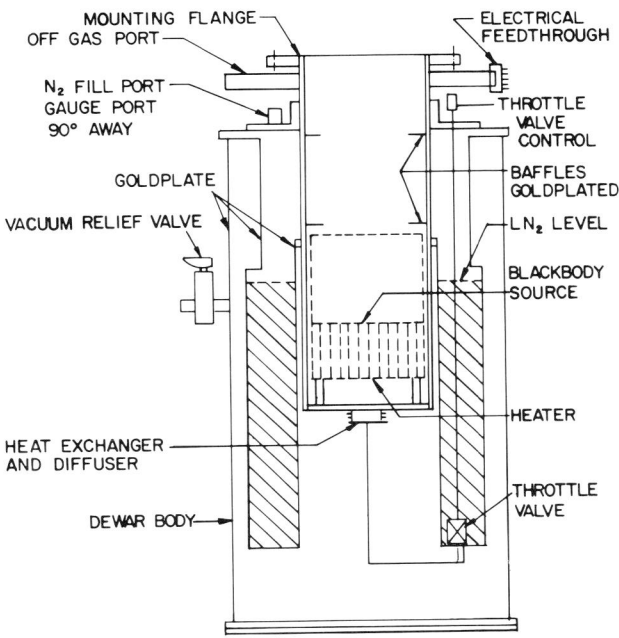

FIG. 7.15. Schematic of a variable low-temperature blackbody (-195 to $+50°C$) employing liquid nitrogen (courtesy of the Eppley Laboratory).

stability of the working standard lamp sources. Future investigations will, undoubtedly, be applied to improving the agreement between the International Pyrheliometric Scale (IPS) and the U.S. National Bureau of Standards radiation scale (NBS).

A challenge which will not go unheeded is a new experimental determination of the Stefan–Boltzmann constant based on (1) new ideas and techniques for realizing near ideal blackbody sources, (2) better temperature definition (IPTS) in terms of the thermodynamic scale (see Chapter 6), and (3) the design and operation of recently developed absolute detectors. Finally, applying modern precision instruments and advanced technology, the ultimate objective in absolute radiometry will be for a careful investigation experimentally verifying the long-accepted Planck distribution equation.

REFERENCES

Bedford, R. E. (1960). A low temperature standard of total radiation. *Can. J. Phys.* **38**, 1256–1278.

Bedford, R. E. (1970). This volume.

Coblentz, W. W. (1914). Constants of spectral radiation of a uniformly heated enclosure, or so-called blackbody, I. *Bull. Bur. Std.* **10**, 1–77.

Coblentz, W. W., and Emerson. W. B. (1916). Studies of instruments for measuring radiant energy in absolute value. *Bull. Bur. Std.* **12**, 503–552.
Coblentz, W. W. (1917). Constants of spectral radiation of a uniformly heated enclosure, or so-called blackbody, II. *Bull. Bur. Std.* **13**, 459–477.
Coblentz, W. W. (1922). Present status of the constants and verification of the laws of thermal radiation of a uniformly heated enclosure. *Sci. Papers Bur. Std.* **17**, 7–48.
de Groot, W. (1950). A note on the constants of radiation. *Physica* **16**, 419–420.
Drummond, A. J. (1970). This volume.
Eppley, M., and Karoli, A. R. (1957). Absolute radiometry based on a change in electrical resistance. *J. Opt. Soc. Am.* **47**, 748–755.
Féry, Ch. (1909). Détermination de la constante de la loi de Stefan, *Compt. rend.* **148**, 915–918.
Gerlach, W. (1912). Eine Methode zur Bestimmung der Strahlung in absolutem Maass und die Konstante des Stefan–Boltzmannschen Strahlungsgesetzes. *Ann. Phys.* **38**, 1–29.
Gerlach, W. (1913a). Zur Kritik der Strahlungsmessungen. I. *Ann. Phys.* **40**, 701–740.
Gerlach, W. (1913b). Zur Kritik der Strahlungsmessungen. II. *Ann. Phys.* **41**, 99–114.
Gerlach, W. (1913c). Zur Kritik der Strahlungsmessungen. III. *Ann. Phys.* **42**, 1163–1166.
Gerlach, W. (1916). Die Konstante des Stefan–Boltzmannschen Strahlungsgesetzes; neue absolute Messungen zwischen 20 und 450°C. *Ann. Phys.* **50**, 259–269.
Gouffé, A. (1945). Corrections d'ouverture des corps-noirs artificiels compte tenu des diffusions multiples internes. *Rev. Optique* **24**, 1–7.
Hilleary, D. T., Heacock, E. L., Morgan, W. A., More, R. H., Mangold, E. C., and Soules, S. D. (1966). Indirect measurements of atmospheric temperature profiles from satellites: III. The spectrometers and experiments. *Monthly Weather Rev.* **94**, 367–377.
Hilleary, D. T., Anderson, S. P., Karoli, A. R., and Hickey, J. R. (1968). The calibration of a satellite infrared spectrometer. *Proc. XVIII Intern. Astronautical Congr., Belgrade*, 1967, **2**, 423–437. Pergamon Press, New York.
Hoare, F. E. (1928). A determination of the Stefan–Boltzmann radiation constant using a Callendar radio balance. *Phil. Mag.* **6**, 828–839.
Hoare, F. E. (1932). A determination of the Stefan–Boltzmann radiation constant using a Callendar radio balance. *Phil. Mag.* **13**, 380–392.
Hoffman, K. (1923). Bestimmung der Strahlungskonstanten nach der Methode von Westphal. *Z. Physik* **14**, 301–315.
Hoffman, E. (1950). On the new value of the second radiation constant. *Z. angew. Phys.* **2**, 88–95.
Jaffey, A. J. (1954). Solid angle subtended by a circular aperture at point and spread sources: formulas and some tables. *Rev. Sci. Instr.* **25**, 349–354.
Jakob, M. (1957). "Heat Transfer," Vol. II, 652 pp. Wiley, New York.
Kahanowicz, M. (1917). The constant σ in the Stefan–Boltzmann law. *Nuovo Cimento* **13**, 142–167.
Karoli, A. R., Hickey, J. R., and Drummond, A. J. (1968). Continuation of experimental investigations of low-temperature radiation sources. Final Report, Phase I, 34 pp. ESSA Contract E-75-68N.
Karoli, A. R., Hickey, J. R., and Nelson, R. E., (1967). An absolute calibration source for laboratory and satellite infrared spectrometers. *Appl. Opt.* **6**, 1183–1188.
Keene, H. B. (1913). A determination of the radiation constant. *Proc. Roy. Soc. (London), Ser. A* **88**, 49–60.

Kurlbaum, F. (1898). Über eine Methode zur Bestimmung der Strahlung in absolutem Maass und die Strahlung des schwarzen Körpers zwischen 0 und 100 Grad. *Ann. Phys.* **65**, 746–760.

Kussman, A. (1924). Bestimmung der Konstanten σ des Stefan-Boltzmannschen Gesetzes. *Z. Physik* **25**, 58–82.

Lummer, O., and Pringsheim, E. (1899). Die Verteilung der Energie im Spectrum des schwarzen Körpers. *Verhandl. Deut. Phys. Ges.* **1**, 23–41.

Lummer, O., and Pringsheim, E. (1900). Über die Strahlung des schwarzen Körpers für lange Wellen. *Verhandl. Deut. Phys. Ges.* **2**, 163–180.

Mendenhall, C. E. (1929). A determination of the Stefan–Boltzmann constant of radiation. *Phys. Rev.* **34**, 502–512.

Müller, C. (1933). Bestimmung der Konstanten σ des Stefan–Boltzmannschen Strahlungsgesetzes. *Z. Physik* **82**, 1–36.

National Academy of Sciences–National Research Council Committee on Fundamental Constants (A. G. McNish, chairman). (1963). New values for the physical constants. *Nat. Bur. Std. U.S. Tech. News Bull.* **47** (No. 10), 175–177.

Paschen, F. (1899a). Über die Verteilung der Energie im Spectrum des schwarzen Körpers bei niederen Temperaturen. *Sitzber. Akad. Wiss., Berlin* **21**, 405–420.

Paschen, F. (1899b). Über die Verteilung der Energie in Spectrum des schwarzen Körpers bei höheren Temperaturen. *Sitzber. Akad. Wiss., Berlin* **21**, 959–976.

Roich, E. (1938). Isochromatic energy curves for a blackbody in the region 2800–3060 Å *Phys. Z. Sowjetunion* **13** (1), 11–22.

Rubens, H. and Kurlbaum, F. (1901). Anwendung der Methode der Restrahlen zur Prüfung des Strahlungsgesetzes. *Ann. Phys.* (4) **4**, 649–666.

Rubens, H., and Michel, G. (1921). Prüfung der Planckschen Strahlungsformel. *Phys. Z.* **22**, 569–577.

Rutgers, G. A. W. (1949). The second radiation constant in Planck's radiation formula. *Physica* **15**, 985–989.

Schumacher, P. E. (1964). A high temperature circular aperture blackbody radiation source. Res. Rept. 64-1, 26 pp. Rocketdyne Division of North American Aviation Canoga Park, California.

Simmons, F. S., DeBell, A. G., and Anderson, Q. S. (1961). A 2000°C slit-aperture blackbody source. *Rev. Sci. Instr.* **32**, 1265–1266.

Sitnik, G. F. (1961a). General principles for the realization of a model of a blackbody at a high temperature. *Soviet Astron.—A. J. (English Trans.)* **4**, 74–82.

Sitnik, G. F. (1961b). A blackbody model operable at high temperature. Soviet Astronomy—A. J. **4**, 1013–1022.

Snyder, N. W., Gier, J. T., and Dunkel, R. V. (1955). Total normal emissivity measurements on a aircraft materials between 100 and 800°F. *Trans. ASME, Ser. C: J. Heat Transfer* **77**, 1011–1019.

Sparrow, E. M. (1965). Radiant emission characteristics of nonisothermal cylindrical cavities. *Appl. Opt.* **4**, 41–43.

Stair, R., Johnston, R. G., and Halbach, E. W. (1960). Standard of spectral radiance for the region of 0.25 to 2.6 microns. *J. Res. Nat. Bur. Std.* A **64**, 291–296.

Valentiner, S. (1910). Vergleichung der Temperaturmessung nach dem Stefan–Boltzmannschen Gesetz mit der Skale des Stickstoffthermometers bis 1600°. *Ann. Phys.* **31**, 275–311.

Van Dusen, M. S., and Dahl, A. I. (1947). Freezing points of Co and Ni and a new determination of Planck's constant c_2. *J. Res. Nat. Bur. Std.* **39**, 291.

Vermeulen, D. (1935). A new determination of the radiation constant c_2. Diss. Univ. Utrecht.

Vul'fson, K. S. (1951). An absolute method of measuring the temperature of a blackbody. *Soviet Phys. JETP* (English Transl. of *Z. Eksper. Teor. Fiz.*) **21**, 507–509.

Warburg, E., and Müller, C. (1915). Über die Konstante c des Wien-Planckschen Strahlungsgesetzes. *Ann. Phys.* (4) **40**, 410–432.

Wien, W., and Lummer, O. (1895). Methode zur Prüfung des Strahlungsgesetzes absolut schwarzer Körper. *Ann. Phys.* (3) **56**, 451–456.

Zaalberg van Zelst, J. J. (1936). Experimental determination of the second radiation constant in Planck's formula. *Diss. Univ. Utrecht.*

—8—
LABORATORY METHODS OF EXPERIMENTAL RADIOMETRY INCLUDING DATA ANALYSIS

J. R. Hickey

8.1. Introduction

The experimental scientist is not always as fortunate in controlling the factors which may affect his measurements adversely as the person working in a standards laboratory. Possibly, we should make a distinction between the radiometric experiments which are conducted expressly for the advancement of scientific knowledge and those which are conducted as an engineering requirement. Some of the other contributors have restricted their discussions to the background of radiometric calibration using both controlled sources and controlled detectors, which have a theoretical basis for their construction and operation and which, in general, are used to calibrate the instruments that, in turn, are used as working standards. The question always arises as to the true reliability of any measurement performed with an instrument under different conditions than those of the calibration.

Transfer methods are probably the most commonly used radiometric measuring techniques. Any discussion of experimental radiometry outside the calibration laboratory must necessarily include transfer methods and, possibly, even double or triple transfers through a series of sources or detectors.

This chapter then is really a discussion of transfer methods. An attempt will be made to cover those experiments which are of general interest. There will also be included an attempt to reconcile the notion of traceability which is inherent in virtually every engineering specification.

Let us attempt to define a "transfer method" and an "instrument of transfer." A transfer method is one in which a regulated radiation source or stable radiation detector is employed as a constant radiometric device, while another device is subjected to varying conditions of environment or positioning with respect to the first. An instrument of transfer is a radiometric source, detector, or filter (here used generically to include instruments which contain an optical system with prisms or gratings), or combinations of these devices, which can be maintained in a stable condition while two other elements are compared with each other by means of the instrument. In this latter definition, the two elements being compared need not be under the same conditions. In practical use, one unit is, quite often, a secondary or working standard or an instrument previously calibrated against such a standard. Seldom, in engineering applications, is a primary standard used as such element. The delicate nature, difficulty in use, and specific environmental or angular field requirements of primary standards usually render them impractical.

In any instance where traceability is required, an accepted standard must be one element of a transfer chain. The measurement may become meaningless if there is no compatible transfer method for comparing a standard to an instrument used in the engineering application.

In the United States, the accepted standards are those traceable to the National Bureau of Standards (NBS), as embodied in the various types of standard lamps. Chapters 3 and 4 of this volume should be consulted, in this regard, in conjunction with this chapter. The universally accepted IPS reference (see Chapter 1) should also be considered in the traceability of total radiometric calibrations of a meteorological or geophysical nature.

To proceed logically, total radiation and spectral radiation methods are treated separately since they are traceable through separate standards. The practical application which will be stressed in this chapter is that of evaluating simulated solar radiation in space environmental test facilities. This has been chosen since it is a current problem and involves extremes in measurement conditions.

8.2. Total Radiation Scales in Practice

8.2.1. NBS Scale

From the previous chapters, one should have a good knowledge of what a radiation scale is. Let us reiterate the method in which the scales are maintained. The NBS scale of total radiation (actually irradiance) was originally embodied in a group of 50-W carbon-filament lamps (Coblentz, 1915). These lamps were calibrated by transfer against a blackbody maintained at NBS. Two types of lamps are shown in Fig. 8.1. When calibrating a detector against such a standard, the detector is placed at a distance of 2 m and the lamp is operated at three current settings. For practical purposes, we are interested in the value of the irradiance available for transfer calibration. For the three currents and for all the lamps, this value is between 40 and 100 $\mu W\ cm^{-2}$. While this is a reasonable level for low-intensity work, it is a low value for someone working in solar simulation work if we consider the solar constant to

8. LABORATORY METHODS OF EXPERIMENTAL RADIOMETRY

FIG. 8.1. Carbon-filament lamps used for standards of total radiation (old lamp is on left).

be 136 mW cm^{-2}, which is 1.36×10^3 times the highest carbon lamp value. Any nonlinearity in the detector's curve of sensitivity versus irradiance, or in the readout system for the detector, will certainly affect a measurement made by an instrument calibrated against this scale unless precautions are taken.

Let us also consider what problems arise from the spectral distribution of this standard. The carbon filament emits energy similar to a blackbody at about 1800°K. But this is limited by the envelope which is glass and may be coated with spalled carbon after aging. It also includes radiation emitted by the heated envelope. To use such a standard absolutely, it is necessary to have a detector which is spectrally flat over a fairly large spectral region. Eppley–Parsons black or 3M black (Chapter 7), properly applied to the receiver will assure us of this.[1] Realizing (1) that most detectors are not coated with these blacks, because the massive receivers required would result in long time constants, and (2) that a window on a thermopile is a necessity for low-intensity measurements under normal laboratory conditions, it behoves the experimentalist to determine how much of the energy is in the region transmitted by quartz which is the most commonly used thermopile window for

---

[1] In order to be certain that the black receiver is indeed flat, the instrument can be calibrated against low-temperature (less than 150°C) and high-temperature (1200°C) blackbodies for checking purposes. Another method is to perform a spectral response investigation employing a cavity detector as reference.

shortwave radiation. Also, it should be pointed out here that most blacks worthy of the name are flat absorbers over the quartz transmission region. To find out how much of the carbon lamp energy is transmitted by a quartz window 1 mm thick, a well-blackened detector is exposed, with and without window, to the standard lamp under stable conditions. The following data are necessary for such an interpretation.

Carbon lamp current		Filter factor[a]	Effective transmission
Old lamps	New lamps (A-19)		
250 mA	300 mA (45 μW cm^{-2})	1.21	0.826
300 mA	250 mA (65 μW cm^{-2})	1.20	0.832
350 mA	400 mA (80 μW cm^{-2})	1.19	0.840

[a] To be defined later in this chapter.

A glance at the last column indicates the change in the radiation distribution with lamp current. In the first instance, it can be seen that 82.6 % of the standardized irradiance of a carbon lamp, at its lowest current setting, is incident on the receiver of a thermopile fitted with a quartz window. Direct current output voltages from laboratory thermopiles will be of the order of 1 to 15 μV (since the average sensitivity is about 0.1 μV μW^{-1} cm^{-2}) during carbon lamp calibration (40–100 μW cm^{-2}).

A newer type of NBS standards of total irradiance are tungsten-filament lamps (Stair et al., 1967). Three references (see Fig. 8.2) are available with type C13 filaments[2]:

1. 100 W   100T 8/½,   120 V
2. 500 W   500T 20, 115–120 V   DNK
3. 1000 W   1000T 20, 115–120 V   DPT

The operating currents are 0.75, 3.60, and 7.70 A, respectively. A typical 100-W lamp produces an irradiance of approximately 520 μW cm^{-2} at 1 m; the 500-W lamp yields about 750 μW cm^{-2} at 2 m; and the 1000-W lamp about 1845 μW cm^{-2} at 2 m. A specified shutter and baffle system (Fig. 8.3) is necessary for use with these lamps. Suitable range ammeters of 0.25 % accuracy should be used to monitor lamp current. The calibration range of these lamps may be extended by the user, employing "inverse square law" extrapolation to other detector distances.

In 1968, NBS announced a new standard of total irradiance which yields

[2] The 100-W lamps are no longer available with C13 filaments; CC13 filaments are now employed.

Fig. 8.2. Tungsten-filament lamps used for standards of total radiation (1000-, 500-, 100-W left to right).

one solar-constant level radiation at a distance of 40 cm. The G.E. type DXW lamp is V-mounted in a ceramic reflector which has a flame-sprayed coating of aluminum oxide on the interior surface (Fig. 8.4). These lamps are also calibrated spectrally, as will be described in a later section.

8.2.2. IPS Scale

As has been discussed in the previous chapters, the International Pyrheliometric Scale is embodied in a group of instruments of which we shall consider the Ångström compensation pyrheliometer (see Chapter 1). This instrument may be used practically in a range extending from about 10 mW cm^{-2} to one solar constant (136 mW cm^{-2}) in its present commercially available form. This limitation is somewhat dependent on the control unit capability which is based on specifications, for the normal usage of the instrument, in measuring natural normal incidence solar radiation. If one were to select a well-known reference irradiance level at which intercomparisons of the highest order were accomplished, this would be in the range of 80 to 100 mW cm^{-2}. Thus, we could say that the higher flux scale is still not completely known at one solar-constant level. However, the instrument would most likely operate

Fig. 8.3. Shutter and baffle unit for use with tungsten-filament standards of total radiation.

just as well at this level. We must remember that the instrument is designed for use, at the ground, in natural sunlight.

8.2.3. Cross Referencing the Scales

It should be apparent that one cannot measure the carbon-filament or NBS tungsten-lamp outputs with an Ångström compensation pyrheliometer. This is because of the low intensity of the older total radiation standards and of geometry problems of the newest type, when viewed by an Ångström pyrheliometer. The two scales can be reconciled using a source of intermediate intensity between these standards and natural sunlight. One such source is a well-regulated 5000-W tungsten-projection lamp in a suitable lamphouse unit (see Fig. 8.5). This unit then becomes an instrument of transfer for this cross

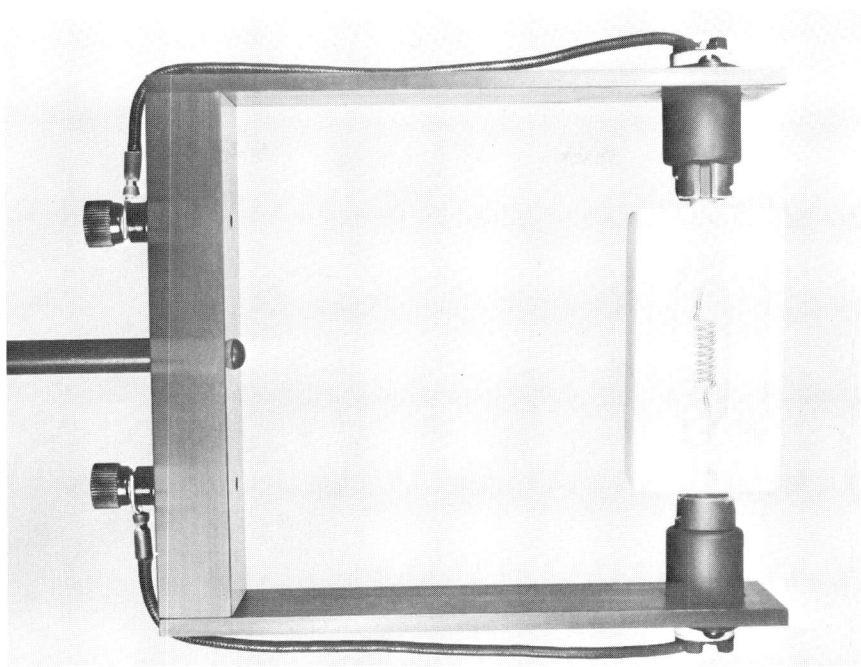

Fig. 8.4. New high-intensity (solar-constant level) standard of total and spectral irradiance.

Fig. 8.5. High-intensity lamphouse unit with 5000-W tungsten filament lamp.

calibration. The energy from the lamp should be limited to the spectral region of approximately 0.3 to 4.0 μm, by use of a cooled quartz window, to avoid problems caused by infrared radiation from the heated glass envelope of the lamp. It would also be possible to use a high-wattage system employing a compact arc lamp, if the constancy of flux could be maintained over the time of transfer. The method, in general, is a double operation. Refer to Fig. 8.6.

(a) Starting from the standard lamp calibration of a thermopile, the procedure is as follows. The thermopile is placed on the optical bench associated with the lamphouse. This bench should be about 3 m long and aligned with the lamphouse. To obtain a thermopile output of the same magnitude as that measured when the thermopile was exposed to the standard lamp, the thermopile is repositioned (or the lamp power may be varied) until the proper irradiance level is found. Then, a quartz window is alternately placed in front of the receiver and removed. If the spectral limiting is sufficient, the ratio of the "quartz out" to "quartz in" readings should be 1.084 ±0.002 for a 1-mm thick crystal quartz slide. The standard lamp level has now been transferred to the high-intensity bench. If a carbon lamp was used for the original calibration, the position will be greater than $2\frac{1}{2}$ m from the lamp (5 kW tungsten). It is reasonable to assume that the "inverse square law" will hold for this system to a distance as close as ten times the filament dimension, if internal reflections in the housing are at a minimum. In practice, this assumption is

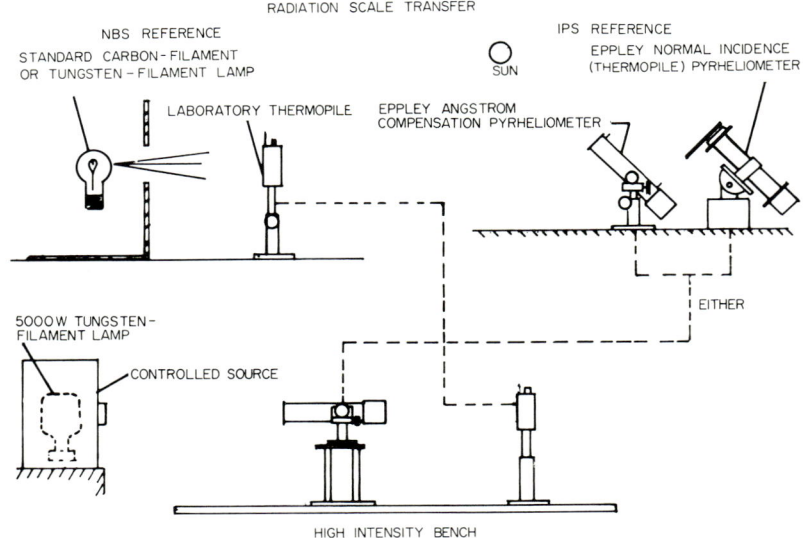

FIG. 8.6. Diagram of radiation scale transfer methods.

high-powered source to enter the tube. Baffling may be necessary if too much scattering or unwanted reflections are encountered. By using the quartz-in–quartz-out method, the linearity is established for air, by varying the lamp power or position sufficiently to cover the range 70–280 mW cm^{-2}. The quartz window on the instrument is then removed (since it is not needed in vacuum and may cause reradiation effects due to the lack of convective cooling). The chamber is evacuated to less than 10^{-4} torr pressure.[6] The lamp settings, for the known irradiance levels as previously determined, are then repeated and the instrument outputs noted. The ratio of the output at a given irradiance to that in air is termed the vacuum-to-air ratio. It should be emphasized that the sensitivity in air is for the instrument with its quartz window in place and that in vacuum, without the window. The vacuum-to-air ratio depends on instrument construction and is a function of intensity. It is always equal to or greater than 1.0. With the instrument in vacuum, a water (jacket) stabilizing system is used to maintain a uniform case temperature. The use of a circulating arrangement, with heaters and coolers, allows changing of the instrument temperature (which is monitored using a thermocouple within the instrument). Thus, at any intensity setting, the temperature may be varied over the limits of temperature of normal tap water (hot and cold) at the installation.[7] This then is the temperature transfer operation. If the instrument is very sensitive to temperature changes, the detector may be compensated for such an effect. These methods are discussed elsewhere (Marchgraber and Drummond, 1960). The outcome of all these transfers is a plot of instrument output versus irradiance under the specified conditions.

8.3.1. Use of the Calibrated Radiometer

Let us now assume that the radiometer is in the space chamber (to continue with our example) and ready for operation. By varying a selectable aperture angle, it has been determined that the source is between 6 and 7° off collimation. The aperture is then set at the larger aperture angle, the water cooling is turned on, and the chamber is pumped down and the walls cooled. It now occurs to the operator that he cannot easily shutter the source but must shutter the instrument. The energy measured by a thermopile is the net energy. The thermopile will receive energy from warmer objects within its field-of-view and radiate to the cooler objects. Thus, if the solar beam covers 6.3° of the field and the cold walls the remaining 0.7°, these are the conditions

---

[6] It has been shown that the change in thermopile sensitivity is negligible below 10^{-4} torr (Karoli *et al.*, 1960).

[7] If the instrument were also intended for use in air at atmospheric pressure, the temperature transfer could be accomplished in a temperature chamber. Water from temperature-regulated circulators could also be used.

of the measurement and must be communicated to those responsible for using the data for thermal design work. The next question is the effect of shuttering. The instrument shutter should be very close in temperature to the radiometer body if the instrument design is good. This means that very little radiative transfer will occur between the shutter and thermopile when the shutter is closed. However, if the shutter is slightly colder than the reciever, the output signal measured with the shutter closed will be negative. (This may also be the case for some filters of a filter radiometer—both the shutter-open and shutter-closed readings may be negative.) The net radiative exchange, as measured under these conditions, should be interpreted as the difference in radiation between a field containing 6.3° of arc, at about 5800°K,[8] coupled with 0.7° of arc at the instrument temperature. Now let us assume that the solar simulator entrance port is blocked by a shutter at 80°K. This is sometimes referred to as determination of the "tare." Now the interpretation is slightly different. Assuming the instrument shutter is left open and the source is shuttered, the difference in radiation interchanged is that between a source as described previously and 7° of arc at 80°K. Since the instrument temperature determined by the water system is about 300°K, a larger negative signal can be expected. Assuming all components to radiate as blackbodies, the former case would yield a signal smaller by a nearly insignificant amount, less than one part per thousand.

Special care must be taken that the radiometer body does not come in good thermal contact with cold lines or heaters within the chamber and that it is thermally isolated from structural members with large thermal inertia. If such is not the case, certain portions of the radiometer body may be preferentially heated or cooled, causing fluctuations with a long time constant. Also, the shutter should not be left open for lengthy periods during measurements, since the aperture assemblies could be heated and reradiate to the receiver.

8.4. Sensitivity Definitions for Thermopiles

Few thermopiles have a unique sensitivity for all applications. The variation with respect to environmental factors has already been discussed in preceding sections. There are also differences in sensitivity over the area of the various thermopile receivers. The discussions in this chapter apply to measurements during which the entire receiver is filled with a uniform radiation flux. Also, we are primarily concerned here with dc measurements.

Let us now consider, in detail, the variations due to spectral differences in source, window or black. A functional relationship of the form

(8.1) $$V = f[\alpha(\lambda), \tau(\lambda), H_\lambda(\lambda)]M$$

---

[8] This will depend on the fidelity of the solar simulation.

exists, where V is the thermopile output, $\alpha(\lambda)$ is the spectral absorptance of the black, $\tau(\lambda)$ is the spectral transmittance of the window, $H_\lambda(\lambda)$ is the spectral irradiance produced by a given source (or sources), and M is a factor relating to thermopile construction. Let us first define a window as a filter which exhibits a flat (or nearly flat) transmittance characteristic over a wide wavelength region, and which is used primarily for protection or stability reasons and not for spectral selection. Among window materials are quartz, lithium fluoride, calcium fluoride, pyrex, silver chloride, cesium iodide, fuzed silica, certain clear glasses, germanium, sodium chloride, thallium bromoiodide, and others. These materials can also be considered filters (i.e., spectral limiting devices) under certain conditions. Blacks are also wavelength dependent. However, they usually exhibit a certain spectral flatness over a given region if applied properly. Thin coatings of blackening material usually are not spectrally flat. When only a thin coating is applied to a receiver, the absorptance properties of the substrate (receiver material) may greatly affect the spectral absorptance properties of the instrument. Common blackening materials are lampblack (carbon), camphor soot, gold black, platinum black, and various paints and lacquers such as Parsons black, 3M black, and many others. For all practical purposes, blacks are flat over the uv, visible, and near infrared regions of the spectrum and vary in their long wavelength cutoff. It should be mentioned that even white paint or a deposit of MgO could be a good black in the infrared region, despite their high reflectivity in the visible. The spectral absorptance characteristics of a black should be investigated prior to its selection as a receiver coating for a given application.

The first consideration in choosing a thermopile should be the nature of the source to be measured. Once this has been reasonably defined, a black which exhibits a flatness over the spectral range of the source should be selected for the receiver. For any thermopile used as the detector of radiation from a source emitting in the spectral region for which the detector is flat, there is a "basic sensitivity," defined as the ratio of the output signal (V_t) to a known total irradiance (H_{true}) at the receiver plane. Note that no window is considered here. Since good windows are characterized by a flat transmittance characteristic over the major portion of their transmission band, the spectral limits of this band may be selected as the usable limits of the window. The thermopile should be flat over a wider band than the window to assure the best measurements. In such a case, and if the source emission is confined to spectral limits of the window, the irradiance at the receiver (H_{rec}) will be equal to the true irradiance (H_{true}) times the main band transmittance (τ_w) of the window. This can be written as

(8.2) $$H_{\text{rec}} = H_{\text{true}} \tau_w$$

Since the basic sensitivity (S_b) is defined as

(8.3) $$S_b \equiv V_t/H_{true}$$

and the sensitivity of such an instrument with the window in place is

(8.4) $$S_w \equiv V_m/H_{true}$$

then,

(8.5) $$S_w = V_m \tau_w/H_{rec}$$

Now, since V_m is a signal produced by a true irradiance H_{rec}, it can be seen that

(8.6) $$S_w = S_b \tau_w$$

It should be mentioned that in checking τ_w, the Fresnel formulas can be used. Reference should also be made to the previous section treating carbon lamp calibration. In that instance, the window did not transmit over the entire source emission region. Sensitivities can be defined for such cases also, as will be described later. For such a case, the quartz is considered a filter.

It is sometimes more desirable to measure power than irradiance. The two functions are simply related if the beam is uniform over the area sampled. The power is then

(8.7) $$P = HA$$

and the power sensitivity is

(8.8) $$S_p = S_h/A$$

where A is the area of the receiver and the subscripts p and h refer to power and irradiance, respectively.

Chapter 2, 3, 5, and 6 should also be consulted with regard to thermopile calibrations and indicative values of various parameters for different types of thermopiles. For a more complete discussion, one should refer to the references (Smith *et al.*, 1968; Nicodemus, 1965).

8.5. Filter Radiometry with Thermopile-Type Detectors

Remembering the qualifications presented in the last section, the subject of filter measurements should now be considered. Even when filter measurements are being made, a window is usually[9] placed in front of the thermopile.[10]

---

[9] A window is not used when the instrument is in vacuum, due to the lack of convective cooling. Stability should not be a problem in vacuum since air currents or pressure changes are the usual causes of instability.

[10] The filter is the first element in the beam—then the windowed thermopile. The window is usually chosen so that it blocks any reradiation from the absorbing-type filters, which tend to heat up, although this is not always possible in infrared studies.

8. LABORATORY METHODS OF EXPERIMENTAL RADIOMETRY 243

The window remains in place as the filters are changed. Filters chosen for use should have spectral limits between those of the window. For the purposes of this discussion, two definitions are required.

Broadband filters[11] include colored (or clear) glasses which have a steep lower wavelength transmittance characteristic, described by a center of lower cutoff (wavelength), and which exhibit a flat transmittance characteristic over a main band (usually covering a portion of the visible, at longer wavelengths than the cutoff, and the near infrared).

Narrowband filters include those which are commonly referred to as "interference filters," combinations of glasses in a "sandwich," and the type of glass which exhibits a wide band of transmittance which is not completely flat.

This last type of filter includes green and blue glasses and such uv filters as Schott UG 11. However, these may be treated as either narrowband or broadband depending on the source.

Mathematically, there are three integrals which should be considered. These are:

$$(1) \quad \int_{\lambda_1}^{\lambda_2} H_\lambda(\lambda) \tau_n(\lambda) \, d\lambda,$$

$$(2) \quad \int_{\lambda_1}^{\lambda_2} H_\lambda(\lambda) \, d\lambda$$

$$(3) \quad \int_0^\infty H_\lambda(\lambda) \, d\lambda.$$

In these integrals $H_\lambda(\lambda)$ is the spectral irradiance of the source and $\tau_n(\lambda)$ is the spectral transmittance of the filter, which is usually designated by a number n when it is one of a set. The limits λ_1 and λ_2 define the band limits of the filter, as will be discussed later. Integral 1 represents the true irradiance behind[12] the filter. Integral 2 represents the true irradiance within the spectral limits of the filter, which constitutes that quantity which is the desired result of the measurement. Integral 3 is the total irradiance of the source at the instrument. Two other definitions are needed for a treatment of filter radiometry. These are:

1. *Partial filter factor*—that factor by which the measured irradiance must be multiplied to obtain the value of true irradiance within the filter band.[13]

---

[11] These filters have been described extensively in the literature. They are manufactured by Schott, Corning, and Chance, etc.

[12] The term "behind" here means the other side of the filter from the source.

[13] This quantity will be referred to as simply the filter factor for the remainder of this chapter.

This will be designated F_n.

2. *Total filter factor*[14]—that factor by which the measured irradiance must be multiplied to obtain the value of total true irradiance. This will be designated F_{tn}. It follows that

(8.9) $$F_n = \frac{\text{(Integral 2)}}{\text{(Integral 1)}} = \frac{\int_{\lambda_1}^{\lambda_2} H_\lambda(\lambda)\, d\lambda}{\int_{\lambda_1}^{\lambda_2} H_\lambda(\lambda)\tau_n(\lambda)\, d\lambda}$$

and

(8.10) $$F_{tn} = \frac{\text{(Integral 3)}}{\text{(Integral 1)}} = \frac{\int_0^\infty H_\lambda(\lambda)\, d\lambda}{\int_{\lambda_1}^{\lambda_2} H_\lambda(\lambda)\tau_n(\lambda)\, d\lambda}$$

As mentioned previously, the quantity desired in any filter radiometric measurement is expressed by Integral 2. Let us define

(8.11) $$H_{\Delta\lambda}^{(n)} = \int_{\lambda_1}^{\lambda_2} H_\lambda(\lambda)\, d\lambda$$

where the superscript n refers to the filter number. If a window is employed with the thermopile,

(8.12) $$H_{\Delta\lambda}^{(n)} = F_n \int_{\lambda_1}^{\lambda_2} H(\lambda)\tau_n\, d\lambda = F_n V_n / S_w$$

where V_n is the output signal when filter n is in place. If there is no auxiliary window,

(8.13) $$H_{\Delta\lambda}^{(n)} = F_n V_n / S_b$$

S_w and S_b have the same meaning as in the last section. It should be obvious that V_n of Eq. (8.12) does not equal V_n of Eq. (8.13). The results obtained can be expressed either in tabular or histogram form. In a table there should be at least three columns headed as shown in the following example.

Filter number	Spectral band	$H_{\Delta\lambda}$
1	380–450 nm	10 mW cm^{-2}

For histogram presentation (see Fig. 8.9), the average spectral irradiance over the band should be plotted as a straight line parallel to the abscissa (wavelength axis between the wavelength limits λ_1 and λ_2). The average

---

[14] See the section concerning carbon lamp calibrations for an example. It is implied that the filter band does not cover the entire source emission spectrum.

8. LABORATORY METHODS OF EXPERIMENTAL RADIOMETRY 245

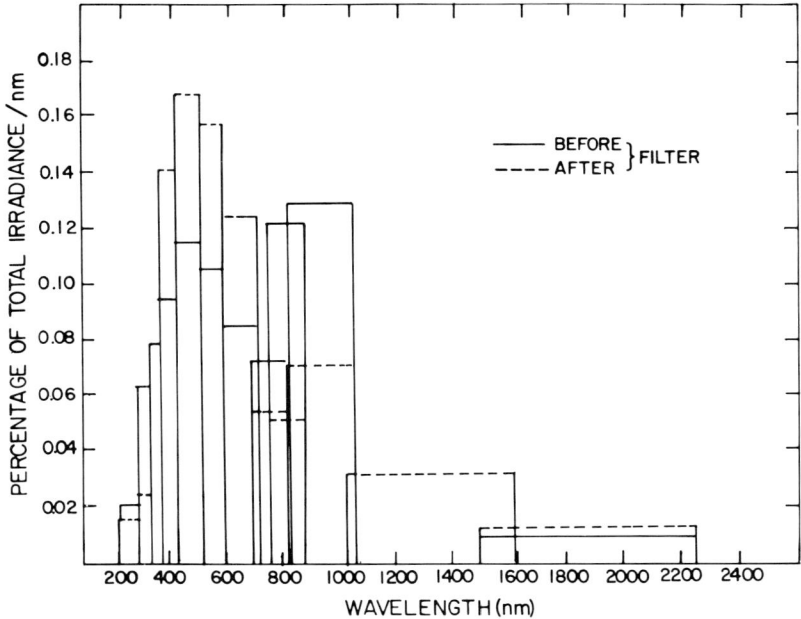

FIG. 8.9. Histogram presentation of filter radiometer results. Data were obtained at CNES (France) for filtered and unfiltered source [after Drummond and Hickey (1967)].

spectral irradiance for the band of filter n is

(8.14) $$H^{(n)}_{\Delta\lambda \, \text{avg}} = H^{(n)}_{\Delta\lambda}/\Delta\lambda$$

where

(8.15) $$\Delta\lambda = \lambda_2 - \lambda_1$$

For the example in the table above

$$H^{(1)}_{\lambda \, \text{avg}} = \frac{10 \quad \text{mW cm}^{-2}}{70 \quad \text{nm}} = 0.143 \quad \text{mW cm}^{-2} \, \text{nm}^{-1}$$

Examples of histogram solutions will be given later. It is customary to distinguish the side of the band to which the irradiance function is weighted. The wavelength of the "center of integrated irradiance"[15] may be marked

---

[15] Not to be confused with what is sometimes called the "center of gravity" or "center of transmittance" of a filter, which is defined as $\bar{\lambda} = \int \tau(\lambda)\lambda \, d\lambda / \int \tau(\lambda) \, d\lambda$.

along the line of the histogram. This wavelength (λ_c) is defined such that

$$(8.16) \qquad \int_{\lambda_1}^{\lambda_c} H_\lambda(\lambda)\, d\lambda = \int_{\lambda_c}^{\lambda_2} H_\lambda(\lambda)\, d\lambda$$

Sometimes, straight-line segments are drawn between consecutive λ_c positions in the histogram. One should be careful to explain the significance of such an apparent source curve to avoid misinterpretation.

Special cases. There are two types of filter that allow a less tedious method of data reduction than that indicated in the previous section. These will be discussed here prior to the problems inherent in calculating more difficult filter factors.

The first of these cases is that of filters exhibiting the ideal characteristic—a rectangular transmittance characteristic. It should be obvious that the filter factor is the reciprocal transmittance. A glance at Eq. (8.9) provides this assurance. No such filter exists in the narrow or broadband class. However, judicious choices can be made which, for most practical purposes, allow this approach to be taken. First, let us consider broadband filters. The lower transmittance cutoff is steep enough to be considered rectangular for many of these glasses. Therefore, if the source emission is less than 1 % of the total in the spectral range beyond the longwave limit of the main band transmittance, Eq. (8.9) yields

$$(8.17) \qquad F_n = 1/\tau_n$$

Also, when measuring discontinuous (line structural) sources, a narrowband filter may be chosen such that it is a flat transmitter over the region of a single emission line or a small band (or group of lines). If comparatively high continuum and line emission are present at the same wavelengths, one should not use this approach.

The second case concerns very narrowband filters, which exhibit a nearly symmetric transmittance function about the wavelength of maximum transmittance (designated λ_{max}). A process of data interpretation, termed "normalization-to-peak analysis,"[16] is used in this instance. Assuming a window is employed, the thermopile system can be considered to have a spectral sensitivity response function

$$(8.18) \qquad R(\lambda) = S_w\, \tau(\lambda)$$

and the sensitivity at λ_{max} is

$$(8.19) \qquad R_{max} = S_w \tau(\lambda_{max}) \qquad [\mu V\ \mu W^{-1}\ cm^{-2}]$$

[16] This method is described in many texts on infrared technology. It usually is restricted to use with spectrally selective detectors.

8. LABORATORY METHODS OF EXPERIMENTAL RADIOMETRY

The irradiance which is normalized-to-peak is

(8.20) $$H \equiv V/R_{\max} \quad [\mu W \, cm^{-2}]$$

The bandwidth for this case is defined as

(8.21) $$\Delta\lambda \equiv \int_{\lambda_1}^{\lambda_2} R(\lambda) \, d\lambda / R_{\max} = \frac{\int_{\lambda_1}^{\lambda_2} \tau(\lambda) \, d\lambda}{\tau(\lambda_{\max})} \quad [nm]$$

The peak-normalized value of *spectral* irradiance assigned to $\lambda_{\max}$ is

(8.22) $$H_\lambda(\lambda_{\max}) = \frac{V}{R_{\max} \Delta\lambda} = \frac{V}{S_w \int_{\lambda_1}^{\lambda_2} \tau(\lambda) \, d\lambda} \quad [\mu W \, cm^{-2} \, nm^{-1}]$$

It can be seen that a large number of filters suitable for this type of analysis, used in conjunction with a thermopile, would yield a spectral irradiance curve like that obtained from monochromator methods. However, such filters usually have too small a bandwidth to allow sufficient energy to reach the receiver, to give a meaningful signal with the thermopile elements presently available.

The filter factor originally defined by Forsythe (1937) included the effect of spectrally selective detectors. Other recent discussions (Schneider *et al.*, 1967; Nicodemus, 1968) treat filter radiometry with selective detectors. The Schneider work also treats monochromator correlations which will be discussed later in this chapter. However, here, our discussions will pertain to flat detectors only.

Sufficiency of total irradiance measurement. Before proceeding to further treatment of the filter techniques, let us consider the amount of spectral information which can be gained from a measurement of total irradiance. There are instances when the relative spectral distribution of sources is fairly well known. Blackbodies, graybodies, tungsten lamps operated at specific settings, and even some of the compact-arc lamps exhibit a known spectral emission function. The percentage of the total energy which is contained in a certain spectral band is, therefore, an easily calculated quantity. However, the total irradiance at a given location is a function of the geometry of the source array and the configuration of source and detector. Thus, it can be seen that if the relative spectral emission is known (or can be assumed), and the total integral irradiance measured, the irradiance in any subordinate spectral region can be inferred by the following approach.

The relative spectral curve is drawn; the area under the curve is assigned a convenient scale in integration units; the number of integration units is set equal to the total measured irradiance as

(8.23) $$H_\lambda = K \int_0^\infty J_\lambda(\lambda) \, d\lambda$$

where $J_\lambda(\lambda)$ is the relative spectral emission and K is the proportionality factor relating integration units to true irradiance. Any subordinate spectral region ($\Delta\lambda$) can now be evaluated using the equation

$$(8.24) \qquad H_{\Delta\lambda} = K \int_{\Delta\lambda} J_\lambda(\lambda)\, d\lambda$$

8.6. Correlation of Broadband and Narrowband Filter Radiometer Measurements

When filter radiometric correlations are being made, one usually works from better to poorer resolution. The implication of this statement will be appreciated after the discussion of monochromator correlation.

A broadband filter properly selected will yield two important results:

(a) the integrated absolute irradiance at longer wavelengths than the center of cutoff (λ_m); and

(b) the integrated absolute irradiance at wavelengths shorter than the center of cutoff.

The latter is found by a differential method using the total irradiance determined from an unfiltered measurement. When two or more broadband filters are used, the integrated irradiance between their centers of cutoff can also be determined. If we assume the radiometer has a proper window, the results can be obtained as follows:

(c) The integrated irradiance from the center of cutoff to the practical source limit is

$$(8.25) \qquad H^{(n)}_{>\lambda_m} = V_n / S_w \tau_n$$

where the n's again refer to the filter number.

(d) The integrated irradiance over the wavelength region shorter than the center of cutoff is

$$(8.26) \qquad \begin{aligned} H^{(n)}_{<\lambda_m} &= (V_{\text{tot}}/S_w) - H^{(n)}_{>\lambda_m} \\ &= [V_{\text{tot}} - (V_n/\tau_n)]/S_w \end{aligned}$$

(e) The integrated irradiance between the centers of cutoffs of filters 1 and 2, assuming filter 1 has the smaller λ_m, is

$$(8.27) \qquad \begin{aligned} H^{(1,2)}_{\Delta\lambda} &= (V_1/\tau_1 - V_2/\tau_2)/S_w \\ &= (F_1 V_1 - F_2 V_2)/S_w \end{aligned}$$

8. EXPERIMENTAL METHODS OF LABORATORY RADIOMETRY

If the set of broadband filters is well matched in the main band, Eq. (8.27) becomes

(8.28) $$H_{\Delta\lambda}^{(1,2)} = F_b(V_1 - V_2)/S_w$$

where F_b is the reciprocal main band transmittance. Differential broadband results can be plotted in histogram form V as described for narrowband filters.

To correlate broad and narrowband data, it is necessary to add up the $H_{\Delta\lambda}$ values of the narrowband filters over either an integrated broadband region, such as in Eqs. (8.25) and (8.26) or a differential region such as in Eqs. (8.27) and (8.28). It is seldom found that wavelength limits will match exactly. Certain approximations must be made in the correlation. Remembering that the histogram solution for narrowband filters gives an average *spectral* irradiance value, a first approximation can be calculated using it. There are essentially two cases to be considered. These are voids and overlaps. The method can best be shown by an example. Consider an integrated broadband value of 30 mW cm^{-2} below the cutoff of a filter for which λ_m is 500 nm. In a certain set of narrowband filters, assume that there are six filters which cover the range below 500 nm and that the values for these filters are as follows:

Filter number	Spectral band (nm)	$H_{\Delta\lambda}^{(n)}$ (mW cm^{-2})	$H_{\lambda\mathrm{avg}}^{(n)}$ (mW cm^{-2}nm^{-1})
1	250–300	1.40	0.028
2	305–340	2.94	0.084
3	340–380	4.48	0.112
4	375–410	4.90	0.140
5	410–450	7.20	0.180
6	450–500	10.00	0.200

Consider 250 nm to be the lower limit of source emission. The region 300–305 nm constitutes a void and the region 375–380 nm constitutes an overlap. The total of the $H_{\Delta\lambda}^{(n)}$ values is 30.92 mW cm^{-2}. However, the amount between 300 and 305 nm must be added to this and the amount between 375 and 380 nm must be subtracted. To a first approximation, this can be accomplished by averaging the two $H_{\lambda\mathrm{avg}}^{(n)}$ values of the filters involved and multiplying by the void or overlap bandwidth; for our example, this would be

$$H_{250-500} \simeq \sum_{n=1}^{6} H_{\Delta\lambda}^{(n)} + \tfrac{5}{2}[H_{\lambda\mathrm{avg}}^{(1)} + H_{\lambda\mathrm{avg}}^{(2)}]$$

$$- \tfrac{5}{2}[H_{\lambda\mathrm{avg}}^{(3)} + H_{\lambda\mathrm{avg}}^{(4)}]$$

$$= 30.92 + 0.280 - 0.630 = 30.57 \quad \mathrm{mW\ cm}^{-2}$$

which is higher than the broadband value by 1.9%.

It is not likely that filter limit matching, as in the example, would be obtained in practice. Also, it is stressed that this is only a first approximation method. If a relative source emission curve is available for the source or a similar source, a more suitable value for $H_{\lambda \text{avg}}$ in the void or overlap region can be obtained. An example of such an analysis is shown in Fig. 8.10.

The results of a filter correlation analysis (Hickey, 1965) are given in Table I. The "R" symbol in the data to the left signifies the summation in the column directly above. Narrowband versus narrowband and broadband versus broadband correlations are shown at the lower right. The source for this example was a solar simulator employing mercury-xenon arcs.

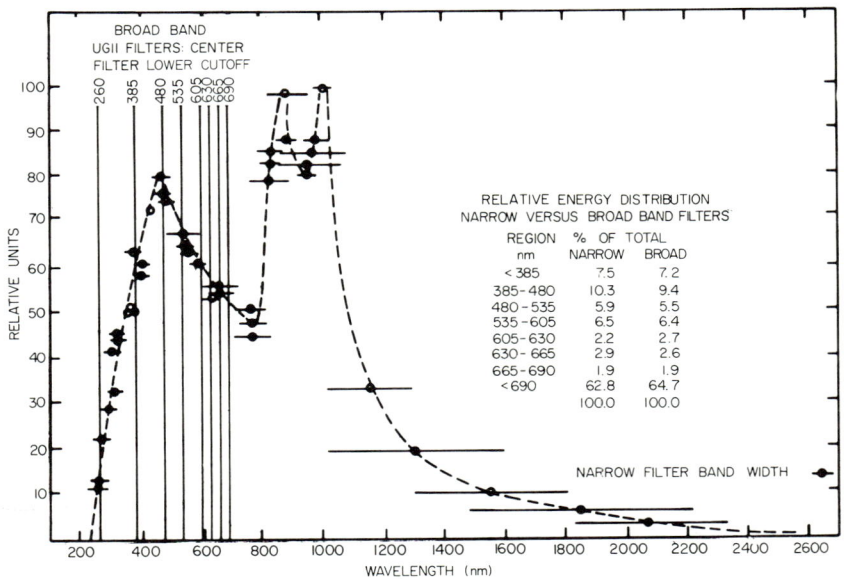

FIG. 8.10. Diagram showing correlation of narrowband and broadband filter radiometer results using the unfiltered CNES solar simulator as source [after Drummond and Hickey (1967)].

8.7. Monochromator Measurements

8.7.1. Status of Spectral Standards

There are two classes of spectral standard available for use on an engineering or working standard basis. These are the standard of spectral radiance and the standard of spectral irradiance. There are two types of spectral radiance

8. LABORATORY METHODS OF EXPERIMENTAL RADIOMETRY   251

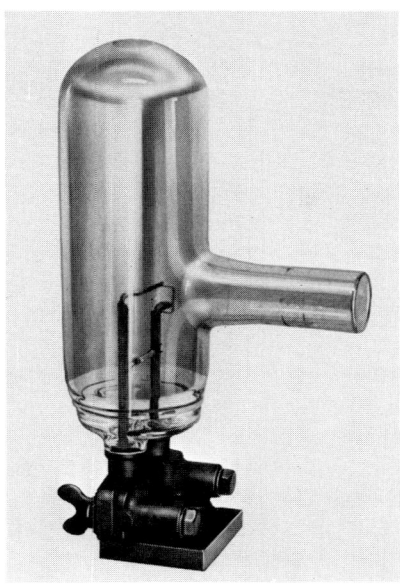

FIG. 8.11(a). G. E. type 30A/T24/17 lamp used as standard of spectral radiance.

standard both employing the same basic strip-filament lamp. These have been treated in Chapters 3 and 4. The traceability (within NBS) is different. The lamp of the latter (Kostkowski *et al.*, to be published) uses a notched filament, G.E. type 30 A/T24/13. The former (Stair *et al.*, 1960) is a G.E. type 30A/T24/17. This is the lamp discussed below. The earlier standards of spectral irradiance are discussed in Chapter 3. The 200-W version (Stair *et al.*, 1963) is a G.E.6.6A/T4/CL lamp, while the later 1000-W version is a G.E. type DXW. The new high-intensity standard of total and spectral irradiance (Schneider *et al.*, 1968) makes use of a DXW lamp in a reflector. This reflector is slip-cast fused silica with a reflecting surface of flame-sprayed aluminum oxide,[17] as shown in Fig. 8.4. The appropriate wavelength ranges for the standards are given below.

Spectral radiance	250–2500 nm
High-accuracy spectral radiance	210–800
Spectral irradiance	250–2500
High-intensity spectral irradiance	300–2500

Figure 8.11(a) shows the radiance lamp for which curves of spectral output

[17] The trade name is Rokide A and is registered to the Norton Co., Worcester, Massachusetts.

TABLE I. Correlation of filter results of data taken at the JPL (Hg-X compact arcs) solar simulator.[a]

Narrowband versus broadband				Narrowband versus broadband			
NB region λ(nm)	Total (%)	BB region λ(nm)	Total (%)	NB region λ(nm)	Total (%)	BB region (λnm)	Total (%)
250-300	2.9	UG 11	16.3	770-890	2.8	RG 8	48.6
300-350	6.6			850-1070	13.0	770-675	(-2.5)
345-380	7.6			890-850	(-2.0)		
350-345	(-0.5)			1050-1600	21.2		
R 250-380	16.6	265-380	16.3	1070-1050	(-0.5)		
265-295	2.0	UG 11	16.3	1500-2200	11.5		
295-345	6.8			1600-1500	(-2.2)		
345-380	7.6			>2200	(0.5)		
R 265-380	16.4	265-380	16.3	R >770	44.3	>770	46.1
390-420	3.8	GG 22	82.8	Narrowband versus narrowband			
420-430	(0.9)	-GG 14	-71.0	NB region λ(nm)	Total %	NB region λ(nm)	Total (%)
430-490	7.7			250-300	2.9	265-295	2.0
490-480	(-0.7)			300-350	6.6	295-345	6.8
R 390-480	11.7	390-480	11.8	345-380	7.6	345-380	7.6
480-490	(0.7)	GG 14	71.0	350-345	(-0.5)		
490-600	18.9	-RG 1	-51.5	R 250-380	16.6	265-380	16.4
600-605	(0.4)						
R 480-605	20.0	480-605	19.5				

8. LABORATORY METHODS OF EXPERIMENTAL RADIOMETRY

535–525	(−1.2)	OG 1	67.8		
525–600	16.7	−RG 1	−51.5		
600–605	(0.4)			600–850	8.7
R 535–605	15.9	535–605	16.3		
565–600	10.1	OG 2	61.6		
600–605	(0.4)	−RG 1	−51.5	R 600–850	8.7
R 565–605	10.5	565–605	10.1		
675–710	(0.9)	RG 8	48.6	600–620	(0.9)
710–810	2.5			620–710	4.0
810–800	(−0.4)			710–810	2.5
800–1200	21.6			810–850	(1.3)
1200–1300	3.0				8.7
1300–1800	17.2				
1800–1750	(−1.5)				
1750–2600	7.0				
R 675–2600	50.3	>675	48.6		

Broadband versus broadband

BB region λ(nm)	Total (%)	BB region λ(nm)	Total (%)
Q·WG 7	2.2	UG 11	16.3
WG 7-GG22	15.0		
390–380	(−1.0)		
R <380	16.2	265–380	16.3

[a] R is the summation of components.

are included in Fig. 8.11(b). The DXW lamp is shown in Fig. 8.12(a). The spectral irradiance curves of the 1000-W DXW lamp, with and without reflector, are given in Fig. 8.12(b).

With the exception of the high-accuracy radiance standards (Chapter 4), the spectral standards have similar traceability. From a practical experimental viewpoint, all measuring systems to be used in comparing other sources to the known standards must be compatible with these standards, so that transfer may be accomplished. It is assumed that these standard lamps will be used in calibration work. The issued lamp should therefore be retained as a shelf standard and used only when necessary. Other lamps of the same or a compatible type should be calibrated as working standards, employing the shelf standard and a suitable instrument for transferring the calibration. The instrument of transfer should be chosen judiciously to assure a meaningful

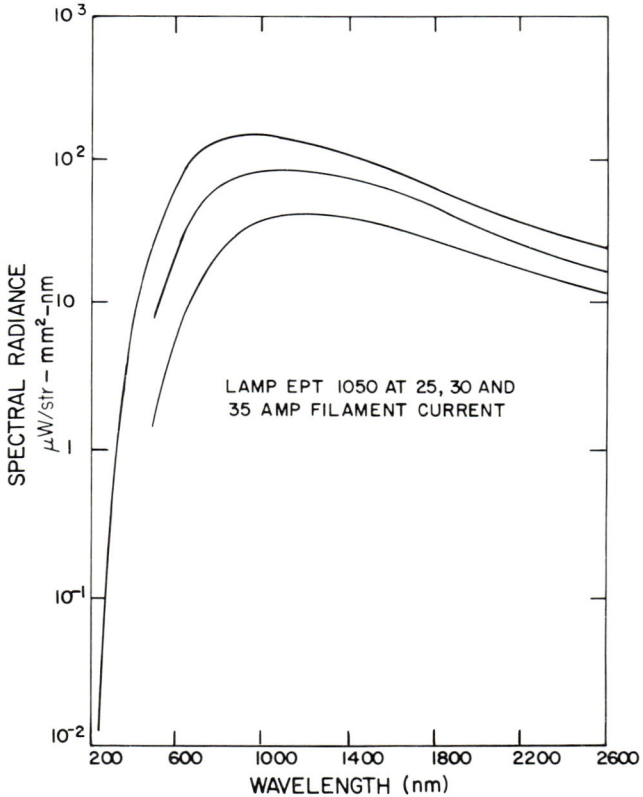

FIG. 8.11(b). Typical spectral radiance distribution curves at three current settings.

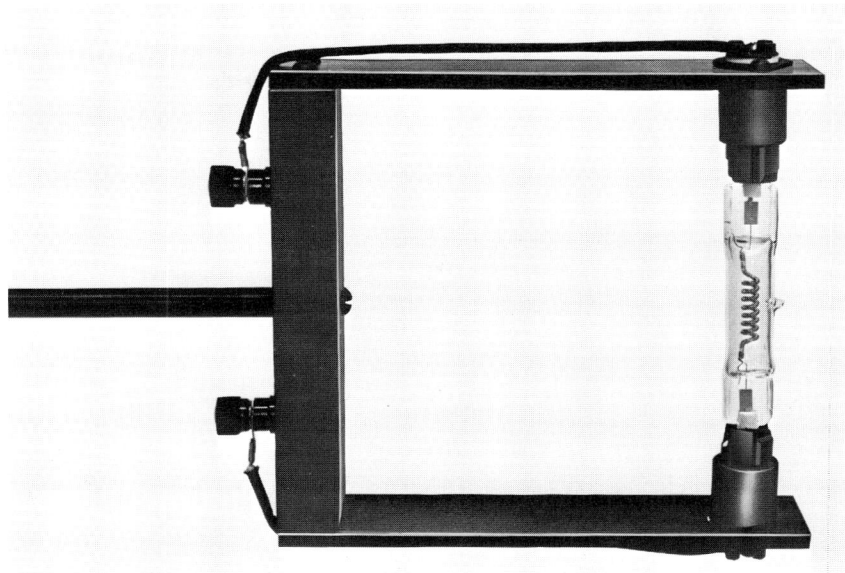

Fig. 8.12(a). G.E. type DXW, 1000-W tungsten-quartz-halogen lamp used as the standard of spectral irradiance.

transfer. Periodic checks of the working standards should be performed on a schedule based on the number of operating hours. Twenty-five hour intervals are usually sufficient.

8.7.2. Basic Compatibility Considerations in Using Spectral Standards with Monochromators

The use of a strip-filament lamp with a monochromator will most likely require imaging techniques. The use of irradiance standards with a monochromator requires diffusion techniques for the purpose of image spoiling at the ultimate detector. To state the problem simply, a source which is a plane can be imaged on a plane and a source which is three dimensional cannot.

The first and foremost problem in using spectral standards is measuring a meaningful output signal from the transfer instrument, especially below 300 nm. The second is to make a meaningful measurement of an unknown source without greatly varying the instrument settings from those used in measuring with the standard.

It cannot be overemphasized that the use of standards in comparing sources should be carefully undertaken. Attention must be afforded to the

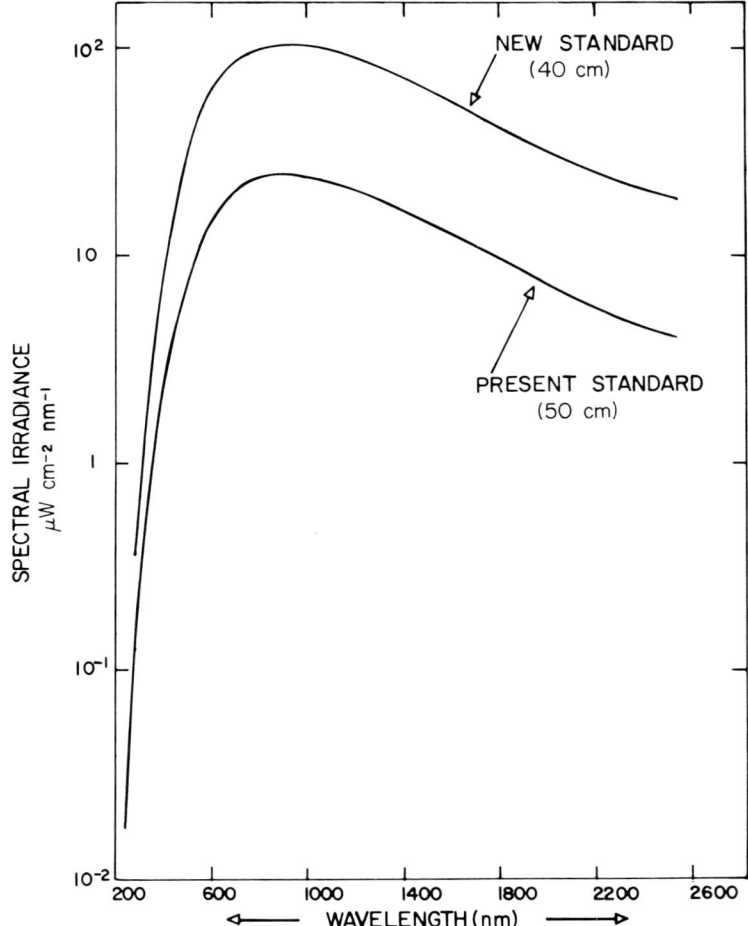

Fig. 8.12(b). Typical spectral irradiance curves for a DXW lamp operated at 8.3 A and at a distance of 50 cm without reflector and 40 cm with reflector.

calculation of slit functions if the sources are significanctly different in emission characteristics. This may be eliminated only if very high resolution can be obtained, such that the bandwidth of the monochromator is considerably less than that of the emission line being measured.

The spectral radiance standards and the 200-W spectral irradiance standards were not satisfactory for use in measuring high-intensity sources (especially below 350 nm). This was due to the great differences in energy, which introduced instrument nonlinearity in transfer measurements. The 1000-W irradiance standard alleviated this situation to some extent. The new reflector lamp further reduces the problem but is not useful below 300 nm.

8.7.3. Measurement of Source Emission by Monochromator Methods

By monochromator methods here is meant that a monochromator is used as the transfer instrument for comparing a known source (such as a spectral standard) with an unknown source. Emission spectroscopy is an old art as well as a science. Numerous texts contain references to the calculation of slit functions, monochromator design, and such related topics as resolution and instrument dispersion. When comparing a standard source and an unknown source, all the conditions for use of the standard must be met. The geometric relationship between the source and the monochromator is second only to instrument (mainly detector and readout) stability. The unknown source must then be set up with relation to the monochromator in the same geometrical fashion—or at least the beam of light from the unknown source must fill the monochromator in the same way. The next important consideration is instrument linearity. If the energy at any wavelength emitted by the unknown is considerably higher or lower than that of the standard, such problems may arise. However, unless very weak sources are being measured, the monochromator limitation will probably be reached because of an insufficient amount of energy from the standard. This is especially true if it is necessary to use a diffuser (e.g., a MgO coated sphere) at the entrance of the monochromator. Some types of attenuators, such as screens or neutral density filters, can usually be used if there is too much energy. These devices introduce more possible errors into the intercomparison.

A rule of thumb in such measurements is to attempt the greatest instrument throughput possible with the highest possible resolution. That is to say, the slits should be closed to the smallest width which allows a meaningful signal to be obtained at the wavelength of lowest source output.

Another important problem to be considered is stray light within the measuring instrument. Stray light can be investigated by allowing light, at other wavelengths from that for which the instrument is set, to enter the slits. This is accomplished by filtering at the entrance slit. Each instrument is different and must be investigated. A generalization could be made here that the cost and quality of components are almost secondary to the operator's knowledge of his instrument's capabilities. A qualified spectroscopist should get reasonable results even with a relatively poor instrument. A monochromator set at any wavelength is little more than a narrowband filter. The distribution of wavelengths at the exit slit is the one distinguishing feature which separates it from a common filter in the interpretation of results.

A word should be mentioned about sources. A source does not drastically change its relative output characteristics. Continuous sources do not become line sources or vice versa. In choosing a transfer instrument, one should consider the sources he desires to measure. An instrument which operates at a constant bandwidth over its spectral range would be the most practical for general purposes. If this is not possible, the bandwidth should be set the same, for any given wavelength, for both the known and unknown source measurements.

When using grating instruments, special care should be taken to eliminate second and higher order wavelengths from the measurement. Suitable filters can be obtained for this purpose.

Examples of results of monochromator measurements are given in Figs. 8.13 and 8.14. These figures depict one method of displaying monochromator results, i.e., normalized-to-peak adjusted ordinate value. In assessing the energy in any spectral region, integration with wavelength is undertaken.

Recently, spectroradiometers such as the Cary model 14 have been developed which allow direct intercomparison of sources. The instrument is fitted with a standard source which is viewed sequentially with the unknown by means of a rotating integrating sphere. A plot of the ratio of unknown to standard, as a function of wavelength, is presented directly. The absolute spectral irradiance, as a function of wavelength in analog or digital form, can be obtained, depending on the accessories which are available with the instrument.

8.8. Use of Monochromator Data in Calculating Filter Factors

The discussion of filter radiometry of previous sections was not really complete since the method of obtaining the $H_\lambda(\lambda)$ or $J_\lambda(\lambda)$ values inherent in the three fundamental integrals was omitted. Such values are obtained by the transfer methods using monochromators in the manner just described. The integrations inherent in the calculation can now be performed with numerical methods. Filter factors vary with the source, as is obvious from the equations developed earlier in this chapter. In Table II, the factors for a set of 12 filters are given for three different sources. It can be seen that the differences are striking. It should be realized that one of the most important procedures is establishing the filter limits. The method used when the filter transmittance has been measured on a spectrophotometer is explained below.

The 0.1% transmittance limits are usually chosen to begin the analysis. It may not be obvious that the 0.001τ filter factor is the logical starting point. In defense of this selection, the following example is relevant. The irradiance sensed at the radiometer is represented by the product function integral. First, assume that the analysis is to be carried out by using a filter which has

8. LABORATORY METHODS OF EXPERIMENTAL RADIOMETRY 259

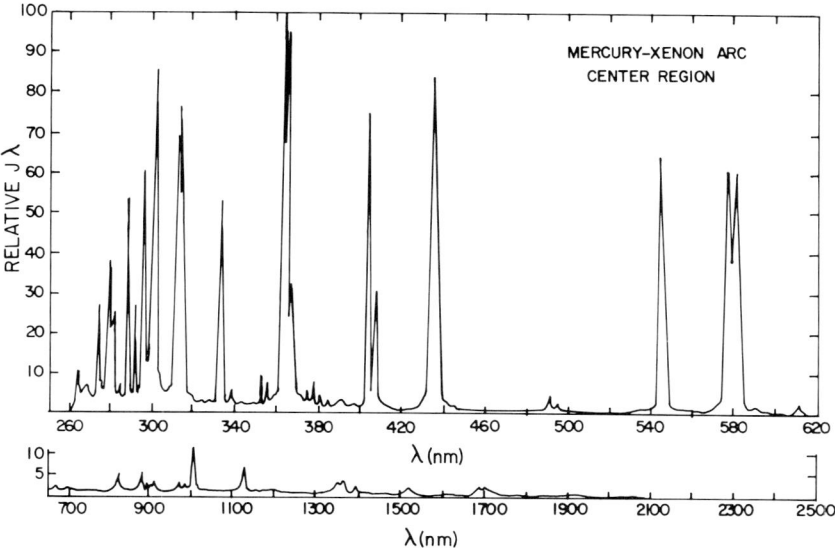

FIG. 8.13. Relative spectral emission of a 2500-W Hanovia mercury-xenon arc lamp.

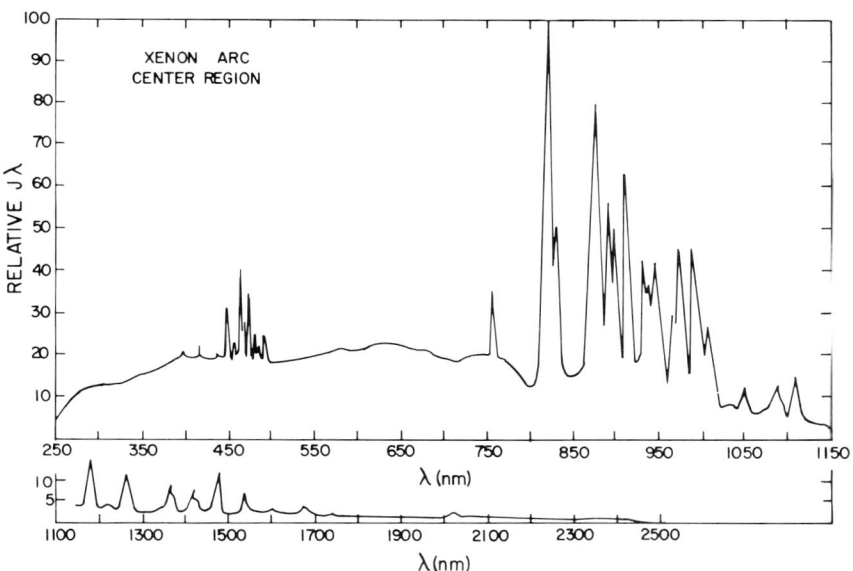

FIG. 8.14. Relative spectral emission of a 2200-W Hanovia xenon arc lamp.

TABLE II. Change in filter factor with source for a typical set of 12 filters.[a]

Filter Number	Wavelength limits (nm)	Filter factors		
		Xenon arc	Mercury-xenon arc	Carbon arc 5800°K
1	242–300	7.31	11.1	10.1
2	275–325	11.7	10.4	13.1
3	280–320	10.9	7.80	12.2
4	323–350	6.86	3.80	6.09
5	350–385	4.70	3.28	4.90
6	386–417	4.22	2.96	4.16
7	415–455	2.91	2.01	2.93
8	515–573	2.39	1.65	2.46
9	555–600	3.52	1.84	3.50
10	600–850	3.75	4.16	3.17
11	800–1300	2.08	2.09	2.18
12	1100–1950	2.32	2.60	2.38

[a] Filters were selected originally for use with mercury-xenon sources.

an effective transmittance of 0.30 within the 0.001τ limited band which is 30 nm wide. As a limiting case, assume a relative source curve (normalized to 100) to have a value of 100 over the range of the filter. The product function integral for this filter would then have a value of 900 units. In order to have 1% of this total included in the side bands beyond the 0.001τ limits, it would be necessary to obtain 9 units of integrated product function requiring a side band of 9000 nm extension, which is greater than the entire emission region of the sources being considered here. Although this is a limiting case, it can be shown that proper selection of filters for use with a given source will yield similar results. In most cases, the 0.001τ band is a well-chosen point from which to start in shrinking the filter limits (Hickey, 1965). As the filter band shrinks, the filter factor becomes smaller (i.e., the effective band transmittance increases). A point that is often overlooked in calculating filter factors is that a solution exists for every set of band limits chosen.

The nature of the radiometer detection should now be analyzed. At signal levels of greater than 50 μV, a thermopile radiometer, with the associated readout system, can distinguish between irradiance levels of about 1%. This corresponds to an error of approximately 5 μW cm^{-2}. Errors of less than 5% are possible for signals of 10 μV. For output signals of about 1 μV an error of about 20% of the measured signal is possible with the system normally employed in these tests. Signals of less than 1 μV are meaningless here, although these could be measured accurately with laboratory precision standard-type potentiometric equipment.

Based on the 1% energy discrimination limitation, and remembering that the reasons for the expanding band limits have been reduced to the absurd, it can be concluded that a shrinking of the product function integral by 0.5% at each end of the band is justified without increasing the uncertainty of a measurement beyond the inherent errors. The computer output contains integral values for very small bands (sometimes less than a nm in the uv), as well as the cumulative totals for the entire band. The new wavelength limits, which are designated the practical limits, are obtained by choosing those wavelengths which are nearest to the integral values of 0.5% and 99.5% of the total integrated product function. The wavelength shifts involved may vary from filter to filter for a given source. Data in Table III show how the filter factor and the band limits change in accordance with this procedure. JO and PFO refer to the measurements outside a space chamber while JC and PFC refer to monochromator curves which have been corrected to inside the chamber.

Direct filter factor measurements. Probably the best method of obtaining a true filter factor is to measure it in the beam of the unknown source. The emission curve of the source, as modified by the filter, is measured in the same manner as the emission curve itself. This is the product function curve and is defined

(8.29) $$PF_\lambda(\lambda) = H_\lambda(\lambda)\tau(\lambda)$$

as measured using the monochromator as the transfer instrument. The method has the advantage of taking into account effects which depend on the combination of filter and unknown source, which is not possible when the filter transmittance is measured separately by a spectrophotometer.

TABLE III. Change of filter factor with change of limits for filter Hg(JPL)5.

Band limits[a] (nm)	Filter factor (from JO and PFO)	Percent of total integral PFO	Filter factor (from JC and PFC)	Percent of total integral PFC
345–397	8.59	100.0	8.59	100.0
350–397	8.37	99.9	8.37	99.9
354–397	7.99	99.6	7.98	99.6
355–397	7.91	99.5	7.90	99.5
355–395	7.72	99.5	7.73	99.5
355–390	7.34	98.8	7.34	98.8

[a] The original 0.001 τ limits are 345–397 nm and the practical limits are 355–390 nm.

In choosing the limits using this method, the integrated PF curve is available. Shrinking of limits can be accomplished simply from the computer output data of the PF integration. All insignificant (usually 0.5%) subdivisions of the total integral are subtracted from each end of the curve integration. The wavelengths of the limits of the central 99% integral are then used in determining the factor.

Performing the tests for determining factors in this manner for a set of filters is a long and tedious job, but is worth the effort if the highest accuracy is desired.

The changing of filter factors as the resolution of the monochromator is changed has been studied using the method, previously described, of measuring both the relative source emission and product function curves directly. These tests were performed using a Perkin-Elmer Spectracord instrument in single-beam mode, with the filter mounted in the sample chamber of the spectrophotometer, not at the entrance slit. Figure 8.15 shows a sample of such an analysis. In general, changes in resolution will not greatly affect the filter factors, within the practical limits of instruments use.

Filter factors are not greatly influenced by small changes in the spectrum of a given source. Each case, however, must be judged on its own conditions.

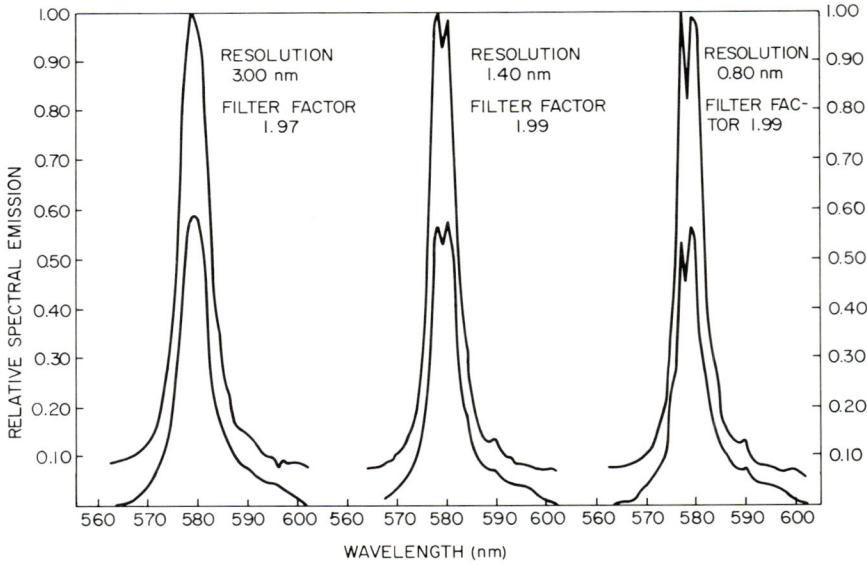

FIG. 8.15. Source emission and product function curves at three monochromator resolution settings showing that the filter factor does not change [Hg-Xe-Zn arc lamp; filter 6 from Hg set—after Hickey (1965)].

Changes in the filter itself are probably of more concern. New filters are being developed to withstand all the conditions of solar simulators better than has been possible previously. One must remember that the filter factor for a given filter is different for different sources. However, a given source does not change its spectral profile radically during short periods of time, with the possible exception of the shortwave region of those described by brightness temperature and having an appropriate Planckian distribution. Since the filter factor is a ratio of the integral of two functions of the source emission, changes in denominator and numerator occur together, tending to cancel any net change in the factor. Assuming that the filter has not changed its transmittance characteristics, spectral changes of the source can be detected by the filter radiometer. In addition, the absolute values of irradiance measured for each band will not be greatly affected.

Other presentations of monochromator data. Very few truly absolute results have been obtained using monochromators for the reasons previously stated. Curves of $H_\lambda(\lambda)$ or $J_\lambda(\lambda)$ are usually plotted and presented as results. The resolution or bandwidth and its variation with wavelength should be given if this type of presentation is chosen. The monochromator results can also be presented in a histogram form, as was explained for the filter radiometer data. An example of such a histogram is given in Fig. 8.16. The integrated $J_\lambda(\lambda)$ data over selected wavelength intervals is divided by the interval $\Delta\lambda$, to give the average $J_\lambda(\lambda)$ over the band. This method may also be used for monochromator–radiometer correlation.

8.9. Correlation of Monochromator and Filter Radiometer Measurement Results

The methods of expressing monochromator results and radiometer results have been discussed separately. The monochromator results histogram may be directly compared with the filter radiometer histogram if both are expressed using the same (practical filter) limits. Figure 8.17 outlines a correlation procedure which is used in comparing the percentage of the total energy in a given filter band. The direct factor method is not shown in the diagram. The blocks designated "monochromator" and "computer", in the upper right-hand corner, could be replaced by a single block in the case of a spectroradiometer.

The point to remember is that (unless reliable absolute monochromator results can be obtained) the methods are mutually dependent. The radiometer method relies on the monochromator method to produce the filter factors, while the monochromator method relies on the radiometer method for absolute irradiance values.

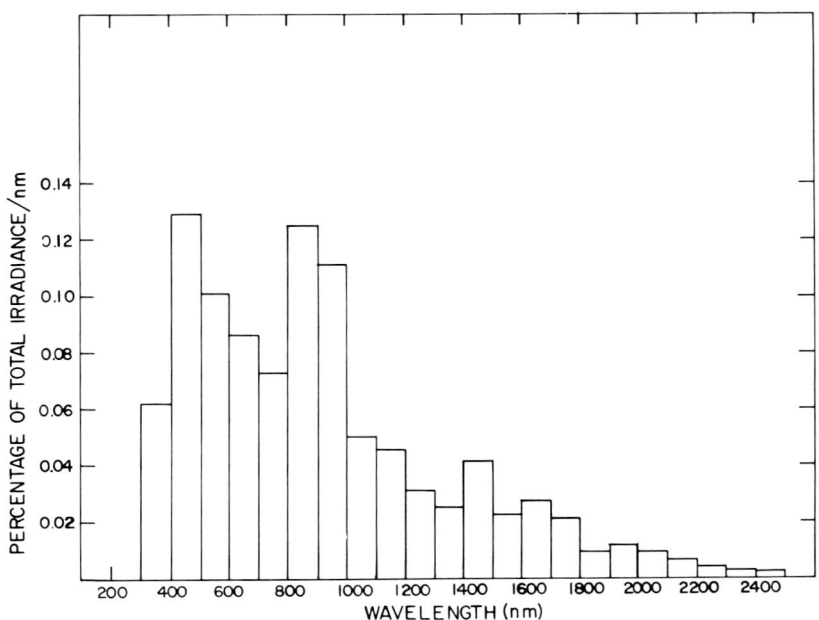

Fig. 8.16. Histogram presentation of monochromator results using results of the CNES solar simulator (unfiltered) as source.

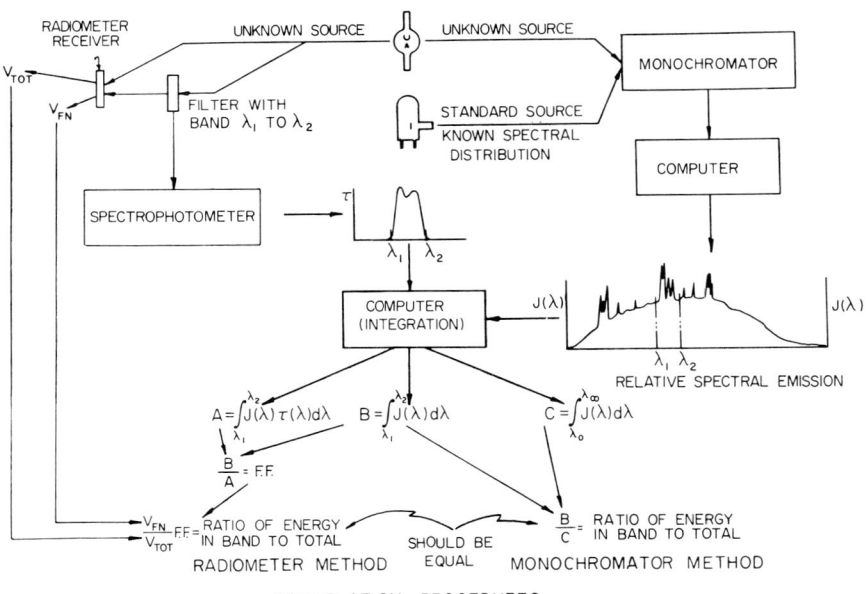

Fig. 8.17. Diagram of monochromator-radiometer correlation procedures [after Hickey (1965)].

8. LABORATORY METHODS OF EXPERIMENTAL RADIOMETRY   265

8.10. Other Uses of Spectral Standards for Calibration Purposes

8.10.1. Use of Spectral Radiance Standards as Spectral Irradiance Standards

A method of accomplishing this has been described (Stair et al., 1965). The method presented here is somewhat different. To obtain extremely low values of spectral irradiance from a strip-filament radiance standard, a suitable set of stops or apertures is placed between the source and the detector. The first aperture should be about 1mm in diameter and should be placed at the front of the fused silica window, which is approximately 4 in. from the filament. The second stop could easily be the detector receiver (e.g. the photocathode of a photomultiplier). The detector ought to be placed far enough from the source so that it cannot see (through the first aperture) more than the central 2 mm of the lamp filament. Since the lamps are uniform emitters within a few per cent over that area of filament, the small aperture becomes the effective source. The radiance of the standard is then the radiance of the aperture considered as source. The spectral irradiance at the detector is

$$(8.30) \quad H_\lambda(\lambda) = N_\lambda(\lambda) A_a \Omega_{ad} / A_d$$

where $N_\lambda(\lambda)$ is the aperture radiance, A_a the area of the effective source, Ω_{ad} the solid angle subtended by the detector, at the source, and A_d the area of the detector. Since

$$(8.31) \quad \Omega_{ad} \simeq A_d / D^2$$

where D is the distance between the aperture and detector,

$$(8.32) \quad H_\lambda(\lambda) = N_\lambda(\lambda) A_a / D^2$$

Figure 8.18 is a diagram showing this procedure.

8.10.2. Use of Irradiance Standards as Radiance Standards

It is necessary sometimes to calibrate such instruments as photometers which have small fields of view (often less than 1%) in terms of radiance sensitivity. This can be accomplished using an irradiance standard. The method involves the viewing of a MgO (or other diffuse) reflecting surface which is illuminated by the light from the spectral irradiance standard. The surface is perpendicular to a line from its center to the standard source when properly oriented with regard to its holder. The spectral irradiance on the surface is known and the reflected radiance is then calculated using the equation

$$(8.33) \quad N_\lambda(\lambda) = H_\lambda(\lambda) \rho(\lambda) / \pi$$

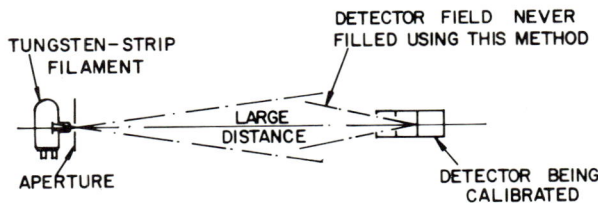

DETECTOR SHOULD SEE APERTURE AS POINT SOURCE
OF KNOWN SPECTRAL INTENSITY OR RADIANCE

Fig. 8.18. Use of a spectral radiance standard as a low-level spectral irradiance standard for the calibration of sensitive detectors.

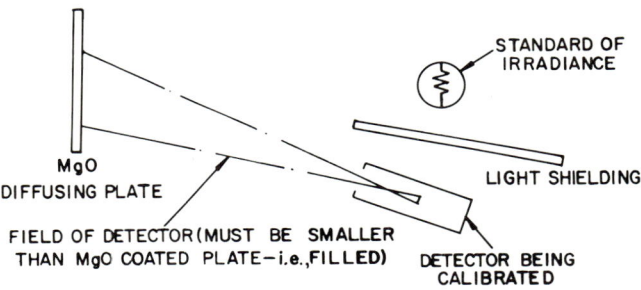

Fig. 8.19. Use of a spectral irradiance standard to obtain a known spectral radiance for the calibration of sensitive detectors.

where $H_\lambda(\lambda)$ is the spectral irradiance at the MgO surface and $\rho(\lambda)$ the spectral reflectance of the surface. The distance from the irradiance standard may be greater than that of the calibration (50 cm for 1000-W lamps). In this case, inverse square law corrections are applied so that

(8.34) $$H_\lambda(\lambda) = H_\lambda^\circ(\lambda)(50/d)^2$$

where $H_\lambda^\circ(\lambda)$ is the standards' known spectral irradiance at 50 cm and d the distance from the standard to the diffusing surface (see Fig. 8.19).

Calibration of detectors using methods 1 and 2. The configurations used in methods 1 and 2 can be used to calibrate very sensitive detectors at low levels of spectral radiance or irradiance (Hickey *et al.*, 1967). Filters with extremely narrow bands (a few nanometers or less) are interposed between the source and detector. The appropriate value of radiance or irradiance in the spectral band of the filter is then calculated. If a number of filters with bands over the sensitivity range of the detector are used, it can be seen that the spectral response function of the detector can be determined. Stray light must be thoroughly eliminated to assure the success of calibrations utilizing these methods.

REFERENCES

Coblentz, W. W. (1915). Measurements on standards of radiation in absolute value. *Bull. Bur. Stds.* **11**, 87–97 (issued November 1914).
Drummond, A. J. (1968). The absolute calibration of thermal radiation detectors. *ISA Trans.* **7**, 197–206.
Drummond, A. J., and Hickey, J. R. (1967). Measurement of the total flux and its spectral components in solar simulation systems with special reference to the extraterrestrial radiation. *Solar Energy* **11**, 1–11.
Forsythe, W. E. (1937). "Measurement of Radiant Energy," p. 101. McGraw-Hill, New York.
Hickey, J. R. (1965). Correlation of monochromator and filter radiometry determinations of the spectral distribution in large solar simulators. *Proc. IES ASTM Intern. Symp. Solar Radiation, Los Angeles, 1965.*
Hickey, J. R., Garvey, J. A., Scholes W. J., Griffin, R. N., and Arvesen, J. (1967). Calibration of the Earth albedo experimental package of the OSO satellite series. *Proc. XVII Intern. Astronautical Congr., Madrid, 1966*, **2**, 237–246. Gordon and Breach, New York.
Karoli, A. R., Ångström, A. K., and Drummond, A. J. (1960). Dependence on atmospheric pressure of the response characteristics of thermopile radiant energy detectors. *J. Opt. Soc. Am.* **50**, 758–763.
Kostkowski, H. J., Erminy, D. E., and Hattenburg, A. T. Higher accuracy spectral radiance measurements. *J. Res. Nat. Bur. Std.* (to be published).
Marchgraber, R. M., and Drummond, A. J. (1960). A precision radiometer for the measurement of total radiation in selected spectral bands. Monograph No. 4, pp. 10–12. Intern. Union of Geodesy and Geophys., Paris.
Nicodemus, F. E. (1965). Radiometry Rept. No. EDL-G324, 98 pp. Sylvania Electronics Systems.
Nicodemus, F. E. (1968). Radiometry with spectrally selective sensors. *Appl. Opt.* **7**, 1649–1652.
Schneider, W. E., Stair, R., and Jackson, J. K. (1967). Spectral irradiances as determined through the use of prism and filter spectroradiometric techniques. *Appl. Opt.* **6**, 1479–1486.
Schneider, W. E., Waters, W. R., and Jackson, J. K. (1968). Development of a high-intensity standard of total and spectral irradiance. NBS Rept. 9899, 81 pp. U.S. Dept. of Commerce.
Smith, R. A., Jones, F. E., and Chasmar, R. P. (1968). "The Detection and Measurement of Infrared Radiation," 2nd ed., 503 pp. Oxford Univ. Press, London and New York.
Stair, R., Johnston, R. G., and Halbach, E. W. (1960). Standard of spectral radiance for the region of 0.25 to 2.6 microns. *J. Res. Nat. Bur. Std. A* **64**, 291–296.
Stair, R., Schneider, W. E., and Jackson, S. K. (1963). A new standard of spectral irradiance. *Appl. Opt.* **2**, 1151–1154.
Stair, R., Fussell, W. B., and Schneider, W. E. (1965). A standard for extremely low values of spectral irradiance. *Appl. Opt.* **4**, 85–89.
Stair, R., Schneider, W. E., and Fussell, W. B. (1967). The new tungsten-filament lamp standards of total irradiance. *Appl. Opt.* **6**, 101–105.

—9—

ON DETERMINATIONS OF THE ATMOSPHERIC TURBIDITY AND THEIR RELATION TO PYRHELIOMETRIC MEASUREMENTS

A. K. Ångström

9.1. Introduction

The question concerning the transmission of solar radiation through the atmosphere cannot be adequately treated without consideration of the radiation from the source itself, viz. from the sun. The sun sends out a whole spectrum of wavelengths, as is demonstrated in Fig. 9.1, which also gives a schematic idea of how this radiation is influenced, for different wavelengths, through extinction in the atmosphere. The integral of the solar radiation over all wavelengths, reduced to the mean distance between earth and sun, is

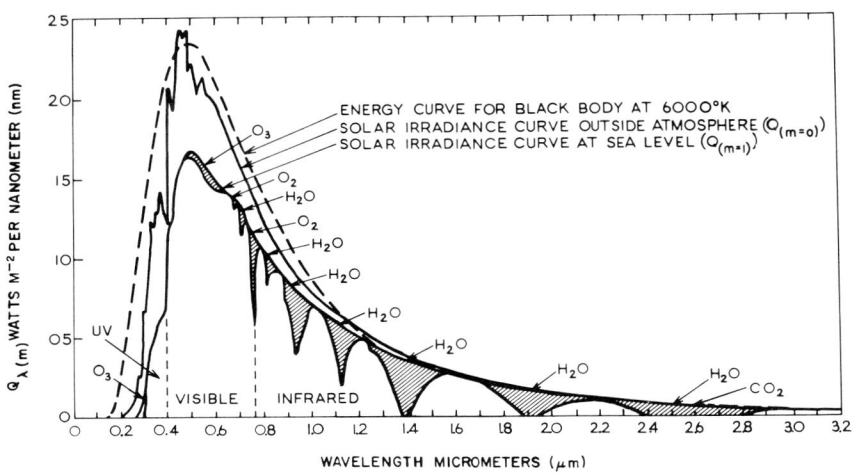

Fig. 9.1. Solar spectral irradiance outside and within the atmosphere (after Gast, 1960).

269

generally denoted the solar constant. Through the elaborate and painstaking work of Abbot and his collaborators, at the Astrophysical Observatory of the Smithsonian Institution, it has been established that this so-called constant, for most meteorological considerations concerned with the energy exchange within the system earth-atmosphere, really is constant (Abbot et al., 1908–1954). It is no exaggeration to say that this result ought to be regarded as one of the most important achievements in geophysics in our century. Different authors treating the very great number of determinations of Abbot have (with application of different corrections for unobserved parts of the spectrum) obtained somewhat different values for the solar constant, the mean falling at about 2.00 cal cm^{-2} min^{-1} (Drummond, 1965, and this volume, Chapter 1). There are suggestions of small variations, perhaps of up to 0.5 % or slightly more, in the value of the solar constant, but controversial opinions have been expressed concerning these apparent variations (and it is not intended to discuss this). What is said here concerning the "integral solar constant" ought not to be applied to "partial solar constants" referred to special wavelength intervals. It seems evident that great variations occur in the extreme ultraviolet, but also that these variations only influence the energy integral to a very small extent over all wavelengths.

A glance at Fig. 9.1 tells us that there are different processes at work producing the extinction within the atmosphere. We have a selective absorption limited to certain lines or bands in the spectrum, where the absorption by the atmospheric gases—especially noteworthy are water vapor, carbon dioxide, oxygen, and ozone—produces marked depressions. However, there is also an extinction of more general character influencing all wavelengths, affecting the short ones much more than the long ones.

9.2. Scattering of Solar Radiation within the Atmosphere

This last-named extinction is found to have a double cause, namely (1) scattering by the air molecules and (2) scattering and absorption by solid and liquid particles within the atmosphere. The existence of such a molecular scattering was first pointed out and investigated on a theoretical basis by Lord Rayleigh. With slight supplements to his theory, especially with regard to the so-called depolarization effect of unsymmetrical molecules, it can be stated that observations in very clear air, in the Antarctic and on high mountains, show close agreement with the theory (Ångström, 1929 and 1930). The Rayleigh scattering, as it is often called, is inversely proportional to the fourth power of the wavelength.

On the other hand, the scattering by the liquid and solid particles of the atmosphere—by what is usually termed aerosol—is highly dependent on the purity of the atmosphere, viz., on the number, sizes, and qualities of the scattering particles.

The basic relation between the different factors causing the extinction of solar radiation and the observed radiation may be written as follows:

$$(9.1) \qquad I = \frac{1}{S} \int_0^\infty I_0(\lambda)\, e^{-A(\lambda)}\, d\lambda$$

where I is the radiation intensity measured by a pyrheliometer and expressed in the International Pyrheliometric Scale, λ the wavelength, $I_0(\lambda)$ the radiation intensity outside the atmosphere at the mean sun distance (in a small wavelength interval $\Delta\lambda$ which must be so chosen that the extinction exponent (see below) can be regarded as constant), $S = R^2/R_m^2$ the reduction factor for mean sun–earth distance (where R is the distance corresponding to the date and R_m the mean value), and e the base of natural logarithms.

$$(9.2) \qquad A(\lambda) = A_R(\lambda) + A_D(\lambda) + A_W(\lambda)$$

is the extinction exponent composed of three parts, each of them dependent on the wavelength.

$$(9.3) \qquad A_R(\lambda) = m a_R(\lambda) = m \cdot 0.00897 \lambda^{-4.09}$$

gives the extinction in clear dry air according to Rayleigh's theory, supplemented by a depolarization factor of 1.061 (Penndorf, 1957), where m is the optical air mass, with the vertical air mass at the earth's surface, at standard pressure (1000 mb), as unit.

$$(9.4) \qquad A_D(\lambda) = m_h a_D(\lambda) = m_h \beta_p \lambda^{-\alpha} = m\beta\lambda^{-\alpha}$$

gives the extinction of haze according to the definitions of the present author.

$$(9.5) \qquad m_h = m \cdot 10^3/p$$

is the relative air mass, where p is the atmospheric pressure (in mb) at the observation site.

$$(9.6) \qquad a_D(\lambda) = \beta_p \cdot \lambda^{-\alpha}$$

is the extinction coefficient due to the haze and is expressed as a function of wavelength.

β_p is the turbidity coefficient of A. Ångström (1961 and 1964a) corresponding to the wavelength 1 μm. It is consequently expressed in the same unit generally used for expressing the Rayleigh scattering.

$$(9.7) \qquad \beta = \beta_p \cdot 10^3/p$$

where β is the turbidity coefficient referred to unit optical air mass.

$A_W(\lambda)$ is the extinction caused by the water vapor contained in the atmosphere.

The relationship (9.6) constitutes an empirical approach founded mainly on the extensive observational material collected by the Smithsonian Institution.

The exponent α has been found to be closely related to the size of the scattering particles and to the frequency of their size distribution. This exponent, as a rule in the case of aerosol, has a value smaller than 4, which would be expected if the particles—like molecules—were very small compared with the wavelength of light. In the case of a clear sky and a turbidity not influenced by volcanic eruptions or forest fires, the value of α is close to 1.0 to 1.5, viz., the coefficient of scattering for various wavelengths is, with some approximation, inversely proportional to the wavelength.

Theoretical investigations provide an interesting background for those empirical results. If we follow the considerations of Mie, in order to get an idea about scattering by small particles, we generally find a rather complicated relationship between scattering and wavelength. It has been shown, however, that if the frequency distribution of the different size particles is given by

$$(9.8) \qquad dN = -K \, dr/r^\gamma$$

where dN is the number of particles within the interval dr of radius r and γ is a constant independent of N and r, then the scattering coefficient due to their combined effect takes the form (9.6), provided that $\gamma = \alpha + 3$. It has been found, in many cases, that the size distribution comes near to what is suggested by Eq. (9.8). Probably, however, the distribution can deviate considerably from the one expressed by this simple rule. Another explanation may be considered through reference to the results derived by Foitzik (1950). He shows, in the case of small scattering water droplets, how the scattering is dependent on the radius of the droplets and on the wavelength. For droplets of a radius between 0.2 and 0.5 μ, we have evidently a wavelength dependence which is characterized by a continuous decrease of the scattering coefficient within the visible spectrum, and it seems rather natural that a given preferred size will also lead to a preferred value for the exponent α.

9.3. Determination of the Ångström Turbidity Coefficient

Further studies of the variation of the values of β and α seem to show that a derivation of these two turbidity parameters provides a basis on which sufficiently detailed ideas, for most purposes, may be obtained with regard to aerosol extinction as a function of wavelength. Evidently, assuming the validity of Eq. (9.6), we may determine β as well as α through measurements of the intensity of the solar radiation at two points of the spectrum where no selective absorption occurs. We then get two equations of the type (9.4), from which both unknown parameters may be derived. The radiation measurements executed within meteorological networks generally cover rather wide wavelength intervals for which we may put

$$(9.9) \qquad \bar{I} = \int_{\lambda a}^{\lambda b} I_\lambda \, d\lambda = \int_{\lambda a}^{\lambda b} I_{0\lambda} \cdot \tau \cdot \exp(-\beta_\lambda \cdot m) \, d\lambda$$

Here $\bar{I}$ is the radiation obtained directly from pyrheliometric measurements and corresponding to an interval, between wavelengths a and b, where practically no selective absorption occurs. The product $I_{0\lambda} \cdot \tau$ is evidently the partial solar constant at the wavelength λ multiplied by the transmission factor τ for the Rayleigh scattering. This product can easily be obtained from appropriate tables or otherwise computed (e.g., Penndorf, 1957). It is now found that under conditions in which the extinction coefficient of the aerosol can be expected to be very low or to be totally eliminated, there is the experimental coincidence:

$$(9.10) \qquad \bar{I}_{\beta=0} = \int_{\lambda a}^{\lambda b} I_{0\lambda} \cdot \tau \, d\lambda$$

The validity of Eq. (9.10) is corroborated especially by the maximum values obtained by the Smithsonian Institution (Abbot and collaborators) on the top of Mount Whitney, 4420 m, by Liljequist (1956) in the Antarctic, and by the U.S. Weather Bureau at Mauna Loa Observatory, 3380 m, Hawaii (treated by Drummond and Ångström, 1965, 1967; Ångström and Drummond, 1966).

If now tables are created giving the right-hand integral of Eq. (9.9) for different values of β, they may be used for determining β after $\bar{I}$ has been found through measurements. This is the most simple way to determine β, but it assumes α to be constant, an assumption which is only approximately true. An analysis of the great mass of transmission values published by Abbot shows, however, that this assumption forms a reasonable approximation whenever the atmosphere is comparatively clear. Table I is an example of this kind of procedure. The turbidity is determined from direct measurements behind the Schott filter RG8. This filter transmits practically all radiation from the sun of wavelength above 700 nm. Subtracting the corresponding measurement from a measurement of the total integral radiation $\left(\int_0^\infty F(\lambda) \, d\lambda\right)$, we obtain (after a slight correction for ozone absorption) a value for the integral $\int_0^{700} F(\lambda) \, d\lambda$, which, with the named correction, can be regarded as free from effects of selective absorption.

TABLE I. Determination of the turbidity coefficient from measurements with the RG8 filter ($\lambda > 700$ nm).

m ($\beta \cdot 10^3$)	0.0	0.5	0.6	0.7	0.8	0.9	1.0	1.25	1.50	1.75	2.0	3.0	4.0	5.0	6.0
				I (mcal cm^{-2} min^{-1}) for $\lambda < 700$ nm										
0	948	852	835	818	802	787	771	742	716	698	662	569	499	436	397
5		848	830	812	795	779	762	732	704						
10		843	824	806	788	771	753	721	691						
15		838	819	800	781	763	744	710	678						
20		834	813	794	773	755	735	699	665						
25		830	808	787	766	746	726	688	652	619	588	479	396	333	282
30		825	802	781	759	738	717	678	640						
35		820	797	775	752	730	709	668	629						
40		815	791	769	745	722	700	658	618						
45		810	786	762	738	714	691	648	607						
50		805	780	756	732	707	682	638	596	559	522	404	321	256	206
75		782	754	726	698	670	642	592	546	505	466	342	260	196	152
100		758	726	695	664	634	604	549	500	455	415	290	208	148	111
125							557	510	460	413	370	246	168	114	81
150							535	475	421	371	330	200	136	90	60
175							503	445	386	337	295	180	110	68	45

N.B. Extraterrestrial flux (total) taken as 1.98 cal cm^{-2} min^{-1}, after Nicolet (1951).

The β-values thus determined provide an idea of the optical clarity of the atmosphere, which for many practical purposes is accurate enough. Thus, Ångström and Drummond (1962a,b), investigating the illumination from sun and sky at Pretoria in South Africa, found that the illumination measured in kilolux could be derived with an accuracy of $\pm 1.5\%$ from the derived β-values, as is shown in Table II. Since β closely reflects the extinction of the atmosphere within the visible solar spectrum, the result is what could be

TABLE II. Global illumination (kilolux) at sea level of a horizontal surface.

β	m								
	0	1	1.25	1.50	1.75	2.0	3.0	4.0	6.0
0.00	142	127	100	82	69	59	36	25	15
0.05		118	90	72	60	51	31	19	11
0.10		110	83	66	54	45	26	18	10
0.15		104	77	61	50	41	23	17	11
0.20		100	73	58	46	38	23	17	11
0.25		98	71	55	44	37	22	17	12
0.30		96	71	55	44	38	24	18	14

expected. It also demonstrates, however, that the assumption of a constant value for α, in this practical case, constitutes an acceptable simplification.

The use of the two simple parameters α and β, or occasionally β alone, must be justified by studies of the correlation existing between these parameters and certain factors of practical or theoretical interest. Thus, as has been shown by Ångström (1929, 1930), Schüepp (1949), and, particularly, Valko (1961), the β-values are closely correlated with the origin of the air masses and may be used for a characterization of the synoptic situation and thus for weather forecasting. Tropical air masses are usually much more turbid than arctic ones.

9.4. Influence of Circumsolar Sky Radiation on Pyrheliometric Measurements

Another problem in which the β-measurements have recently been found highly useful concerns the influence of the aureole radiation—the circumsolar sky radiation—upon ordinary pyrheliometric measurements (Ångström and Rodhe, 1966). It is clear that the aureole around the sun has its cause in the scattering within the atmosphere, and β, as we have seen, is intended to be a measure of this scattering. What we measure with our pyrheliometer is not only the direct radiation from the sun but also an additional radiation dependent on the aperture of the instrument and on the amount and kind of scattering particles within the atmosphere. We have a number of measurements by different investigators concerning the falling off of sky radiation just beyond the solar disk. Based on a synthesis of a great number of such measurements, we arrive at the result shown by Fig. 9.2. As parameters we have chosen the product βm, where m is the absolute air mass and φ the aperture. From the diagram, it is concluded that two pyrheliometers with different apertures must show differences under ordinary conditions, dependent on the turbidity of the sky, of which β is a measure. First, through the introduction of β into our considerations, a simple way of taking account of the turbidity in pyrheliometric comparisons has been found (see Fig. 9.3). For specified theoretical studies, a more detailed idea of the scattering for different wavelengths may be necessary, but for ordinary comparative pyrheliometric work, the simplification introduced by the use of the easily accessible parameters α and β seems justified.

Especially for the different models of the Ångström electrical compensation pyrheliometer (which has rectangular diaphragms), Ångström and Drummond (Ångström and Rohde, 1966) have introduced the term "effective aperture." This aperture is defined as the full angle which, in the case of a circular opening, will give the same radiation to the central part of the receiver as does the actual rectangular aperture. The radiation reaching the center of

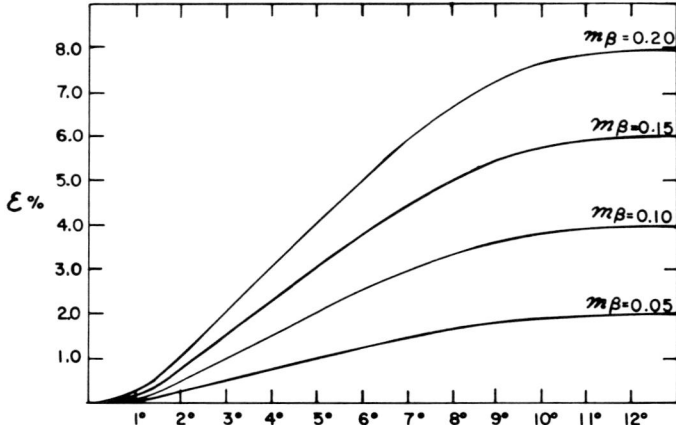

Fig. 9.2. Radiation from the aureole given in percentage of the direct solar radiation simultaneously measured. The figure shows ε in its dependence on the turbidity parameter $m\beta$ and on the aperture of the measuring instrument (pyrheliometer).

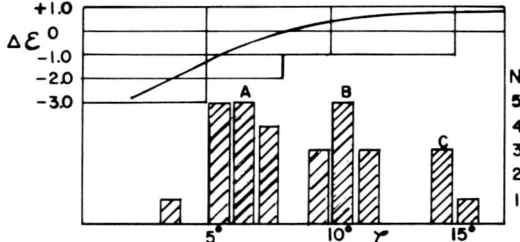

Fig. 9.3. The figure demonstrates the frequency of the apertures of the pyrheliometers represented at the WMO (Davos) international comparisons in September 1964 (lower part of figure). If all these pyrheliometers are assumed to be normalized so as to yield values equal to the Davos standard Ångström number 210 ($\psi = 8°$), under conditions where the turbidity parameter $m\beta$ has a given value, we get (from the upper part of the figure) the differences from the standard which are expected to occur when the measurements are repeated with the turbidity parameter 0.10 higher.

the strips was determined through graphical integrations. These computations made it evident that the effective aperture, in general, is practically independent of the turbidity. At present, the aperture of pyrheliometers, with rectangular openings, is defined by the two angles, α_s and α_l, corresponding to the short and long sides of the opening, respectively. As the intensity of the radiation from the aureole decreases rapidly with the distance from the sun, an increase of the larger angle α_l above a certain limit value produces no appreciable increase in the effective aperture. In fact, for rings of the sky, with distances above about 5° (aperture > 10°), an overcompensation of the aureole radiation takes place, in that from such a ring more radiation is falling on the screened than on the exposed strip. As a good rule of thumb, valid when $\alpha_l > 2.5\alpha_s$, we may apply

$$(9.11) \qquad \varphi = (4/3 \pm 0.05)\alpha_s$$

For an aperture of, for instance $6 \times 24°$, we then have $\varphi = 8.0 \pm 0.3°$. The effective aperture was thus computed for each of the compensation pyrheliometers compared at Davos (World Meteorological Organization, September 1964 International Pyrheliometric Comparisons). The frequency distribution of the different apertures of these pyrheliometers is graphically demonstrated in Fig. 9.3. We find that they fall chiefly within three main intervals which are denoted A, B, and C.

The A group which contains 14 of the 30 pyrheliometers (viz., 40 %) is characterized by apertures between 5 and 8°; the B group (11 instruments) represents apertures between 9 and 12°; and the C group (4 instruments) apertures between 14 and 16°. The mean apertures within the three main groups are: A—6.5°, B—10.7°, and C—15.0°. Let us now imagine a case which is of some interest. Assume all the instruments to have been compared under turbidity conditions defined by a given mean value for the turbidity parameter $m\beta$ (about 0.075 to 0.10) and apply corrections to their earlier constants in order that they all read equal to a given standard, e.g., Å 210 with an effective aperture of 8.0°. Let us now expose the instruments under conditions characterized by a value of $m\beta$ which is 0.10 higher, viz., about 0.20, a value which often must occur at places like Washington, Kew Observatory, and Potsdam, at relative air masses between 1.2 and 2.0. The curve in the upper part of Fig. 9.3 shows how the different instruments may now be expected to deviate from the assumed standard. The instruments of groups B and C will read about 0.5 to 0.7 % higher; the instruments A between 0.2 and 1.2 % lower. A little group represented by the smaller aperture instruments of the Smithsonian Institution can be expected to read about 2.0 % lower than the assumed standard. In the extreme cases (apertures of about 4° on the one hand and about 15° on the other), instruments which were adjusted to give equal readings for the low atmospheric turbidity at Davos can be expected to differ by about 3 % under the turbidity conditions often prevailing at low-altitude stations.

9.5. Absorption of Solar Radiation by Atmospheric Water Vapor

A third problem, where the introduction of the parameter β has introduced a revision of popular ideas, concerns the absorption by water vapor in the atmosphere. This absorption, as can be easily shown, is dependent not only on the optical path within the water vapor but also, to some extent, on the turbidity of the atmosphere. If we try to determine the atmospheric water-vapor content, from measurements of the absorption in the direct solar beam (as have often been made), we arrive at faulty results, as pointed out in special cases by, for instance, Foitzik and Hinzpeter (1958). The error is in considerable part due to absence of proper consideration of the atmospheric turbidity. Figure 9.4 shows how the water-vapor absorption can be graphically expressed as a function of the parameters um and βm, where u is the precipitable water in the vertical column and the product βm is the parameter discussed earlier (Ångström, 1964b).

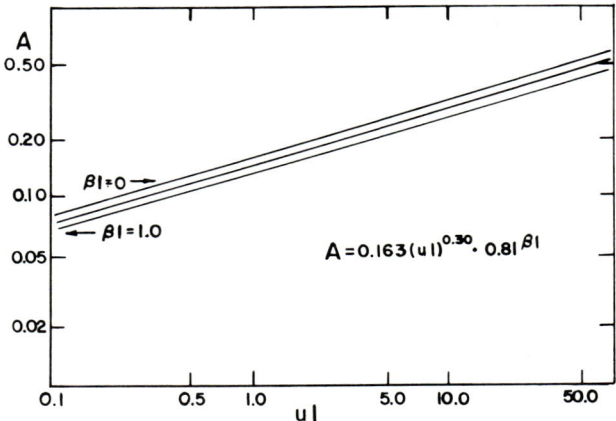

FIG. 9.4. Absorption by water vapor as a function of the parameters $m\beta$ and um (denoted here by βl and μl, where l is the path length). [Extracted from Ångström (1964b).]

Especially with high turbidity—great scattering of the atmosphere—a determination of the water content disregarding the scattering properties may easily introduce errors of more than 30 % in the evaluation of the water content. It seems probable that with proper regard to scattering, we may attain results which, at any rate, must be a valuable control of aerological data and which, in many cases, may supersede such data from the viewpoint of accuracy. How the water-vapor absorption and the influencing moisture content of the atmosphere can be determined from turbidity measurements obtained in pyrheliometric operations is demonstrated by the diagrams of Fig. 9.5.

These were used for deriving such estimates for the atmosphere above Mauna Loa (Drummond and Ångström, 1965 and 1967). In this figure, the turbidity coefficient β is first arrived at by evaluating the radiation below

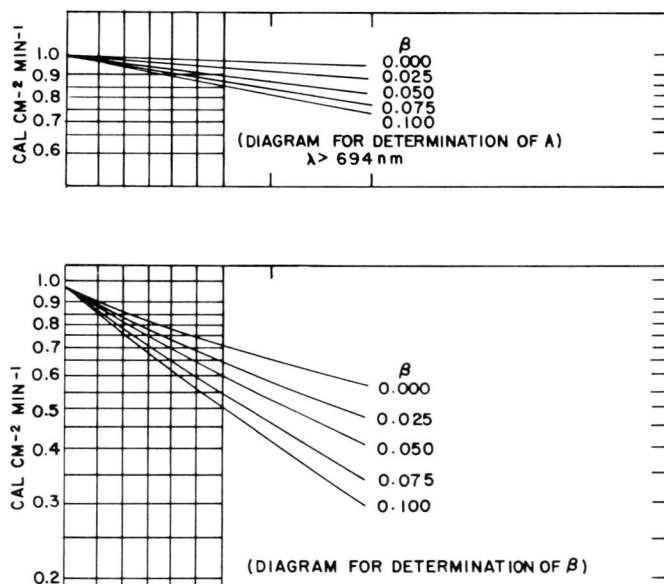

FIG. 9.5. Graphical representation for evaluating the turbidity coefficient β and the water-vapor absorption A, from values of the air mass (absolute) and measurements of radiation.

694 nm, corresponding to given observations of the two groups of radiation:

$$\int_{694}^{\infty} F(\lambda)\, d\lambda - A \quad \text{and} \quad \int_{0}^{\infty} F(\lambda)\, d\lambda - A$$

(A is the selective absorption by water vapor and CO_2). On subtraction, we obtain

$$\int_{0}^{694} F(\lambda)\, d\lambda$$

With this value, we use the curves of Fig. 9.5 (lower diagram) to derive β.

Next, applying this value of β to the upper diagram of the figure, we determine the value of the integral $\int_{694}^{\infty} F(\lambda)\, d\lambda$, viz., the radiation which ought to have been found through measurements behind the RG8 filter, for the case where no water-vapor absorption had occurred. The difference between this value and the actually observed value gives us A. From A we readily obtain the precipitable water above (see Fig. 9.4).

In applying this procedure to the measurements on high mountains, e.g., Mount Everest, 5710 m (Bishop et al., 1966), and Mauna Loa, 3380 m (Ångström and Drummond, 1966), we derived the following average values for the turbidity (β), the water-vapor absorption (A), and the precipitable water (W) at these stations:

	β	A	W
Mount Everest (5710 m)	0.048	0.04 cal	0.1 mm
Mauna Loa (3380 m)	0.017	0.13	5.0

The turbidity of the atmosphere is occasionally subjected to changes which are not caused by climatic changes; this does not mean, however, that climatic changes may not be caused by changes in the average turbidity. It seems probable, as has been emphasized by several authors (e.g., Abbot, Kimball, Humphreys), that the volcanic outbreaks and intense forest fires, which give rise to an increased turbidity, may cause marked climatic changes, especially concerning temperature.

9.6. Influence of Atmospheric Turbidity on the Planetary Albedo of the Earth

An idea of the influence of turbidity on reflection of the sun's radiation out to space, viz., on the *albedo of the earth*, may now be gained from measurements and registrations of the illumination corresponding to different values of the turbidity (Ångström, 1962). Dependable measurements of this kind are still comparatively rare. They have, however, been carried out in a systematic way at Kew Observatory, London, as reported by Robinson (1956), and also by Drummond (1956, 1958) at Pretoria (South Africa). The instrumental measurements of the illumination, from sun alone as well as from sky and sun, are supplemented by very extensive filter measurements, which allow an effective control of the illumination values. The measurements are especially valuable as they refer to conditions where the parameters involved have varied within a large range; for example, turbidity of $\beta = 0.05$ to 0.35 and absolute air mass (m) penetrated by the solar rays of 0.85 to 4.0 (Ångström and Drummond, 1962a,b).

I have given, earlier in this chapter, an idea about the dependence of illumination on the turbidity coefficient β. Here, I will briefly mention another important application. In considering the *"loss* in illumination" caused by a given turbidity, we may form an idea about the reflection out to space under different conditions of turbidity. From there, we may go a step further and inquire into the influence of turbidity on the planetary albedo.

The whole procedure involves a number of considerations which are of such a character that the absolute values arrived at may be rather doubtful. For details, I must here refer to my original paper (1962). On the other hand, the changes produced through changes in the turbidity are probably rather well reflected in the derived values which are presented in the adjacent Table III.

As we see from the table, an increase in the turbidity coefficient from 0.10 to 0.20—a change easily produced during a period of violent volcanic activity —produces a change in the albedo from 0.335 to 0.38, which would mean a change in the effective temperature of the earth of more than 4°C. The matter is, consequently, well worth our attention.

Here, I have tried to give a general survey concerning our attempts to measure the turbidity of the atmosphere and to express the results in a form accessible through rather simple pyrheliometric measurements. I hope that I have been able to show that the simplifications introduced in our concepts are justified from a practical point of view, and that our meteorological radiation measurements, through the methods described or suggested, can reach an increased value and usefulness.

Finally, in this connection, I call attention to Figs. 9.6 and 9.7, which show the annual variation of the turbidity at some representative stations.

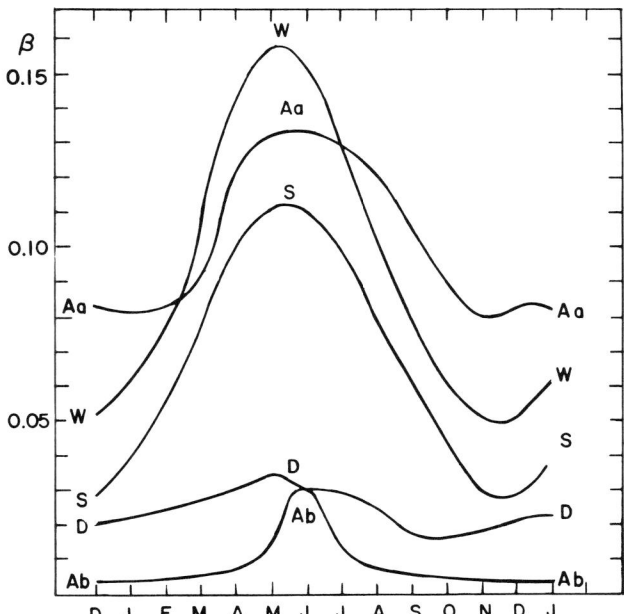

FIG. 9.6. Annual variation of the turbidity coefficient β for Aachen (Aa), Abisko (Ab), Davos (D), Stockholm (S), and Washington (W), according to Hoelper (1935) and Tryselius and Ångström (1934); from Tryselius (1936).

TABLE III. Planetary albedo of the earth for different values of the turbidity.

β	0.00	0.05	0.10	0.15	0.20	0.25
a	0.29	0.31	0.335	0.36	0.38	0.40

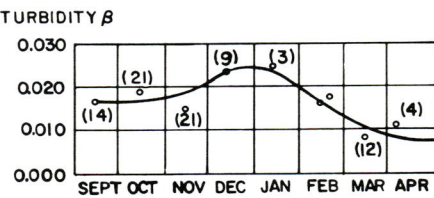

FIG. 9.7. Monthly means of turbidity β at Maudheim, according to the Ångström–Hoelper method; the figures within parentheses give the number of days from which the means are computed; from Liljequist (1956).

APPENDIX

After the previous text was written, some new works were described and partly published, the results of which have an important bearing upon the review presented in this chapter. First of all, the joint Jet Propulsion Laboratory–Eppley Laboratory experiment, at altitudes above the ozonosphere (Drummond *et al.*, 1968; see also Chapter 1, this volume), has confirmed a result obtained by these laboratories from measurements at lower levels: The solar radiation at wavelengths below about 607 nm is considerably lower (about 7 %) than was derived from the measurements of the Smithsonian Institution. If further investigations should confirm this result, it would mean that the β- and α-values, computed on the basis of the earlier ideas concerning the spectral energy distribution within the extraterrestrial solar radiation, need a correction which, especially in the case of the lower β-values, may be considerable. A more detailed consideration of the effect of such errors in the spectral energy distribution on some parameters derived from pyrheliometric measurements will be reported elsewhere.

Another important aspect refers to some results concerning the influence on solar radiation, in India, by dust in the atmosphere. In a recent article (Mani, 1969), it is shown that the large variations of the character of the dust within this region result in highly variable α-values ($0.0 < \alpha < 3.5$). Useful information concerning the spectral solar radiation, under these conditions, can be gained first through a determination of β as well as α.

References

Abbot, C. G., Fowle, F. E., Aldrich, L. B., and Hoover, W. H. (1908–1954). Annals of the Astrophysical Observatory of the Smithsonian Institution II–VII (see especially III, 99 and 113, and VII, 176), Washington, D.C.

Ångström, A. (1929). On the atmospheric transmission of sun radiation and on dust in the air. *Geografis. Ann.* **H2**, 156–166; (1930). On the atmospheric transmission of sun radiation II. *Geografis. Ann.* **H2-3**, 130–159.

Ångström, A. (1961). Techniques of determining the turbidity of the atmosphere. *Tellus* **13**, 214–223.

Ångström, A. (1962). Atmospheric turbidity, global illumination and planetary albedo of the earth. *Tellus* **14**, 435–450.

Ångström, A. (1964a). The parameters of atmospheric turbidity. *Tellus* **16**, 64–75.

Ångström, A. (1964b). On the absorption of solar radiation by atmospheric water vapor II. *Arkiv. Geofysik*, **4**, 503–512.

Ångström, A. K., and Drummond, A. J. (1962a). Fundamental principles and methods for the calibration of radiometers for photometric use. *Appl. Opt.* **1**, 455–464.

Ångström, A. K., and Drummond, A. J. (1962b). On the evaluation of natural illumination from radiometric measurements of solar radiation. *Arch. Meteorol. Geophys. Biokl.* **B12**, 41–46.

Ångström, A. K., and Drummond, A. J. (1966). Note on solar radiation in mountain regions at high altitude. *Tellus* **18**, 801–805.

Ångström, A. K., and Rodhe, B. (1966). Pyrheliometric measurements with special regard to the circumsolar sky radiation. *Tellus* **18**, 25–33.

Bishop, B. C., Ångström, A. K., Drummond, A. J., and Roche, J. J. (1966). Solar radiation measurements in the high Himalayas (Everest region). *J. Appl. Meteorol.* **5**, 94–104.

Drummond, A. J. (1956). Notes on the measurement of natural illumination: I Some characteristics of illumination recorders. *Arch. Meteorol. Geophys. Biokl.* **B7**, 437–465.

Drummond, A. J. (1958). Notes on the measurement of natural illumination: II Daylight and skylight at Pretoria; the luminous efficiency of daylight. *Arch. Meteorol. Geophys. Biokl.* **B9**, 149–163.

Drummond, A. J. (1965). The extraterrestrial solar spectrum. *Proc. IES-ASTM Intern. Symp. Solar Radiation Simulation, Los Angeles*, pp 55–64. Inst. of Environmental Sciences, Mount Prospect, Ill.

Drummond, A. J. (1970). This volume, Chapter 1.

Drummond, A. J., and Ångström, A. K. (1965). Analysis of the solar radiation measurements made in Hawaii (Mauna Loa and Hilo). 1 March 1961–30 June 1962. Final Rept. U.S. Army Contract Da-19-129-AMC-231 (N).

Drummond, A. J., and Ångström, A. K. (1967). Solar radiation measurements on Mauna Loa (Hawaii) and their bearing on atmospheric transmission. *Solar Energy* **11**, 1–9.

Drummond, A. J., Hickey, J. R., Scholes, W. J., and Laue, E. G. (1968). New value for the solar constant of radiation. *Nature* **218**, 259–261.

Foitzik, L. (1950). Zur meteorologischen Optik von Dunst und Nebel. *Z. Meteorol.* **4**, 289–297 and 321–329.

Foitzik, L., and Hinzpeter, H. (1958). "Sonnenstrahlung und Lufttrübung," 309 pp. Akad. Verlagsges, Leipzig.

Gast, P. R. (1960). "Handbook of Geophysics," rev. ed., p. 16 (15), U.S.A.F. Macmillan Co., New York.

Liljequist, G. H. (1956). Energy exchange of an Antarctic snow-field: short-wave radiation. Norwegian-British-Swedish Antarctic Exped. 1949–1952. *Sci. Results* **2**.

Mani, A. (1969). *Tellus* (in press).

Nicolet, M. (1951). Sur la détermination de flux énergétique du rayonnement extraterrestre du soleil. *Arch. Meteorol. Geophys. Biokl.* **B3**, 209–219.

Penndorf, R. (1957). Tables of the refractive index for standard air and the Rayleigh scattering coefficient for the spectral region between 0.2 and 20.0 μ and their application to atmospheric optics. *J. Opt. Soc. Am.* **47**, 176–182.

Robinson, G. D. (1956). The use of surface observations to estimate the local energy balance of the atmosphere. *Proc. Roy. Soc. (London)* **A236**, 160–171.

Schüepp, W. (1949). Die Bestimmung der Komponenten der atmosphärischen Trübung. *Arch. Meteorol. Geophys. Biokl.* **B1**, 257–346.

Tryselius, O. (1936). On the turbidity of polar air. *Statens Meteorol. Hydrogr. Anst.* (Sweden), Communications Series No. 7.

Valko, P. (1961). Untersuchung über die vertikale Trübungs-Schichtung der Atmosphäre. *Arch. Meteorol. Geophys. Biokl.* **B11**, 143–210.

— 10 —

SOME METEOROLOGICAL ASPECTS OF RADIATION AND RADIATION MEASUREMENT

G. D. Robinson

10.1. General Introduction: Magnitudes

I take my task within this series of articles to be to give a broad survey of radiation and radiation measurement within the earth's atmosphere—why measurements and computations of radiation are made, what is their accuracy, and to what extent they are adequate. Space will allow only the most superficial treatment of many topics. It is useful to begin with an examination of the magnitude of the energy transfer and transformation involved. The sun is the only appreciable energy source. We can take the solar constant to be 1.4 kW m^{-2}. About 40 % of this energy is returned unchanged to space, i.e., the albedo of the earth as a whole averaged over the whole solar spectrum is about 0.4.[1] The remaining 60 % is absorbed and after various transformations is re-emitted to space as terrestrial radiation, since, in the long term, the planet is neither warming nor cooling. Putting in the geometrical factors, we find that if the earth were to radiate as a blackbody, its temperature would be about 245°K and the major part of the emitted energy would be at wavelengths between 5 and 50 μm.

Of the unreflected solar radiation, about 15 % is absorbed in the atmosphere, some 3 or 4 % at levels higher than 20 km, which leaves about 45 % to be absorbed at the earth's surface, an average value over the globe and over the year of 160 W m^{-2}. Of this, there is a net loss of about 60 W m^{-2} in exchange of terrestrial radiation with the atmosphere and space, leaving 100 W m^{-2} to be used for evaporation of water or to be lost by convective transfer to the atmosphere. The mean annual rainfall total for the globe is about 100 cm (10^3 kg m^{-2}); this must equal the mean evaporation and requires a mean expenditure at the earth's surface of 70 W m^{-2}, which is later released in the

---

[1] The latest (1967) estimate of the albedo of the earth from measurements on artificial earth satellites is 0.3. It seems likely that the true value is around 0.35.

atmosphere as latent heat of condensation. The 100 W m^{-2} transferred in these ways from the surface is eventually radiated to space by the atmosphere. We can make a very rough estimate of the rate of production—and therefore of dissipation—of the kinetic energy of all scales of atmospheric motion, by noting that the mean difference in temperature between the regions of heating and cooling in the atmosphere is about 30°K, allowing a thermodynamic efficiency of about 10% and kinetic energy production of about 10 W m^{-2}.

Figure 10.1 summarizes the mean annual budget of all radiation for the earth; it is, of course, a balanced budget, and not necessarily appropriate to any specific place. We would, in general, expect regions in low latitudes to have a net gain of energy by radiation and high latitudes to show a net loss. Figure 10.2 shows estimates, based on measurements at the ground, of the annual radiation budget for Southern England, 52° 30′ N, (Robinson, 1956). There is there, over the year, a mean net loss to space of 50 W m^{-2}, which must be made up by the advection of energy by atmospheric motions.

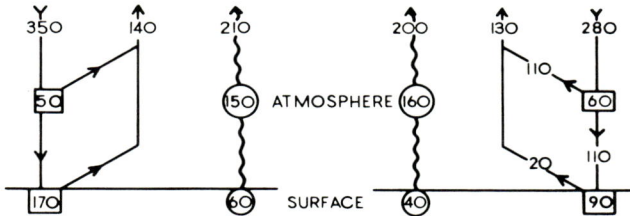

FIG. 10.1. Annual means for the whole earth.

FIG. 10.2. Annual means for Southern England.

Terrestrial radiation budget, unit Wm^{-2}. Straight lines: solar radiation, mainly 0.3–3.0 μm. Square boxes: absorption of solar radiation. Wavy lines: terrestrial radiation, mainly 5–50 μm. Circular boxes: emission of terrestrial radiation.

There are various ways in which we can set a rough time-scale to the broadest features of atmospheric motion. The simplest does not involve radiation directly, but merely compares the total water content of the atmosphere with the turnover rate implied by the mean precipitation and evaporation. The average residence time is 10 days. The theoretical meteorologist cannot hope to understand the cause and nature of atmospheric circulations without a knowledge of the radiation field (not only the gross mean field but its changes with height and with season), and the weather forecaster cannot hope for success with numerical methods for periods comparable with the 10-day time-scale unless he can incorporate radiative processes in his forecast scheme (I do not imply that this is a sufficient condition for success).

This outlines the place of radiation measurement and computation in theoretical meteorology. Oceanographers have a similar interest. The applied meteorologist has other requirements; he has to answer a variety of questions posed by farmers and biologists, architects, engineers, and manufacturers of any materials likely to be exposed to solar and terrestrial radiation at the earth's surface, within the atmosphere and in near space. He is more directly concerned with the spectral distribution of solar radiation than is the dynamical meteorologist. For most of his problems, he can accept a rather lower precision and accuracy in energy measurements, but in one problem in particular—the heat balance of objects in space—he requires just as high a standard in these respects as does the dynamical meteorologist.

10.2. Computation of Radiative Transfer in the Atmosphere

I have referred to computation and shall later suggest that, in some problems of terrestrial radiation, computation gives results as reliable as the best measurements made at present. I can clearly do no more now than indicate some of the general principles involved.

If we consider the intensity I of radiation of frequency ν propagating in direction s at a point P, the integral form of the radiative transfer equation may be written

$$(10.1) \qquad I(\nu, P, s) = \int_0^\infty S(\nu, P', s) T(\nu, P, P', s) \, dx$$

where $S \, dx$, the source function, is the intensity of the radiation of frequency ν in direction s originating in an element dx around a point P' distance x in the direction s from P, and T, the transmission function, is the fraction of this radiation surviving the journey between P' and P. This physically simple equation rapidly assumes a fearsome mathematical complexity when applied to some atmospheric problems, even if we limit consideration to uniformly stratified conditions. Goody (1964) gives a discussion sufficiently detailed for practically all meteorological purposes and certainly for all energetically important processes. We see that if we are concerned with total irradiance at P, we have, as well as the x integration, integrations over frequency and direction. In addition, the path is not in general homogeneous and the monochromatic transmission function T involves an inner integration over distance; and T and S may be frequency, position, and direction dependent.

Figure 10.3 is a much-used diagram of atmospheric transmission/emission properties, which demonstrates how we can distinguish three useful special forms of the transfer equation in the cloudless atmosphere. In the first place, we note that solar and terrestrial radiation can be treated separately, the

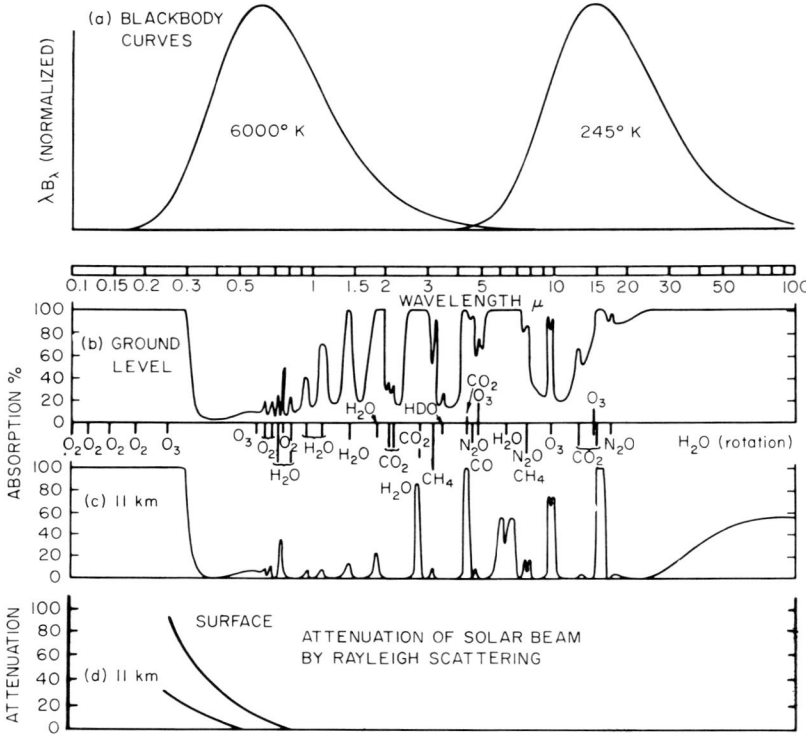

Fig. 10.3. (a) Blackbody emission for 6000°K and 245°K, being approximate emission spectra of the sun and earth, respectively (since inward and outward radiation must balance, the curves have been drawn with equal areas—though in fact 40% of solar radiation is reflected unchanged); (b) atmospheric absorption spectrum for a solar beam reaching the ground; (c) the same for a beam reaching the tropopause in temperate latitudes; (d) attenuation of the solar beam by Rayleigh scattering, at the ground and at the temperate tropopause.

division being made at a wavelength about 4.5 μm. Secondly, we note that absorption of energy by O_2 and O_3 is practically complete at the tropopause and that depletion of the direct solar beam by Rayleigh scattering, is energetically important only in those spectral regions not subject to strong absorption. If we are concerned only with solar irradiance, we can take out radiation absorbed by O_2 and O_3 before computing Rayleigh scattering and ignore scattered radiation in computing tropospheric absorption by H_2O and CO_2. Errors introduced by this procedure will certainly be less than the uncertainty of the solar constant. The three special forms of the transfer equation will be considered in Subsections 10.2.1–10.2.3.

10.2.1. Transmission of Direct Solar Radiation in a Clean Cloudless Atmosphere

Scattering into the beam is energetically unimportant. There is no source function within the atmosphere and the monochromatic problem reduces to Lambert's law.

10.2.2. Diffuse Solar Radiation in a Clean Cloudless Atmosphere

This is a scattering problem, and multiple scattering, polarization, and surface reflection effects are important if details of the field of scattered radiation are required. The source function expressing scattering into the beam is complicated and there is a large amount of literature on the solution of the equation; for example, various papers by Deirmendjian and Sekera (1954) and Coulsen et al. (1960) should be consulted. If the interest is only in total diffuse irradiance of various surfaces, the assumption of isotropic single scattering is sufficiently accurate for use in most practical problems, e.g., Robinson (1962).

10.2.3. Transfer of Terrestrial Radiation in a Clean Cloudless Atmosphere

We can assume local thermodynamic equilibrium and neglect scattering in the atmosphere (though not necessarily at bounding surfaces). The transfer equation takes the form

$$(10.2) \quad I(\nu, P, s) = I(\nu, P^0, s) \exp[-\tau(\nu, P^0, P)] \\ + \int_0^{\tau(\nu, P^0, P)} B(\nu, P', s) \exp[-\tau(\nu, P', P)] \, d\tau(\nu, P', P)$$

where τ is the optical path in direction $-s$, $d\tau(\nu) = k(\nu, x)\rho(x) \, dx$ where k is an absorption coefficient and ρ the density of absorber at x, P^0 is a point at the boundary of the atmospheric slab concerned, $I(P^0)$ the intensity there, and B is the Planck function for frequency ν and temperature appropriate to the position ($P's$). The major difficulty in handling this equation is the rapid, quasi-discontinuous nature of the variation of τ with ν. Other difficulties are caused by the variation of $\tau(\nu)$ with pressure and temperature (i.e., with P') even in an atmosphere of constant composition, and the fact that in a real atmosphere the water-vapor distribution, the major control on $\tau(\nu)$, cannot be accurately expressed analytically. Solution of the equation is still a major field of meteorological research.

Various mathematical transformations of Eq. (10.2) have been used for computational convenience. One which is particularly useful in the inversion problem (Section 10.5) is

$$(10.3) \quad I(\nu, z) = B(\nu, 0)T(\nu, 0, z) + \int_0^z B(\nu, z') \frac{d}{dz'} T(\nu, z, z') \, dz'$$

where we have considered intensity in the upward vertical direction at a height z above a black boundary.

More complex forms of the transfer equation are required to deal with the more general atmospheric problems, examples being as follows in Subsections 10.2.4. and 10.2.5.

10.2.4. Solar Radiation in a Cloudy Atmosphere

This is a multiple scattering problem made considerably more complicated than the Rayleigh atmosphere problem by two factors—firstly, the nature of the scattering from an individual element and, secondly, the existence of absorption. The distribution of scattered radiation with angle (referred to as the scattering function, the phase function, or, in the Russian literature, the scattering indicatrix) is calculable for spherical droplets. It is strongly peaked in the forward direction; its analytical expression is an awkward slowly convergent series. The problem has attracted the attention of some distinguished applied mathematicians and much computing time has been devoted to it. A considerable effort by Russian workers is reported by Feigel'son (1964). The practical value of this elaborate and sometimes elegant work is limited by the enormous diversity of natural clouds, in respect of homogeneity and of the nature and number of particle content. For most practical purposes, the simplified treatments such as those of Neiburger (1949) and Korb *et al.* (1957) are adequate.

10.2.5. Terrestrial Radiation in a Cloudy Atmosphere

There is no direct beam, absorption is very high, and there is a corresponding thermal source function.

The practically important problem concerns the albedo and transmission of clouds for terrestrial radiation. It is customary to treat thick cloud—i.e., cloud through which the sun's disk cannot be seen—as a blackbody with no internal radiative heat transfer and black surfaces. This is clearly unsatisfactory for thin cloud. Feigel'son (1964) has made some computations of cloud albedo from the transfer equation for certain concentrations and size ranges of droplets, as have Yamamoto *et al.* (1966). This work is valuable in giving a qualitative picture of the type of effect to be expected, but the diversity of natural clouds—and the fact that many thin clouds are ice-clouds with large nonspherical particles—limits its quantitative value. Lack of knowledge of the properties of cloud in the terrestrial infrared, in fact, sets the limit to the accuracy of computations of radiative exchange in the atmosphere and of the terrestrial radiation to space.

10.2.6. Radiation Computations as an Adjunct to Dynamical Meteorology

Radiative processes must be incorporated in numerical models of the general circulation of the atmosphere and in numerical systems for weather forecasting for periods beyond about 3 days. There is no great difficulty in principle in applying computations based on the transfer equation, but if this is done by the most rigorous methods yet developed, the computing time required far exceeds that needed for the dynamical computations. It is therefore necessary to investigate simpler and less precise methods, economical in time and matched in precision to the dynamical models. This can only be done satisfactorily by empirical trials, but it seems possible that a suitable method will be to consider only " cooling to space." The radiative exchange at any one level in the atmosphere may be considered as composed of three parts—exchange with the ground or nearest underlying cloud layer, exchange with other layers of the atmosphere, and exchange with (i.e., loss to) space. The first is normally a heating term, the second normally a cooling term, though it may occasionally cause local heating, and the third a cooling term. It is simply obtained from the transfer equation by integrating from the top of the atmosphere to the level under consideration and has the form, for monochromatic radiation,

$$(10.4) \quad \frac{1}{\rho c_p} B(\nu, z) T(\nu, Z, z) \frac{d\tau}{dz} \quad \text{or} \quad \frac{1}{\rho c_p} B(\nu, z) \frac{d}{dz} T(\nu, Z, z)$$

where z is the height of the level concerned and Z is arbitrarily chosen "outside the atmosphere": ρ is now air density and c_p specific heat at constant pressure. We are thus concerned with only a single value of T for each level. If the distribution of absorber in the atmosphere does not vary, the cooling to space depends only on the local value of B and may be approximately considered linearly dependent on local temperature. A recent paper by Rodgers and Walshaw (1966) contains an excellent discussion of cooling rate calculations.

The radiative effects of cloud can, in principle, be incorporated into any dynamical model in which cloud is introduced and forecast—few models include cloud at present, but it is a foreseen and essential development. Haze, however, is a different matter. It is not conditioned by the dynamics in the way that clouds are and is therefore not forecast by the models. If it is radiatively important—and it is in regard to solar radiation though its interaction with terrestrial radiation is less well established—it may prove a limitation on prediction of the state of the atmosphere.

10.3. Measurements of Solar Radiation

10.3.1. Available Measurements

I shall be concerned here with what data are measured and their reliability and availability rather than with details of measuring technique. The dynamical meteorologist is concerned with radiation mainly to know atmospheric heating and cooling rates. These require measurement of the net flux of all radiation at the surface and at various levels. Strictly, divergence of the net flux is required but its direct measurement is impracticable. Particularly in the case of solar radiation, it is convenient to measure the upward and downward streams separately, so that the required flux divergence emerges as the algebraic sum of four irradiance determinations. The most frequently made meteorological measurement is of the solar irradiance of a horizontal plane near the earth's surface; the direct and diffuse components are often measured separately. The upward flux of radiation is less frequently measured. The irradiance of surfaces orientated in other directions in space is regularly measured at only a few meteorological observatories, as is the spectral distribution of solar radiation.

Publication of the measurements in standard form has now been arranged by the World Meteorological Organization. The volumes, entitled "Solar radiation and radiation balance data (The world network)," are issued monthly from the A. I. Voeikov Main Geophysical Observatory, Leningrad, USSR. They contain tables of daily and monthly solar irradiation and, for selected stations, mean hourly values of solar irradiation for the month. The issue for December 1965 contains information from about 300 stations. This information is not all of equal value, but the type of instrument used at each station is stated and reference to the IGY Instruction Manual (CSAGI, 1957) will give an indication of the degree of reliability. I will give an outline of the procedures adopted at stations controlled by the U.K. Meteorological Office, which I believe are reliable stations.

Commercial thermoelectric solarimeters (i.e., pyranometers) are used as detectors. The primary standardization is against Ångström pyrheliometers, by shading direct sunlight. Standardization checks, however, are frequently made against traveling substandards, at the field stations, and in an integrating sphere at the central observatory. Field stations have at least two solarimeters in use—measuring total and diffuse radiation—so that regular consistency checks can be made. Recording is digital minute-by-minute on punched paper tape, with a standby potentiometric record. The punched paper tape is being replaced by $\frac{1}{4}$-in. magnetic tape, the tape punch having proved the most vulnerable link in the system. Tape is processed centrally. The archive medium is magnetic tape, but before archiving a print-out is examined for error in two stages. There is a program for correcting the archive tape in case of retrospectively discovered changes in the calibration constants of thermopiles—which is particularly necessary with cooperating stations not under the full control of the central observatory.

Changes of less than 2 % are not normally made except in data from the central observatory. Data for the WMO publication are printed from the archive tape. (Examples of tabulations are shown in Figs. 10.4 and 10.5.)

About 20 years ago—when recording of solar radiation was just becoming a common activity among meteorologists—Nicolet estimated that an accuracy of long-term irradiation within $\pm 5\%$ represented good and careful work. It is probable that many of the stations publishing in the WMO Journal do not reach this standard. Stations using thermoelectric detectors with digital recording and exercising careful control—for example, the U.K. stations— are probably within $\pm 3\%$ of the correct value of long-term irradiation, but there may still be greater errors—random and systematic—in irradiation in low-sun conditions. Lack of axial symmetry in the sensitivity of the thermopile in use is thought to be the main residual cause of error. Our experience also suggests that there may be an inverse correlation between accuracy of data and promptness of publication.

There has been no systematic attempt to measure solar irradiance as a climatic element away from the ground. I know of no attempt to construct a balloon-borne solar radiation sonde and measurements from aircraft have been sporadic. Roach (1961a, 1964) and I (1958, 1966) have reported work by the British Meteorological Research Flight. Fritz made a few flights in the USA. There has been rather more activity of this type in the USSR, but one gets an impression of rather divergent results by different workers there. In all this work, solarimeters of the type used for surface measurements were employed. Uncertain leveling and uncertain instrument temperature are added to the causes of error. There have been a few successful attempts at precision measurement from aircraft and large balloons, e.g., Drummond *et al.* (1967, 1968) and Kondratiev *et al.* (1967)—see also Chapter 1, this volume.

Measurement of the spectral distribution of solar radiation at the ground has not been systematically developed. The main difficulties arise with the diffuse component. The usual method is to focus light from a diffusing surface within an integrating sphere open to the whole sky (Larché-Kugel in the German literature) on a detector through a grating spectrometer or interference filters. The direct solar beam is more readily investigated and there is a long tradition of measurement in broad spectral regions defined by standard glass filters. These turbidity measurements are described by Ångström in this volume (Chapter 9). They still provide an easily made relative assessment of atmospheric haze, but the more quantitative aspects of this type of measurement can now be better achieved by spectroscopic and

Station Name = Bracknell   Date = 20/03/67   LAT = Standard Time−9

Time	I	T	D	L	F
0	0.0(0)	0.0(0)	0.0(0)	0.0(0)	0.0(0)
1	0.0(0)	0.0(0)	0.0(0)	0.0(0)	0.0(0)
2	0.0(0)	−0.0(22)	0.0(22)	0.0(22)	−0.0(22)
3	0.0(0)	0.1(58)	0.0(58)	0.0(58)	0.0(58)
4	0.0(0)	−0.0(58)	0.0(58)	0.0(58)	0.0(58)
5	0.0(0)	0.3(60)	0.3(60)	0.2(60)	0.2(60)
6	0.0(0)	4.6(60)	3.6(60)	5.2(60)	4.4(60)
7	0.0(0)	8.2(60)	7.6(60)	9.9(60)	8.9(60)
8	0.0(0)	30.1(60)	12.5(60)	33.7(60)	15.9(60)
9	0.0(0)	47.1(60)	16.2(60)	50.3(60)	19.8(60)
10	0.0(0)	51.3(60)	19.5(60)	57.4(60)	23.8(60)
11	0.0(0)	56.1(60)	17.0(60)	67.4(60)	21.2(60)
12	0.0(0)	60.3(60)	22.6(60)	69.1(60)	27.4(60)

Time	B	N	S	E	W
0	0.0(0)	0.0(0)	0.0(0)	0.0(0)	0.0(0)
1	0.0(0)	0.0(0)	0.0(0)	0.0(0)	0.0(0)
2	0.0(0)	0.0(22)	0.0(22)	0.0(22)	0.0(22)
3	0.0(0)	0.1(58)	0.0(58)	0.0(58)	0.0(58)
4	0.0(0)	0.1(58)	0.0(58)	0.0(58)	0.0(58)
5	0.0(0)	0.2(60)	0.1(60)	0.2(60)	0.1(60)
6	0.0(0)	1.7(60)	2.7(60)	6.3(60)	1.7(60)
7	0.0(0)	3.5(60)	4.9(60)	7.7(60)	3.5(60)
8	0.0(0)	5.9(60)	28.5(60)	42.3(60)	5.9(60)
9	0.0(0)	6.5(60)	46.5(58)	47.1(60)	7.0(60)
10	0.0(0)	9.4(60)	48.5(60)	32.7(60)	9.1(60)
11	0.0(0)	8.0(60)	58.5(60)	19.3(60)	9.1(60)
12	0.0(0)	9.9(60)	54.3(60)	11.2(60)	18.1(60)

FIG. 10.4. Print-out of hourly data for Bracknell March 26, 1967 (extract from daily form).

Key
- I   Intensity of direct solar beam (not recorded)
- T   Solar irradiation, horizontal surface
- D   Diffuse solar irradiation, horizontal surface
- L   Illumination, horizontal surface
- F   Diffuse illumination, horizontal surface
- B   Net radiative flux (not recorded)
- N   Solar irradiation, N-facing vertical surface
- S   Solar irradiation, S-facing vertical surface
- E   Solar irradiation, E-facing vertical surface
- W   Solar irradiation, W-facing vertical surface

Units Irradiation mW hr cm^{-2}
Illumination klx h
Number in brackets is number of samples used in forming hourly mean. Equipment switched out between 0137 and 2044 LAT.

Latitude	N 12 40
Longitude	E 45 02
Elevation	3 meters
June 1965	(40507) Khormaksar (Aden), South Arabia

Hourly Totals

Date	00–01	01–02	02–03	03–04	04–05	05–06	06–07	07–08	08–09	09–10	10–11	11–12	12–13
1	−10.0	−10.0	−10.0	−8.0	−8.0	−8.0	−3.0	11.0	25.0	37.0	46.0	53.0	56.0
2	−6.0	−6.0	−6.0	−6.0	−8.0	−8.0	−2.0	10.0	23.0	36.0	45.0	52.0	52.0
3	−8.0	−8.0	−8.0	−6.0	−6.0	−4.0	2.0	15.0	33.0	45.0	54.0	64.0	52.0
4	−4.0	−4.0	−3.0	−3.0	−4.0	−3.0	3.0	17.0	31.0	42.0	52.0	56.0	56.0
5	−4.0	−6.0	−6.0	−4.0	−4.0	−3.0	2.0	15.0	29.0	40.0	50.0	54.0	50.0
6	−6.0	−6.0	−6.0	−6.0	−6.0	−4.0	0.0	11.0	25.0	36.0	46.0	54.0	56.0
7	−6.0	−6.0	−6.0	−6.0	−6.0	−6.0	0.0	12.0	27.0	39.0	48.0	54.0	56.0
8	−6.0	−4.0	−4.0	−6.0	−6.0	−4.0	0.0	12.0	27.0	37.0	46.0	53.0	53.0
9	−8.0	−8.0	−8.0	−8.0	−6.0	−6.0	0.0	12.0	25.0	37.0	48.0	52.0	50.0
10	−6.0	−6.0	−6.0	−6.0	−6.0	−4.0	2.0	12.0	25.0	37.0	46.0	52.0	52.0
11	−6.0	−6.0	−6.0	−6.0	−6.0	−4.0	3.0	15.0	29.0	39.0	46.0	52.0	52.0
12	−8.0	−6.0	−6.0	−6.0	−4.0	−4.0	2.0	14.0	28.0	39.0	48.0	53.0	52.0
13	−4.0	−4.0	−6.0	−8.0	−6.0	−6.0	0.0	14.0	27.0	40.0	48.0	52.0	53.0
14	−8.0	−10.0	−10.0	−10.0	−8.0	−8.0	0.0	12.0	25.0	39.0	50.0	52.0	50.0
15	−8.0	−8.0	−8.0	−6.0	−6.0	−4.0	0.0	12.0	25.0	39.0	46.0	52.0	50.0

FIG. 10.5. Print-out of hourly totals of net radiative flux in form required for WMO publication. Khormaksar (Aden). June 1967 (extract).

Key
B(GD) (Kew Pattern)—Radiation balance meter; Gier and Dunkle type, manufactured at Kew Observatory.
Unit
mW hr cm^{-2}

interference filter techniques. Figure 10.6 (see also Fig. 9.1 for dust-free conditions) illustrates qualitatively the extinction of the direct solar beam by the cloudless atmosphere. The major extinction of the radiation which penetrates the stratospheric ozone is by scattering and by absorption by water vapor, with a significant contribution by CO_2. The O_2 absorption at 762 nm is in a very narrow band and is less important energetically. Figure 10.7 illustrates the possibilities of modern techniques. It is a record of a portion of a solar spectrum obtained at Munich by Quenzel (1966). Superposed on the spectrum are the transmission curves of some of the interference filters used by him for the determination of total atmospheric water vapor.

10.3.2. *Atmospheric Haze and Solar Radiation*

The diffuse and total irradiance at the base of a clean cloudless atmosphere of known ozone and water content, above a surface of known albedo, can be calculated with a precision certainly of 1 or 2 %. When such computations are compared with measured values, it is found that there is normally a considerable excess of diffuse radiation and a deficit in total radiation, which can only be caused by nonselective absorption or upward scattering. I examined this effect in the records of a number of stations and found that only in the Antarctic was the diffuse downward radiation, in cloudless conditions, not significantly greater than computed Rayleigh scattering.

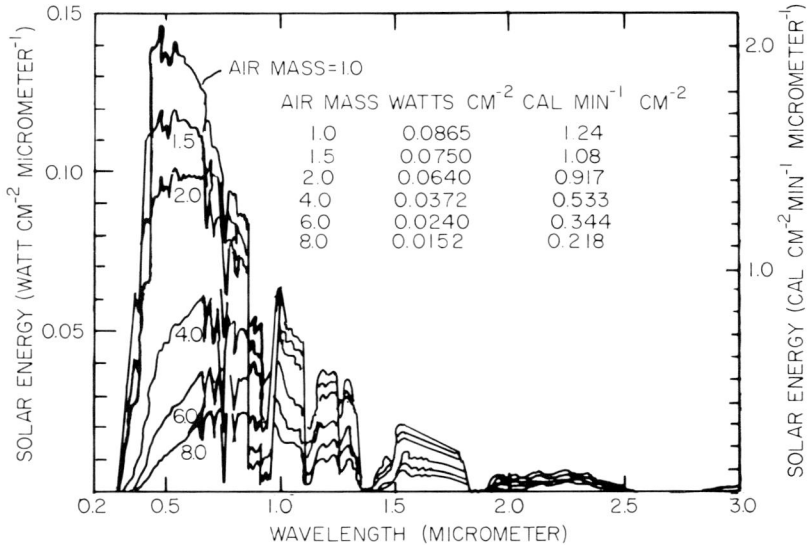

Fig. 10.6. Broad characteristics of atmospheric extinction of the solar beam for different air mass values [after Gates (1966)].

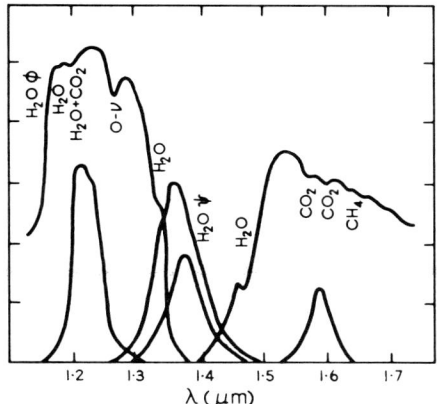

FIG. 10.7. Part of a solar spectrum in the infrared recorded at Munich, 26 September 1962, by Quenzel. Also shown are the pass-bands of interference filters used by Quenzel in determination of total atmospheric water vapor.

In other locations, including rural high-level stations in Africa, Malta, and the Shetland Islands, there is extra downward scattering and absorption of between 2 and 5 % per air mass. These deductions have been confirmed by direct measurement from aircraft (Robinson, 1966). This extinction of the direct solar beam can only be attributed to aerosol—which might include tenuous high cloud although occasions when this was actually reported were excluded in the investigation I have mentioned. The extra extinction by scattering is comparable with that by Rayleigh scattering; the absorption in apparently unpolluted air is 25 % or more of the gaseous absorption and is considerably greater than this in heavily polluted atmospheres.

A practical aspect of the considerable range of the extinction by aerosol is that if we are concerned with solar irradiance, in real atmospheres in a given frequency range, there is little point in making computations of great refinement; for example, the difference between the most elaborate computations of Rayleigh scattering, such as that of Deirmendjian and Sekera (1954), and a simple approximation assuming isotropic single scattering is less than the day-to-day variability of aerosol scattering in an apparently unpolluted atmosphere. Similarly, computation of water vapor and CO_2 absorption by physically sound but elaborate methods, such as that of Roach (1961b), give results which differ from those read off from curves, like that due to Hoelper (see Smithsonian Meteorological Tables), by much less than the uncertainty in the nonselective aerosol absorption. We can, for this practical purpose, dispense with the use of a computer, if that is of any advantage nowadays.

The classic work on the transmission properties of heavily polluted atmospheres is that of Waldram, based mainly on photometry of searchlight beams, carried out in southern England during World War II and reported by him (1945). Waldram measured total extinction of the direct beam and separated the scattering and absorption coefficients by integrating an observed scattering function. To the best of my knowledge, this difficult observation has never been repeated. In what he described as "industrial haze," Waldram's median extinction coefficient was 1 km^{-1} with extremes of about 0.2 and 5 km^{-1}. His absorption coefficients range from very small values to almost 1 km^{-1}. with a median of 0.3 km^{-1}. If we assume a scale height of 1 km for the pollution, Waldram's results indicate for high sun a median extinction of the direct solar beam of about 60 %, with extremes of about 15 and 99 % and a corresponding median absorption of 25 %. These results are entirely consistent with observations of solar radiation in heavily polluted conditions. On cloudless days at Kew Observatory (in a suburb of London), in 1948–1949, with high sun (sin h 0.7 to 0.8), I found the median attenuation attributable to aerosol of the solar beam at about 600 nm to be about 25 % with extremes of 0 to 60 %, and indications that the extinction was roughly equally divided between absorption and scattering.

10.3.3. Volcanic Dust and Solar Radiation

There has long been speculation on the effects of the dust veil from major volcanic eruptions on the transmission of solar radiation. At the time of the eruption of Krakatoa there were no reliable measurements, and there were very few at the time of the eruption of Katmai just before World War I. A significant decrease in the intensity of the direct solar beam following the Katmai eruption seems to be established, but there is no reliable indication of a change in the total (direct and scattered) radiation reaching the ground. The effect on solar radiation at the surface of the recent (March 1963) major eruption of Agung on Bali will be much better documented, but examination of the results is not complete. Results published in Australia indicate significant reduction of the direct solar beam but not of the total radiation. Budyko has examined the records submitted to the Leningrad Observatory, in connection with the WMO publication of solar radiation measurements, and reported (to the 1967 COSPAR/IQSY symposium in London) a world-wide reduction of about 5 % in intensity of the solar beam at the surface. Miller (1967) has extensive and fairly precise measurements of extinction of the direct solar beam between 20 and 40 km made on satellite Ariel 2 (1964–15A), together with a few rocket observations made before the Bali eruption. It now seems certain that the direct solar beam was significantly attenuated by the Bali dust, but the effect on total solar radiation received at the surface is not yet clear.

10.3.4. Cloud and Solar Radiation

The largest single contribution to extinction of solar radiation comes from cloud. I have already referred to the difficulty of computing the optical properties of real clouds, and the results of direct observation from aircraft suggest a very wide range of values of the albedo and absorption of cloud. The best that can be said is that measurements of cloud albedo are not inconsistent with computations for model clouds. Figure 10.8 illustrates this for one series of observations (by the U.K. Meteorological Research Flight) of the albedo of nonprecipitating low-layer cloud. The comparison is made

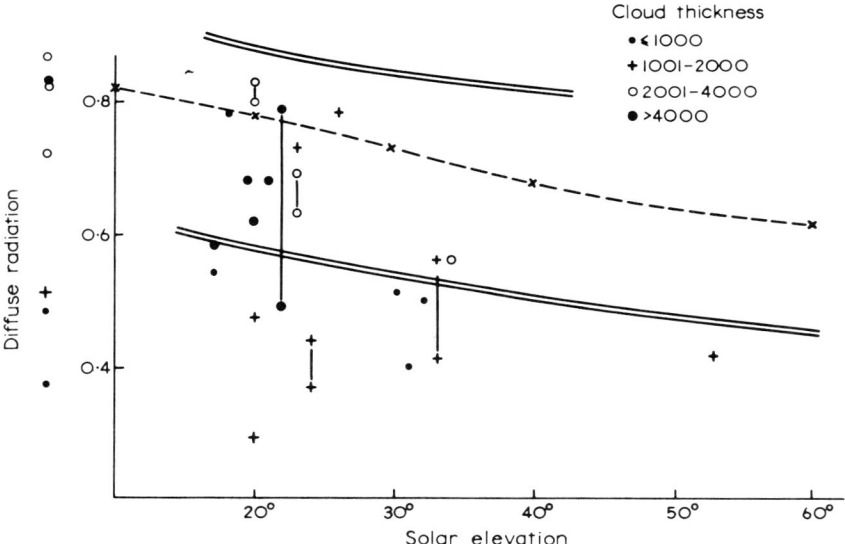

FIG. 10.8. Albedo of stratiform cloud. The points are measurements made on aircraft of the U.K. Meteorological Research Flight. The dotted curve is the theoretical value of Korb et al. (1957) for a layer of stratocumulus 500 m thick. The double lines are theoretical values of Feigel'son and Shiffrin for clouds of optical depths 10 and 50.

with computations by Feigel'son (1964) and by Korb et al.(1957). The observations illustrate not only the variation of the albedo of the same type of cloud from day to day, but also the wide range of values obtained from measurements in flight over extensive sheets of otherwise apparently uniform cloud. When we examine cloud absorption measurements, however, we find that values are generally considerably higher than those indicated by computation. Both Fritz and I have found evidence of clouds absorbing as much as 25 to 30 % of the radiation incident on them—computed absorptions rarely exceed 5 %. Russian observations show a less clear-cut picture;

some agree with the theoretical computation whilst others show larger absorption. The observations can readily be criticized; for example, they depend heavily on interpolation in time, but on balance I think they should be accepted.

10.3.5. Solar Radiation and the Earth's Energy Balance

If we gather together the results of solar radiation observations and look at them from the point of view of the theoretical meteorologist interested in computing the dynamical and thermodynamical development of model atmospheres, we find that the gross albedo of the earth is little different from that suggested by computations of the scattering properties of clear air and cloud. The absorption of radiation at the surface is less than that expected by about 5 % on average, and absorption in the lower atmosphere, due to aerosol and cloud, is correspondingly about 5 % more than theories suggest. The overall energy budget balances at about the value expected, but the distribution of absorption is a little different, the difference being in the direction which would produce a stabilization of the lower atmosphere. The gross albedo of the earth is, of course, a matter of concern to spacecraft engineers as well as to the theoretical meteorologist. Fritz (1949) showed how it might be estimated if the mean cloud albedo were known. It follows also from the solar constant and the net flux of solar radiation at the ground, if atmospheric absorption is known, so that the uncertainty in its value is not greater than the combined uncertainties of these values, which we may put at about 5 %. Actual measurement of the gross albedo naturally requires some form of extraterrestrial observation, but its measurement from satellites has proved unexpectedly difficult. The latest available estimate from satellite measurements—about 0.30—seems to be below the lowest possible indirect estimate. The classical method of Danjon involves photometric comparison of sunlit and earthlit areas of the moon, but results do not seem to be too consistent and may be adversely affected by the airglow and by scattering of moonlight by unobserved cirrus cloud and aerosol.

10.4. Measurements of Terrestrial Radiation

Routine measurements of terrestrial radiation are much less common than those of solar radiation because no weatherproof measuring instrument giving a record which can be automatically tabulated is readily available. Instruments which have been used fall into three types. The most robust, and most commonly used, is a freely exposed black surface. This exchanges radiation with its surroundings; it has a temperature excess or deficit over the ambient air, which tends to vary as incident radiation varies. Ventilation rate, i.e., wind speed, also controls the temperature difference; this effect is either calculated as a correction, as is the practice in USSR, or is made constant by provision of strong artificial ventilation as in the radiometer of Gier and Dunkle. In the second type of instrument, the black receiver is protected by a window transparent to terrestrial radiation, i.e., in the wavelength range 5 to 50 μm. The trouble is that no perfectly transparent window has been found. Crystals such as KRS-5, unless carefully maintained, are not suitable for long-term continuous exposure in all weathers and are generally too expensive for widespread use; the only practicable alternatives are synthetic polymers, e.g., commercial polyethylene film, which have awkwardly placed though not very intense absorption bands. Because of this absorption, it is necessary to use very thin films which are not really weatherproof. The third type of radiometer is based on the same compensation principle as the Ångström pyrheliometer; the greatest practical difficulty in realizing it is to find a surface which is a good mirror in the terrestrial infrared and which remains so in routine exposure to the atmosphere. The first and third types are not suited to automatic recording without human intervention, as they are subject to gross errors when any of the receiving surfaces are wet by rain. Calibration of terrestrial radiation recorders is ultimately by reference to low-temperature blackbody sources, but in practice it is usual first to establish that the surface coating has the same absorptivity for solar and terrestrial radiation, and then to calibrate using solar radiation or lamps in the laboratory.

10.4.1. Measurements of Terrestrial Radiation at the Earth's Surface

The commonly made routine measurement, at the surface, is of the radiation balance or net vertical flux of all radiation—solar and terrestrial. This quantity is of considerable interest to dynamical meteorologists and is useful to agriculture and hydrology. It would be preferable to record, separately, upward and downward components, but the difficulties, and therefore the errors, are greater. Figure 10.5 is a tabulation, in the standard form advocated by WMO, of results from a surface radiation balance meter. Forty stations measured radiation balance during the IGY and IGC. The number of reports has not greatly increased since then, and there is still a complete lack of radiation balance information in the areas where it is of most theoretical interest—the tropical and subtropical oceans. The results are not, in general, so accurate as those for solar radiation. At stations where calibration is carefully controlled, this is mainly because of the need for estimation of the radiation balance during precipitation (when it is small). Most published long-term totals or means should be within ± 10 % of the true value.

10.4.2. Radiative Flux Divergence in the Free Atmosphere

The divergence of the net vertical radiation flux in the free atmosphere is, of course, a direct indication of the local radiative cooling and is, therefore, of the greatest interest to dynamical meteorologists. A few measurements have been made from aircraft—e.g., by A. W. Brewer and J. T. Houghton— and two radiometersondes have been described—a compensation type due to Pohl and used by H. G. Müller of Munich and a polyethylene-shielded black plate, introduced by Suomi and Kuhn and described by Kuhn in this volume (Chapter 12). This is the only radiometersonde which has been extensively used.

We can readily estimate the precision and accuracy required from a radiometersonde. Radiative temperature changes are of order 10^{-5} °C sec^{-1}. For the purpose of dynamical models, we are interested in radiative flux divergence in atmospheric layers not thicker than 100 mb, so we must be able to measure differences in net radiative flux of 10 W m^{-2}, or of 1 W m^{-2} for an accuracy of 10 %. This is, in fact, just the accuracy claimed by Kuhn for a radiometersonde ascent when random errors of individual points have been reduced to the maximum possible extent by smoothing. It is less clear that all possible systematic errors have been identified, and there are, in fact, some peculiarities of radiometersonde results which remain unexplained. They are most apparent in the upper levels of ascents, above 500 mb. The net upward flux of terrestrial radiation in apparently cloudless atmospheres is less, the upward component less and the downward component greater than the values computed. There is considerable confidence in the computation for a purely gaseous atmosphere, and Kuhn rejects the possibility of a sufficiently large systematic error in the measurements, attributing the effect to atmospheric aerosol, which is a plausible explanation but one not positively established. We might note, in this connection, the observation of Williamson and Houghton (1965) who, using a radiometer sensitive between 5.5 and 8 μm, found a considerable downward radiation flux (4 % of blackbody at local temperature), at a height of 30 km, to which they could not attribute a known origin.

However, in any practical application to meteorological problems the most important radiator, energetically, is cloud, and the distribution of cloud and its emissive properties are the biggest uncertainties in any computation of radiative cooling. Feigel'son (1964) presents an approximate theoretical treatment of cloud emissivity, and radiometersonde results can be used for the same purpose if there are sufficiently reliable simultaneous observations of cloud height and temperature. Kuhn reports some interesting and important results in this volume (Chapter 12), estimating the emissivity of complete cloud layers to be

High cloud   0.45;   Medium cloud   0.65;   Low cloud   0.90.

Feigel'son's theoretical values are for cloud models corresponding to the "low" cloud of Kuhn, and she suggests an emissivity of 0.85 to 0.95 for the entire range of terrestrial radiation, but 0.70 to 0.85 in the 8 to 12 μm window. There is thus an impressive agreement between one theoretical determination and deductions from observation, in spite of approximations in the one and the danger of systematic error in the other. We must, however, recall that Yamamoto *et al.* (1966) have also computed the emissivity of clouds in the atmospheric window and find much higher values—0.95 to 0.98—than does Feigel'son. The question of cloud emissivity is particularly important in interpreting the readings of satellite-borne radiometers, and it would seem reasonable to experiment with the values suggested by Kuhn.

10.5. INDIRECT SENSING OF ATMOSPHERIC PROPERTIES BY OBSERVATIONS OF RADIATION

The degree and variation with wavelength of the extinction of the solar beam and the nature of the atmospheric emission spectrum clearly give us qualitative information concerning the atmosphere. To extract useful quantitative information has proved surprisingly difficult. The simplest problem concerns the total amount of a substance scattering solar radiation with a coefficient varying slowly with wavelength. Examples which have been mentioned earlier are measurements of atmospheric "turbidity," using glass filters, and Miller's (1967) recent estimates of the quantity and distribution, with height, of volcanic dust in the stratosphere from observations from rockets and a satellite, of the extinction of the solar beam in the visible region.

Next, in order of difficulty, is determination of the total amount of an absorber in the atmosphere—for example, the spectrophotometric determination of total ozone by Dobson's well-known method, which involves comparison of the intensity in two regions of the spectrum having different extinction coefficients. Quenzel's (1966) recent work on total water-vapor determination from measurements in the near infrared (Fig. 10.7) is another example.

Methods for determination not only of the total amount of an absorber or scatterer, but also its distribution with height—or of the variation of atmospheric temperature with height, given the height distribution of an absorber—are of more general interest. They imply inversion of the transfer equation. The first such method to come into general use in meteorology was, in fact, a rather complex case—the so-called "Umkehr" method for determination of the height distribution of ozone. In this, the ratio of intensities of light scattered from the zenith in two narrow spectral bands, with different absorption coefficients, is measured for different (large) solar zenith angles. The method had been in use many years before difficulties experienced in other experiments involving inversion of the transfer equation led to a critical examination of the information content of a standard set of observations. It is now realized that the results are only marginally significant.

The transfer equation in the form

$$(10.5) \qquad I(\Delta\nu) = \int dz \int_{\nu}^{\nu+\Delta\nu} B(\nu, z) \frac{d}{dz} T(\nu, z) \, d\nu$$

demonstrates the possibility of determination of B as a function of height if T is known as a function of height, or of T if B is known. For a given spectral interval, B is a function of temperature alone. The transmission function T is a rapidly varying, but calculable, function of ν, a slowly varying function of temperature and pressure, and a function of the concentration of absorber. Thus, if we know the variation of absorber concentration with height, if we know enough about its spectrum to make the integration over $\Delta\nu$, and if we have an approximate temperature–pressure–height function (for example, a standard atmosphere), we can in theory recover the actual temperature–height function from intensity measurements in different spectral intervals $\Delta\nu$, or at different zenith distances for the same $\Delta\nu$, each measured intensity providing a differently weighted mean of the temperature over the whole atmosphere. Similarly, if we know the exact temperature–height relation, e.g., from a radiosonde, we have the possibility of computing the height distribution of the absorber. Both these exercises have been attempted. The difficulty is to get sufficient independent pieces of information to give meaningful height distributions. An early attempt to use the inversion method observed ozone emission in the 9.6 μm band at various zenith distances, but the pressure broadening of individual lines is such that emission from the relatively small amount of ozone, at high tropospheric pressures, outweighed that from the larger amount of stratospheric ozone. In principle, there is a great advantage in working in very narrow wavelength bands, at high spectral resolution, when using the inversion technique, but in practice, a prohibitive degree of experimental error or noise-to-signal ratio is quickly attained if the observing point is in a high-pressure region. It is for this reason that the technique has attracted the attention of those concerned with satellite instrumentation; the observation is then made from the low-pressure end of the path. Both logistic and theoretical reasons restrict the number of independent observations which can be made to 5 to 10 for each profile, whereas the meteorologist can use, and may need, many more points on each profile. Prior knowledge of the atmosphere—mean values and variance of the element under investigation—must therefore be fed in during the inversion process if the result is to be of the required form. The question is treated in more detail by Hanel in this volume (Chapter 13).

REFERENCES

Ångström, A., (1969). This volume, Chapter 9.
Coulson, K. L., Dave, J. V., and Sekera, Z. (1960). "Tables Related to the Radiation Emerging from the Planetary Atmosphere with Rayleigh Scattering." Univ of California Press, Berkeley, California.
CSAGI (1957). I.G.Y. Instruction Manual. Pt VI Radiation instruments and measurements. *Ann. Intern. Geophys. Yr.* **IV**, 367–466. Pergamon Press, London.
Deirmendjian, D., and Sekera, Z. (1954). Global radiation resulting from multiple scattering in a Rayleigh atmosphere. *Tellus* **4**, 382–398.
Drummond, A. J., Hickey, J. R., Scholes, W. J., and Laue, E. G. (1967). Multichannel measurement of solar irradiance. *J. Spacecraft and Rockets* **4**, 1200–1206.
Drummond, A. J., Hickey, J. R., Scholes, W. J., and Laue, E. G. (1968). New value for the solar constant of radiation. *Nature* **218**, 259–261.
Feigel'son, E. M. (1964). "Light and Heat Radiation in Stratus Clouds." Moscow. (Translation 1966; Israel program for scientific translations, Jerusalem.)
Fritz, S. (1949). The albedo of the planet earth and of clouds. *J. Meteorol.* **6**, 277–282.
Gates, D. M. (1966). Spectral distribution of solar radiation at the Earth's surface. *Science* **151**, 523–529.
Goody, R. M. (1964). "Atmospheric Radiation." Oxford Univ. Press, London.
Hanel, R. (1970). This volume, Chapter 13.
Kondratiev, K. Ya., Nikolsky, G. A., Badinov, I. Ya., and Andreev, S. D. (1967). Direct solar radiation up to 30 km and stratification of attenuation components in the stratosphere. *Appl. Opt.* **6**, 197–207.
Korb, G., Michalowsky, J., and Möller, F. (1957). Die Absorption der Sonnenstrahlung in der wolkenfreien und bewölkten Atmosphäre. *Beitr. Physik Atmosphäre* **30**, 64–77.
Kuhn, P. M. (1970). This volume, Chapter 12.
Miller, D. E. (1967). Stratospheric attenuation in the near ultraviolet. *Proc. Roy. Soc. (London)* **A301**, 57–75.
Neiburger, M. (1949). Reflection, absorption, and transmission of insolation by stratus cloud. *J. Meteorol.* **6**, 98–104.
Quenzel, H. (1966). Ein Interferenz-Filter Aktinograph zur optischen Bestimmung des atmosphärischen besamtwasser Dampfgehaltes. *Beitr. Physik Atmosphäre* **39**, 112–144.
Roach, W. T. (1961a). Some aircraft observations of fluxes of solar radiation in the atmosphere. *Quart. J. Roy. Meteorol. Soc.* **87**, 346–363.
Roach, W. T. (1961b). The absorption of solar radiation by water vapour and carbon dioxide in a cloudless atmosphere. *Quart. J. Roy. Meteorol. Soc.* **87**, 364–373.
Roach, W. T. (1964). Some aircraft observations of the fluxes of solar radiation in the atmosphere. *Ann. Intern. Geophys. Yr.* **32**, 63–70.
Robinson, G. D. (1956). The use of surface observations to estimate the local energy balance of the atmosphere. *Proc. Roy. Soc. (London)* **A236**, 160–171.
Robinson, G. D. (1958). Some observations from aircraft of surface albedo and the albedo and absorption of cloud. *Arch. Meteorol. Geophys. Biokl.* **B9**, 28–41.

Robinson, G. D. (1962). Absorption of solar radiation by atmospheric aerosol, as revealed by measurements at the ground. *Arch. Meteorol. Geophys. Biokl.* **B12**, 19–40.

Robinson, G. D. (1966). Some determinations of atmospheric absorption by measurement of solar radiation from aircraft and at the surface. *Quart. J. Roy. Meteorol. Soc.* **92**, 263–269.

Rodgers, C. D., and Walshaw, C. D. (1966). The computation of infrared cooling rate in planetary atmospheres. *Quart. J. Roy. Meteorol. Soc.* **92**, 67–92.

Waldram, J. M. (1945). Measurements of the photometric properties of the upper atmosphere. *Quart. J. Roy. Meteorol. Soc.* **71**, 319–336.

Williamson, G. J., and Houghton, J. T. (1965). Radiometric measurements of emission from stratospheric water vapour. *Quart. J. Roy. Meteorol. Soc.* **91**, 330–338.

Yamamoto, G., Tanaka, M., and Kumitani, K. (1966). Radiative transfer in water clouds in the 10-micron window region. *J. Atmospheric. Sci.* **23**, 305–313.

— 11 —

MEASUREMENT OF RADIATION FLUX AND EQUIVALENT RADIATION TEMPERATURE IN THE ATMOSPHERE

J. Strong

11.1. INTRODUCTION

Simple, single-valued parameters such as *temperature, radiation flux, absorption* and *emission coefficients*, and *atmospheric transmission*, are inadequate for representing the physical situation of an object that is exposed in the open air. Under such conditions, one needs to specify two temperatures at the very minimum, viz. the *kinetic air temperature* and the *mean radiation temperature*. And as for the other parameters, one needs to divide flux into a minimum of two wavelength bands: (1) the visible wavelengths together with the very near infrared, such wavelengths as are involved with solar radiation, and (2) the far infrared wavelengths which comprise the intrinsic thermal emission of bodies at terrestrial temperatures. Such a simplistic "two color" spectroscopy certainly needs two absorption coefficients—one for the solar- and another for the terrestrial-type fluxes. We use coefficients with subscripts to designate these two radiation types, such as $a_\odot$ and $e_\oplus$.

A scientific purist would use the term "temperature" only in situations that are characterized by thermodynamic equilibrium. Then, the measurement of any physical property whatever that varies, will yield the same temperature as any other and the kinetic and radiation temperatures are equal.

As a nonequilibrium example, consider a temperature measurement in the open air on a clear, calm, sunny day. If one observed in this situation with two mercury-bulb thermometers—identical except that one had its bulb painted black and the other its bulb painted white—one would get two entirely different temperatures. This is because thermodynamic equilibrium does not obtain.

Meteorologists define and measure the air temperature with a thermometer that is shaded from incoming, warming solar radiation and from outgoing, chilling sky radiation. A standard shade around the thermometer allows free air circulation.

In order to make a first approximation in defining the physical situation in the open air, one specifies the two temperatures—the *kinetic air temperature* and the *equivalent mean radiation temperature*. We define these temperatures by invoking idealized experiments with a polished silver sphere and a black-painted silver sphere. Consider the first one to be coated so that its surface reflects all solar and terrestrial radiation completely. Such a sphere would come to the kinetic air temperature. Consider the second sphere to be coated with a paint so that its surface absorbs all solar and terrestrial radiation completely and, further, consider it enclosed in a vacuum by a perfectly transparent envelope. Such a sphere would come to the mean radiation temperature. Apart from the unattainable vacuum envelope, one cannot hope to obtain silver and real black paints that will satisfy these ideals.

The absorption and emission varies with angle as well as with wavelength; therefore, in a real situation, experiments with these spheres in thermal contact with air involve hemispherical coefficients. Spheres present all angles of inclination to all points of the surroundings.

As for wavelength dependence, Butler *et al.* (1964) give the absorption coefficient as 0.10 for polished silver in sunlight (as simulated by a high-intensity carbon-arc source). They found the emission coefficient of this silver, for room temperature radiation, to be 0.025—making $a_\odot/e_\oplus = 4$.

Figure 11.1 presents Ramsey's (1964) measurements of the black spray paint Krylon 1602. From the laws of radiation, we know at each wavelength that $a_\lambda = e_\lambda = 1 - r_\lambda$. We would accordingly expect $a_\odot/e_\oplus$ to be nearly unity for this paint, as contrasted with the silver value of $a_\odot/e_\oplus = 4$.

Brasefield's (1948) white paint for coating balloon-sonde thermistors is another case in point. For absorption of solar insolation he gives $a_\odot = 0.06$. For many white and colored paints, the $e_\oplus$ in the infrared approaches unity; and for many black paints, the $a_\odot$ for sunlight is approximately equal to $e_\oplus$ for the infrared.

In the 1930's several candidate paints were proposed for covering the 200-in. telescope dome. These and others were studied, yielding the conclusion that Tom Sawyer's whitewash would be superior, thermally, to aluminum paint. Brasefield's paint would perhaps have been even better thermally. However, for durability, aluminum paint was finally used. The $e_\oplus$ values of aluminum paints were measured and found to be 0.5 or less, making $a_\odot/e_\oplus$ about 0.5 as contrasted with an $e_\odot/e_\oplus$ value of about 1/17 for Brasefield's paint.

Also, some experiments that were conducted at that time, including one with reflecting olive-drab camouflage paint, are notable. This olive paint had a visual and near-infrared albedo simulating that of average green leaves, and

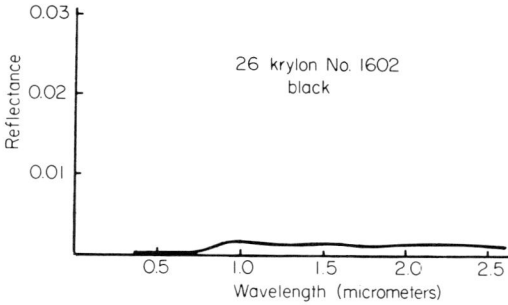
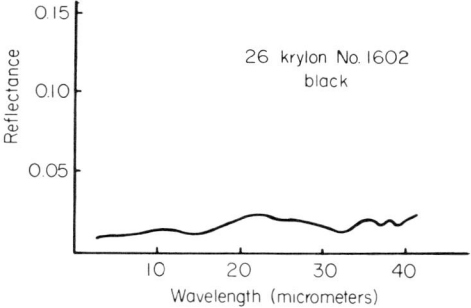

FIG. 11.1. Reflectance of black spray paint.

its infrared $e_\oplus$ was high. The following table gives the observed temperature rises of several painted (on the sunny side only) glass plates exposed normal to the high sun in the open air on a clear day.

Paint	$(T_G - T_A)$ °C
Whitewash	$-1°$
Camouflage	$9°$
Aluminum	$18°$
Black	$27°$

These values illustrate the influence of $a_\odot/e_\oplus$ ratios. The $a_\odot/e_\oplus$ ratio for the black paint was probably nearly unity. As for the camouflage paint, one recalls that half the solar constant is to be found at wavelengths greater than $\lambda = 0.72$ μm. This paint had the high albedo of green leaves for this half.

Figure 11.2 illustrates a method for determining the hemispherical coefficients appropriate for spheres, using a test hemisphere provided with a deeply grooved, copper diameter plate, as shown. This test hemisphere was filled with water and warmed to slightly above room temperature. Then, as it cooled it was placed on insulated points above an uncovered radiation thermopile—first with the grooved and blackened diameter facing down, then with the hemisphere side down. A series of thermopile observations was made during cooling for two cases: first, with the hemisphere side naked and polished, and second, painted black with Krylon 1602 paint. In each case, the alternating thermopile readings were plotted against time as the water-filled hemisphere cooled to room temperature, giving two cooling curves—an upper curve for the diameter radiation (taken as black) and a lower curve for the hemisphere radiation. The ratio of ordinates of these curves yielded $e_\oplus = 0.935$ for the Krylon-painted hemisphere and $e_\oplus = 0.026$ for the case when its polished silver surface was naked.

An absolute method of measuring the total emissivity of flat surfaces, such as painted squares of flat glass, is described and analyzed in a later section.

The differentiation of temperature (determining a kinetic air temperature and an equivalent mean radiation temperature) allows meteorologists,

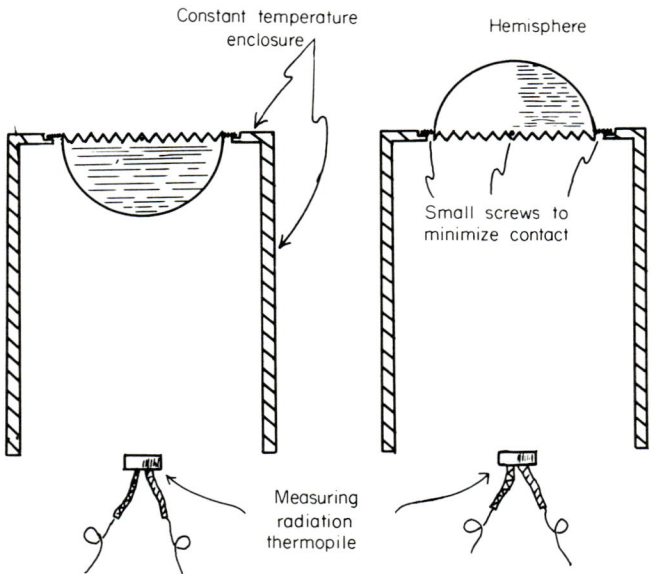

FIG. 11.2. Representation of procedure for measuring emissivity of a hemispherical surface.

ventilation engineers, or physiologists to divide total gains and losses of heat into a radiation fraction and an air conduction fraction. Such partitions are of interest to meteorologists because they are a prime parameter of climate—as when clear-sky radiation induces frost. Indoors, such partitions are important parameters of human comfort and are accordingly of interest to ventilation engineers and physiologists. Situations where the mean radiation temperature is higher or lower than the air temperature are familiar. For comfort, exposure to chilling air must be compensated by concurrent exposure to warming radiation from the sun or from a radiant heater. In the contrasting situation, for comfort in a room where the walls are cold, the air must be heated to compensate for radiation chilling. The first situation feels pleasant; the contrasting one feels dank. The author suggested a special chamber (illustrated in in Fig. 11.3) that provides for the differentiation of radiation and air temperature. This, together with humidity control, defines a variable physical situation under which individual human responses may be studied. This is called partitional calorimetry (Winslow et al., 1936).

There is now considerable interest in the measurement, by balloon-sonde, of equivalent radiation temperatures in the upper air. It would be desirable,

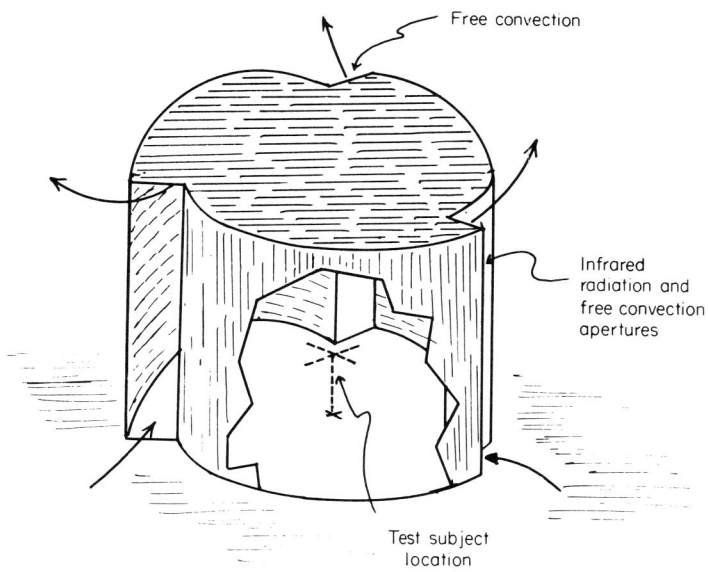

Fig. 11.3. Copper enclosure for control of radiation environment: the copper walls are peened to provide diffuse reflection of infrared radiation entering through three apertures.

for such measurements, to have a meter that is immune to variation of air ventilation. Such a meter is described below. We call it a mean-radiation-temperature meter, or MRT meter.

11.2. A New MRT Meter

Figure 11.4 shows two hollow, silver spheres with thermojunctions attached to their inside wall surfaces and with compensating junctions connected,

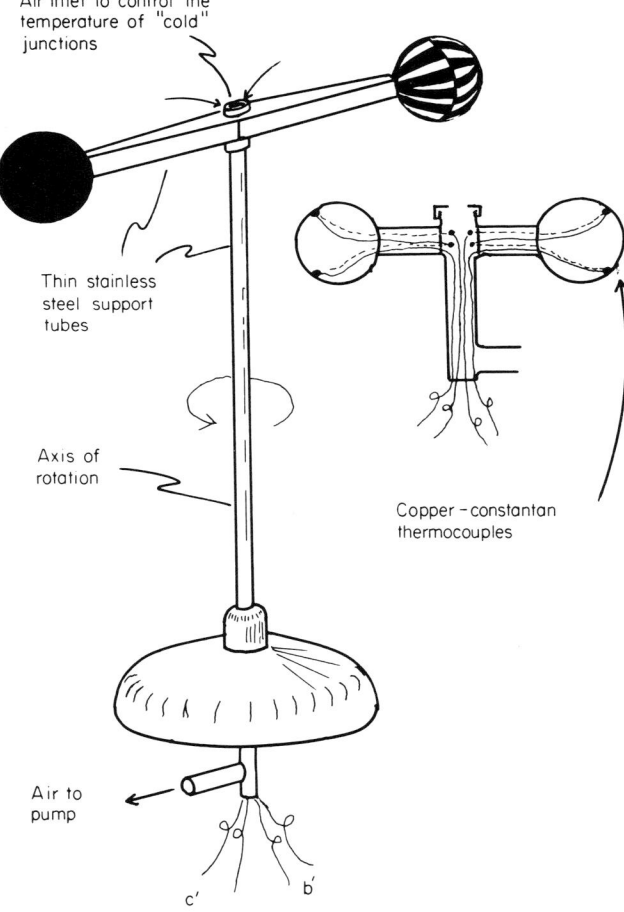

Fig. 11.4. Mean-radiation-temperature meter. By reading two emf's the meter yields a temperature whose value is independent of the degree of air cooling of sensor spheres.

through a mounting tube, and exposed to an induced air stream in a radiation-shielded central section. The sphere to the left is painted black; the sphere to the right is zebra-painted over its polished surface, as shown.

Consider either of these spheres, of radius r, as being isolated (as a test sphere) in air of temperature T_A, and surrounded by a large, concentric, black, hollow sphere of radius R, as shown in Fig. 11.5. The latter represents the radiation environment by which the test sphere is surrounded. Except for a small solid angle (ω), the surroundings are taken to be at temperature T_A. Within ω, the surround temperature is taken to be T_0.

Let $a = e = \beta$ for the first case—the black sphere, and let $a = e = 1.0$ for the surroundings. If f is the fraction of radiation that an elementary area (dS) of the surroundings emits, striking the test sphere and being absorbed by it, then, since in equilibrium as much radiation leaves the test sphere as is absorbed by it, $f = \beta r^2 / R^2$.

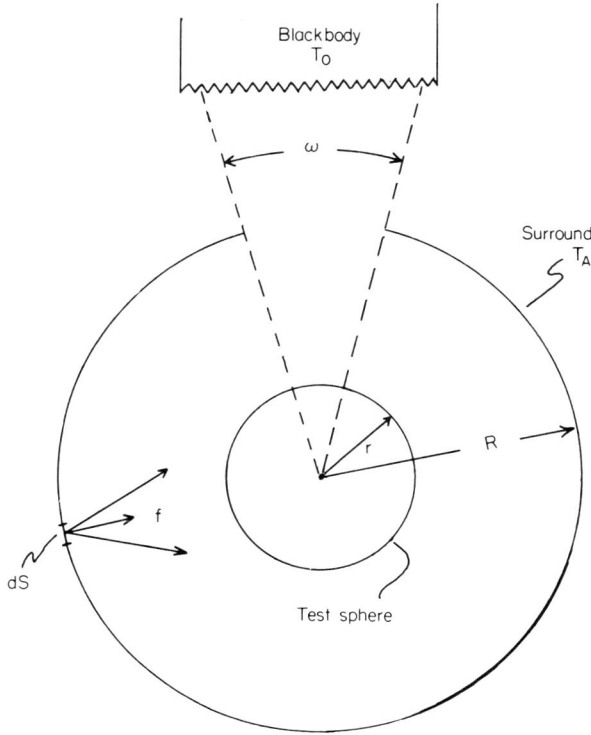

FIG. 11.5. Geometrical model for calculations—see text.

Let N_c be the conduction fraction of the coefficient in Newton's law of cooling—the average loss of heat per unit area that is suffered by the test sphere due to air conduction and ventilation.

The test sphere will come to an equilibrium temperature T_B. Equating heat losses and gains of the test sphere, we get

$$4\pi r^2 \beta \sigma T_B^4 + 4\pi r^2 N_c(T_B - T_A) = 4\pi R^2 f\sigma\{\omega/(4\pi)T_0^4 + [1 - (\omega/4\pi)]T_A^4\}$$

We can simplify this equation, and then reduce it, by invoking the mathematical approximation

$$T_0^4 - T_A^4 = (T_A + t)^4 - T_A^4 \approx 4T_A^3[1 + 1.5(t/T_A) + \cdots]$$

From the second term in the brackets, we see that the first term of this approximation, when $T_A = 300°K$ and $t = 2°C$, is good to 1%. Thus, the simplified equation becomes

$$\beta(T_B^4 - T_A^4) + (N_c/\sigma)(T_B - T_A) = \beta(\omega/4\pi)(T_\theta^4 - T_A^4)$$

And, on applying our approximation, dividing by $4T^3$, letting $b = (T_B - T_A)$ and $n = N_c/4\sigma T^3$, the equation reduces to

$$\beta b + nb = \beta\left[\frac{\omega}{16\pi}\frac{(T_\theta^4 - T_A^4)}{T_A^3}\right]$$

Figure 11.6 shows the manner of achieving a prescribed mean radiation temperature. We have employed this arrangement for laboratory testing and calibration; and an arrangement, similar in principle, can be used for irradiating the whole system of Fig. 11.4.

The expression in brackets above defines an equivalent mean-radiation-temperature increment, θ. This simple case can easily be elaborated to one where a second element of solid angle is at a second different temperature, etc., to represent complex surroundings.

Now, consider the zebra-painted sphere as a test sphere. If the black paint used is the same as before, say $\beta = 0.935$, and if the area that this paint covers, say, is one half of the surface, then one would estimate $e = (0.935 + 0.026)/2 = 0.48$ for the gray (G) sphere. Assuming $a = e = \gamma$ for the gray sphere and $(T_G - T_A) = g$, we may proceed as before. Thus, for the B and G spheres, we will have

(11.1) $$\beta b + nb = \beta\theta$$

and

(11.2) $$\gamma g + ng = \gamma\theta$$

If the apparatus shown in Fig. 11.4, to which these expressions apply, is rotated or oscillated about its vertical stem so that both spheres are equally

11. MEASUREMENT OF RADIATION FLUX

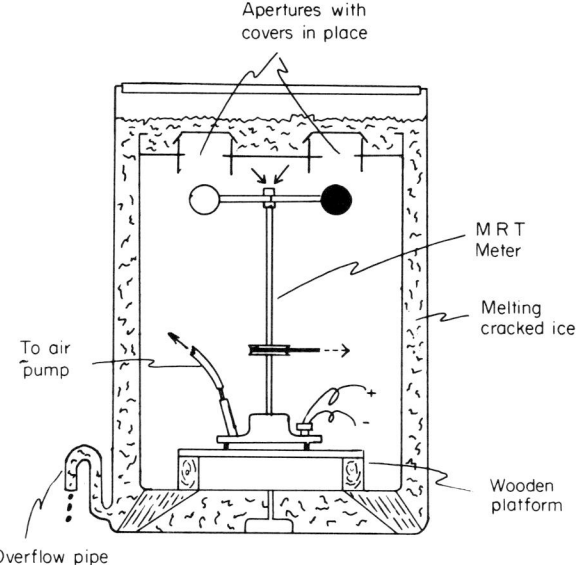

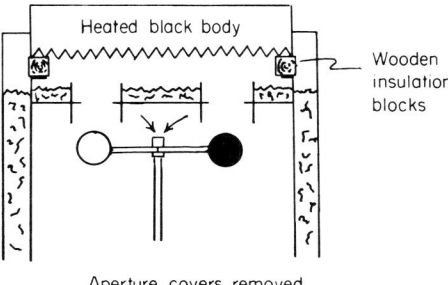

Fig. 11.6. Chamber for studies of MRT meter under controlled conditions.

ventilated, then we can solve the simultaneous equations, Eqs, (11.1) and (11.2) above, and eliminate n. This yields

$$(11.3) \quad \frac{(\beta/\gamma) - 1}{\theta} = \frac{\beta/\gamma}{b} - \frac{1}{g}$$

The validity of this expression has been tested: Fig. 11.7 gives consecutive emf readings which are proportional to b and g. The values of b and g were progressively reduced, even though θ did not vary, as the apparatus was

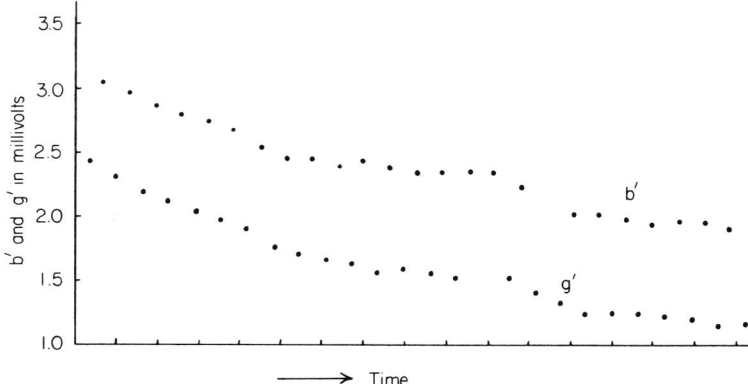

Fig. 11.7. Plot of black and gray sphere emf's with constant irradiation and increasing ventilation (increasing with time).

oscillated more and more vigorously to increase the degree of ventilation. Constant irradiation was achieved by exposure to the high noon sun on an azure, cloudless day, the tests being conducted under glass in the author's greenhouse.

To determine the value of θ from measured values of b and g, we need to know the ratio β/γ. This is the slope of a plot; $1/g$ as ordinate against $1/b$ as abscissa. Such a plot is a straight line because the left-hand side of Eq. (11.3) did not vary. Figure 11.8 is such a plot for the data of Fig. 11.7; Fig. 11.8 yielded $\beta/\gamma = 2.27$. Figure 11.9 shows that the values of θ are constant to $\pm 3\%$. To get this manifest freedom from dependence of θ on ventilation, one must make two temperature measurements rather than just one single temperature measurement. This is to be compared with the alternatives that have been variously proposed (Vernon, 1932; Winslow and Greenberg, 1935; Winslow et al., 1935) that involve measurement of the degree of ventilation with an anemometer and use of calibration data, or employment of servomechanisms, etc.

One can calculate an expected θ, in this instance, using the value of the equivalent solar temperature, the angle that the sun subtends ω, and taking into account reasonable values of the atmospheric and glass transmission for sunlight. Such a calculation gave a value of θ that did not agree with the measured θ. This was due to the fact that the polished silver on the zebrapainted sphere had a value of $a/e = 4$ rather than unity and because, even after taking this into account, we could not properly account for discrepancies arising from reflected sunlight and from the unknown infrared nonsolar irradiation emitted from the remainder of the surroundings.

11. MEASUREMENT OF RADIATION FLUX 317

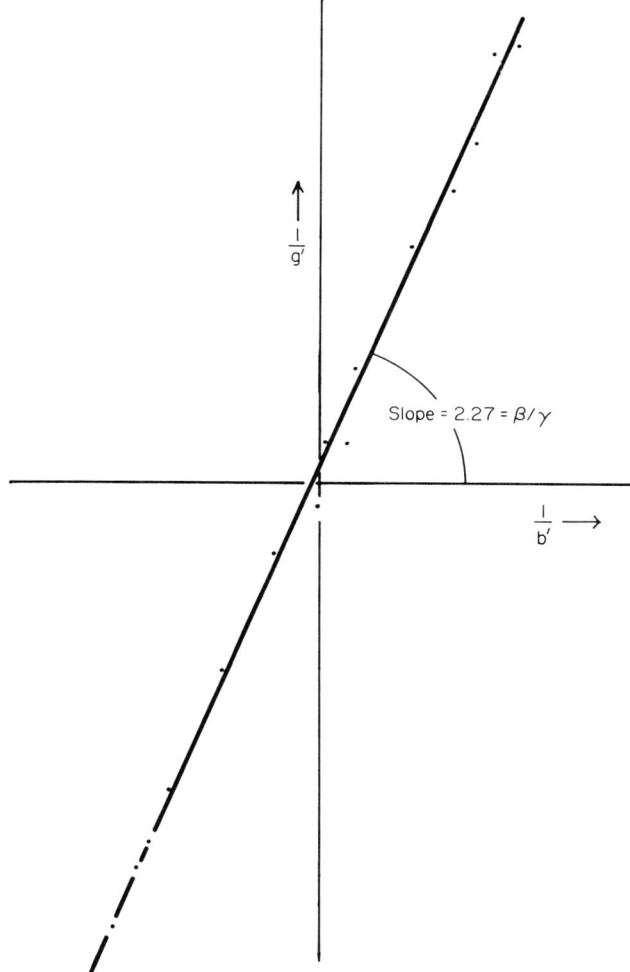

FIG. 11.8. Plot of reciprocals of emf's from Fig. 7.

If we plot the term

$$\left[\frac{\beta/\gamma}{b^2} - \frac{1}{g^2}\right]$$

versus degree of ventilation (or in the case of Fig. 11.7—time), we see that there is a degree of ventilation for which the term vanishes. The significance

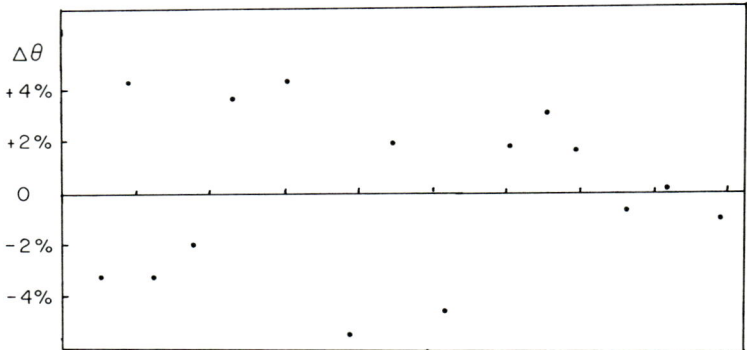

Fig. 11.9. Variance of temperature read-out for data displayed in Fig. 7.

of this vanishing becomes evident on differentiating Eq. (11.3):

$$\left(\frac{-\beta/\gamma - 1}{\theta^2}\right) \Delta\theta = \frac{\beta/\gamma}{b^2} \Delta b = \frac{1}{g^2} \Delta g$$

If $\Delta b = \Delta g$ then $\Delta\theta$ vanishes, or nearly vanishes at the lowest velocities; no error in θ is produced.

This circumstance, coupled with considerations of manufacture and other factors, led us to the design of a different MRT meter described below.

11.3 Four-Sphere MRT Meter

Figure 11.10 shows a four-sphere model. Here, both the black and gray spheres have associated polished silver spheres that serve to measure air temperature, and Δb and Δg have equal air temperature errors. These P spheres substitute for the central hyperventilated "cold-junctions" shelter of the previous model of Fig. 11.4.

Full rotation about the vertical stem at variable speed is provided in this model, rather than the former oscillation—slip rings bringing out the two independent emf's.

The eight hemispheres of this model each have a set of internal thermojunctions, with associated compensating junctions in a central section. Wires from two mating hemispheres, together, pass through a thin Invar supporting tube to the associated compensating junctions. These latter are attached internally to a quadrant of a heavy central copper cylinder. Thus, each pair of hemispheres and their common quadrant form a mechanical and thermoelectrical element. Connection wires from each set of (compensated) junctions, from each hemisphere, pass down the hollow central stem to terminals in a

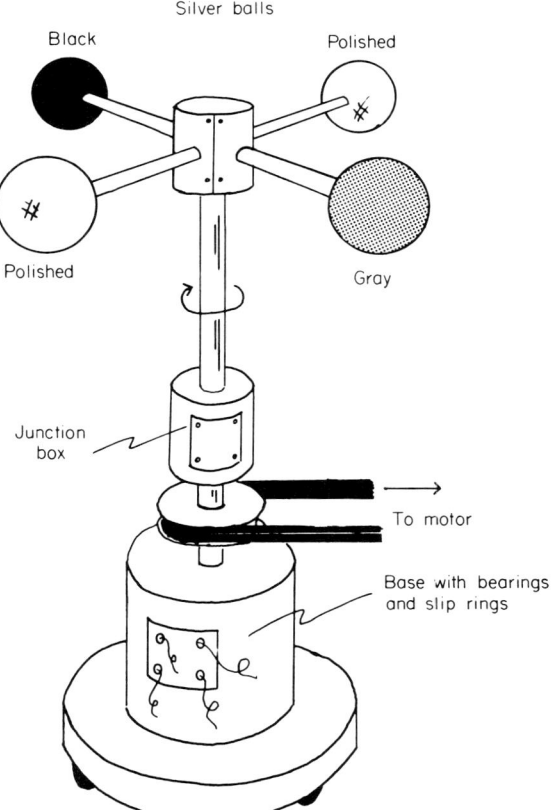

Fig. 11.10. Four-sphere MRT meter—see text.

junction box, after the four quadrants have been integrated mechanically and thermally. The terminals in the junction box may be connected with those of other hemispheres as desired. The appropriate thermoelectric emf's are led to the slip rings. The hemispheres will ordinarily be wired in such a manner that emf's generated by the junctions of the several isothermal quadrant junctions associated with one sphere will cancel those associated with another; thus, the quadrants' temperature is insignificant.

Geometrical and electrical considerations dictated that 20 junctions would be a convenient number to attach to each sphere, since a regular dodecahedron of 20 equal pentagons fits a sphere. These 20 pentagons are easily laid out and they provide regular points for thermojunction locations—10 in each hemisphere. The junctions are thermally in contact with the hemisphere and with the quadrants, but they are electrically insulated by means of a thin sheet of mica that is cemented under each junction. Copper-constantan junctions were employed, yielding in aggregate about 0.8 mV for 1°C difference in temperature between sphere and quadrant. The resistance of each hemisphere set was just under 40 Ω. Due to the high and rapid conduction of heat by silver, the spheres are practically isothermal on the scale of 120-sec time constants (measurements described later).

11.4. Observations with the Four-Sphere MRT Meter

With the central copper cylinder cool (cooled by an ice pack) and all the spheres bright and polished and in vigorous rotation, to bring them to a common air temperature, we compared thermocouple emf's. Shunts of about 4 kΩ across three of the four hemispheres were added to produce the same response for all spheres. When the hemispheres were connected in the manner in which they were intended to be used, each sphere generated an emf that was equal to any other to within one part in a thousand.

The time constant for a sphere and its measurement are not simple (see also section 11.8). However, taking three sets of responses separated by two intervals of time—each of 50-sec duration—we get for the B, G, and P spheres: 2.4, 2.8, and 2.4 for the ratio of responses, $y_1/y_2 = y_2/y_3$, at a certain high ventilation rate, and at a lower degree of ventilation: 2.0, 2.3, and 2.1, respectively. This corresponds to an average time constant of about 2 min. It is desirable to have the time constants of all spheres approximately equal so that the meter will be relatively free from spurious transient responses when air or radiation temperatures fluctuate. Approximate equalities were achieved by using 0.032-in. silver sheet for constructing the B sphere; 0.016-in silver sheet for the P spheres; and 0.032-in. for one hemisphere of the G spheres and 0.016-in. material for the other. This choice arose from consideration that the radiation heat loss N_R and the corresponding air convection loss N_C are approximately equal at low ventilation rates. The heat capacities of the sphere walls were adjusted so that they were proportional to $N = N_C + N_R$.

11.5. Experiments with the Four-Sphere Model

The MRT meter was primarily developed for use in an environmental test chamber such as that shown in Fig. 11.3. In this application, only invisible infrared radiations are involved and absorption coefficients are adequately equal to the emission coefficients. But other experiments can be carried out with the apparatus. For example, tests of the blackness of various paints have been made. One paint was put on one P sphere and its temperature compared with the other P sphere covered with another paint, both being illuminated with a *Sun Gun* lamp. After that, the paints were reversed to cancel possible instrumental error. It was found that the spray paint Krylon 1602 gave the largest emf; the test results of the other four, relative to it, were as follows:

Parsons	0.990
Mautz	0.975
Fuller	0.970
Sicon	0.970

Further experiments are being conducted with both black and white paints that will separate the effects of $a_\odot$ and $e_\oplus$ here.

11.5.1. Case Where $a/e \neq 1.0$

When a and e are not equal, as when sunlight is involved, the starting equation for a sphere is not so simple. We may commence as follows:
$$eT_s^4 + N_c(T_s - T_A) = aT_\theta^4$$
Proceeding mathematically, as before, we get

(11.4) $$et + nt = a\theta + (a-e)(T_A/4)$$

where $t = (T_s - T_A)$. Experiments conducted with the four-sphere model are to be interpreted by application of Eq. (11.4).

11.5.2. Thermistor MRT Meter

Many variations of design have not yet been exploited. For example, we are just now applying the developments of Ney *et al.* (1961), who studied the advantageous influences of small detector size and shape in minimizing errors due to the equivalent radiation temperature. They were concerned with freeing the measurement of air temperature, at the low atmospheric pressures of high altitudes, from radiation effects,

For measurements of effective radiation temperature in the upper air, we are now exploiting a model that uses thermistor sensors. It is equally adapted for measurements in the environmental chamber of Fig. 11.3. One thermistor is attached to the B sphere and another to a G sphere. They measure the temperatures T_B and T_G, of these two spheres. T_A is measured with one or two other small thermistors painted white with Brasefield's paint or silvered. They may be exposed to the air, as in the model of Fig. 11.4, or openly, as the P spheres of Fig. 11.10 are. But by their small size, the developments of Ney *et al.* are appropriately invoked to minimize radiation errors in the air temperature measurement.

11.6. Equivalent Radiation Temperature in the Upper Air

Considerable attention has been given to the measurement of infrared fluxes and equivalent radiation temperatures—both in the lower and upper atmosphere. To mention only a few studies: Measurements from ground stations have been made (Ångström, 1901; Dines and Dines, 1927; Strong, 1941; Richards *et al.*, 1951) although they have never become routine meteorological parameters. Measurements of the upper air from balloon stations include the pioneering black-ball, balloon-sonde temperatures of Gergen (1956, 1957), spectral measurements of upward and downward fluxes (Strong, 1957), and the measurements of the divergence of flux (Strong, 1959). Some of these last measurements are reviewed below. (See also Chapter 12, this volume.)

Figure 11.11 shows wavelength response to the upward flowing infrared flux, as measured with an infrared spectrometer from a station at an altitude of 93,000 ft. The spectrometer was held there by balloon on the night of October 4, 1956. This flux is represented here by the ac thermopile response that is produced when the downward-looking spectrometer had its entrance beam chopped by a thin mica chopper. The chopper was presumably at the local air temperature, 230°K.

A contemporaneous temperature profile has been inserted in this figure to indicate the interpretation of the response at various places in the spectrum. Response at each wavelength is proportional, of course, to the difference between the emission of the mica shutter at a temperature of about 230°K (at 93,000 ft altitude) and the weighted, effective atmospheric emission from the absorbing atmosphere below (or the radiation from ground temperature at wavelengths where the atmosphere is transparent, as between 8 and 13 μm).

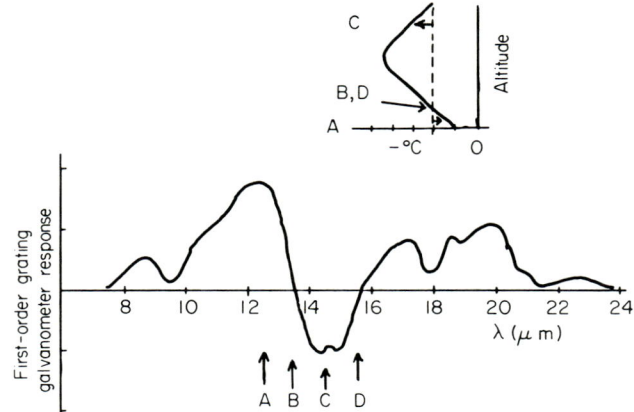

Fig. 11.11. Up-welling radiation spectrum as observed from 93,000 ft altitude.

Near 14 and 16 μm, on either side of the 15 μm absorption band of CO_2, the optical depth, indicated in the insert, is such as to make the effective emission temperature the same as the shutter temperature. Thus, the response at these wavelengths (B and D) is zero. Between these wavelengths around 15 μm where the CO_2 is strongly absorbing, the optical depth is shorter. Hence, the effective emission temperature of the atmosphere at these wavelengths (C) is less than that of the shutter—and the response is negative. Outside these limits, the optical depth reaches to warmer regions of the atmosphere (or to the tops of the Sacramento Mountains over which the balloon floated as these spectra were observed), and at these wavelengths (A) the response is positive.

Let us refer to Fig. 13.10 of Hanel's contribution (this volume) which is a recent profile of equivalent atmospheric "color" temperatures. This curve was obtained from balloon observations, but with much higher resolving power. The spectral components there have been calibrated, in absolute measure, so that each is interpretable as a color temperature. The interpretation is much the same as for the lower resolving power of Fig. 11.11.

Figure 11.12 shows the measured up-and-down flux difference at various altitudes in the atmosphere (as measured at $\pm 30°$ to the horizon). This flux

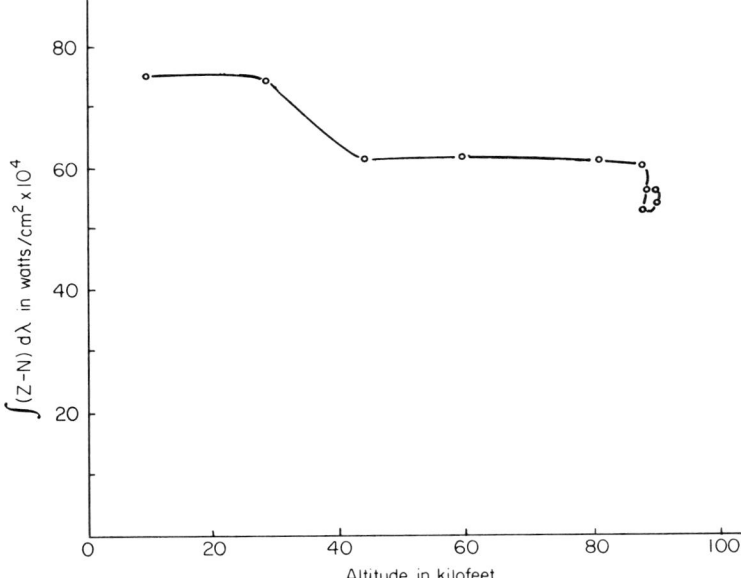

FIG. 11.12. Variation with height of radiation, including the 9.6 μm O_3 absorptive band.

is integrated over the filtered, first-order, bell-shaped response band of a grating spectrometer. Although the equipment yielded considerably higher spectral resolution, the balloon ascension rate was so rapid that it was necessary to integrate over the whole first grating order to get a significant result. The blaze of this order was peaked at 10.6 μm, with half power points at 9 and 12 μm. The plot shows the influence of the atmospheric ozone layer as the balloon rose into it. It is different from the previous spectra in that the instrument looks up through the super-natant atmosphere toward the outer *cosmic cold*, in its off-phase, half-cycle rather than at an ambient-temperature mica shutter. And the rate at which this differential flux varies, with altitude change, measures the divergence of flux.

11.7. Future Problems

Observations over the wavelength range, from 4 out to 30 μm, are not adequate, as some have alleged, to measure the total outgoing terrestrial flux (from high-balloon or satellite stations). This is because a fraction of the emission spectrum of the earth lies at very long wavelengths: $\lambda > 30$ μm. At any temperature, we know that the fraction of the blackbody radiation that falls beyond λ_{max} is 75 %, the fraction beyond $2\lambda_{max}$ is 28 %, and beyond $3\lambda_{max}$ is 12 %. This is because the blackbody curve runs up to its maximum rapidly with wavelength but it descends very slowly—a circumstance that is often not appreciated. At the longer wavelengths, the outgoing terrestrial flux is dominated by the water-vapor emission at upper air-temperatures. If these long wavelength fluxes may be characterized by a 245°K blackbody, 20 % of the total radiation (σT^4) lies at $\lambda > 30$ μm. A substantial error can therefore be made unless the radiation receivers are black at wavelengths longer than 30 μm, and one cannot be confident that such blacks are now within the "state-of-the-art." This is one problem for the future—the problem of blacks for submillimeter wavelengths.

Radiometric observations for determining horizontal temperature gradients in the upper air have been proposed. Because of the possibilities of observing from high-altitude aircraft (such as U-2's) with complete altitude control (as contrasted to balloons), both horizontal gradient observations and vertical measurement of the divergence of flux, wavelength by wavelength, stand out as ready to be exploited to the benefit of synoptic meteorology.

11.8. Absolute Emissivity Measurement of Flat Surfaces

The cooling of the object shown in Fig. 11.13 is measured to determine the absolute emissivities of the flat surfaces A_s. The cooling of the object is carried out in the glass bell jar of a high vacuum pumping system such as is used for thermal evaporation to produce metal mirrors and dielectric films. (Strong, 1938). Details of the method are described hereunder. We assume that the bell jar's interior surfaces, and the base plate, are both effectively black, $e = a = 1$; or that they may be coated black with a paint to make them so. Then the emissivities of the Fig. 11.13 assembly determine the cooling rates.

For simplicity, we describe a particular application of the method that is concerned with the determination of the hemispherical emissivity of the outer surfaces of two glass plates. The glass plates of exposed area A_g are attached

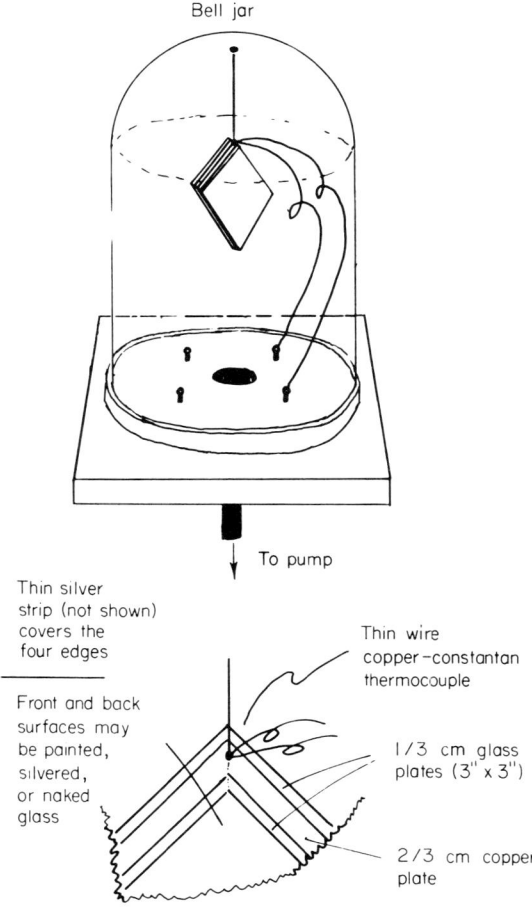

FIG. 11.13. Representation of a cooling-rate method of determining surface emissivities.

by a thin cementing film of beeswax and rosin to a copper plate, to form a "sandwich"—the edge area A_s of this composite being covered with a silver "picture frame" (not shown). A thin wire thermocouple, of constantan and copper, is attached to a corner of the copper plate to support the sandwich at the center of the bell jar. Heat loss along these thermocouple wires is negligible. If the sandwich is 7.1 cm², then $A_g = 2 \times 50$ cm². For this case, the specific heats per unit area of exposed glass surface are $C_g = 0.17$ cal cm^{-2}/°C for the glass and $C_c = 0.27$ cal cm^{-2}/°C for the copper, for 3 and 6 mm thicknesses.

The sandwich is heated warmer than the bell jar which, for simplicity, we assume to be at 300°K, and then the latter is evacuated. Its cooling curve can be conveniently taken by means of a strip-chart recorder. If the sandwich is heated to $\theta = 25$°C above the bell-jar temperature, the thermoelectric emf at the start of the cooling will be about 1 mV.

To determine the time constant τ, we read off the recorder pen deflections y_i, y_{i+i}, etc., at various times that are separated by fixed observation intervals of, say, $\Delta t = 60$ sec.

Now one of the uncertainties of measuring the time constant of the sandwich results from the circumstance that the bell-jar temperature may vary slowly. We cope with this circumstance by assuming that it has a constant temperature during two successive fixed observation intervals. A constant correction δy, for three successive recorder responses y, can be determined by applying the exponential cooling law:

$$\frac{y_2 + \delta y}{y_1 + y} = \frac{y_3 + \delta y}{y_2 + y} = \exp\left(\frac{-\Delta t}{\tau}\right), \quad \text{giving} \quad \delta y = \frac{y_2^2 - y_1 y_3}{y_1 + y_3 - 2 y_2}$$

On applying this correction, one gets an operational value of the time constant τ_θ, corresponding to θ, the temperature elevation of the sandwich above the bell jar's effective temperature. The value of τ_θ thus obtained varies with θ—going asymptotically to a value τ_0 when θ is very small. In this limiting case, on equating heat conduction through the glass at its outer surface with net radiation loss from that surface to the bell-jar walls, we get

$$(C_g + C_c)(d\theta/dt) = \sigma(T_g^4 - T_0^4)(e_g A_g + e_s A_s)$$

Setting $\theta = T_g - T_0$, and applying the approximation cited earlier,

$$(C_g + C_c)(d\theta/dt) = 4\sigma T_0^3(e_g A_g + e_s A_s)\theta$$

which gives

(11.5)
$$\tau_0 = \frac{(C_g + C_c) 4\sigma T_0^3}{(e_g A_g + e_s A_s)}$$

All factors for determining e_g by this equation are known, $e_s = 0.026$ as cited earlier.

Now, in instances such as when $\theta = 25°C$, where the temperature rise cannot be considered to be "very small," the net radiation loss can be corrected by the second-order approximation mentioned earlier, viz.,

$$4\sigma T_0^3 \left(1 + \frac{1.5\theta}{T} + \cdots \right)$$

Also, the glass in this test is slightly nonisothermal: One can calculate correction for the effects of this from its temperature gradients. These gradients, in turn, can be calculated, approximately, from the heat conductivity $k = 25 \times 10^{-4}$ cal sec^{-1} cm^{-1} and from $4\sigma T_0^3 = 1.5 \times 10^{-4}$ cal sec^{-1} cm^{-2}/°C; taking $e_g = 1.0$,

$$k\left(\frac{d\theta}{dx}\right)_{out} = 4\sigma T_0^3 \theta; \qquad \left(\frac{d\theta}{dx}\right)_{out} = \frac{1.5 \times 10^{-4}}{25 \times 10^{-4}} \theta = 0.06\theta$$

At the inside boundary, the copper will be isothermal and the stored heat flowing through this boundary will be lower than at the outside boundary:

$$k\left(\frac{d\theta}{dx}\right)_{in} = \frac{C_c}{C_c + C_g}(4\sigma T_0^3 \theta); \qquad \left(\frac{d\theta}{dx}\right)_{in} = \left(\frac{0.27}{0.27 + 0.17}\right) 0.06\theta = 0.04\theta$$

The average temperature gradient in the glass is 0.05θ, making its average temperature

$$\theta(1 + 0.05 \times \tfrac{1}{6}) \simeq \theta(1 + 0.01)$$

and the temperature of the copper is

$$\theta(+ 0.05 \times \tfrac{1}{3}) \simeq \theta(1 + 0.02)$$

Since these corrections are small, we may apply them after integration. For each value of τ_θ, the factor $[1 + (1.5\theta/T)] = (1 + 0.005\theta)$ is applied as a correction to $4\theta T_0^3$, and the factors $(1 + 0.01)$ and $(1 + 0.02)$ are applied as corrections to C_g and C_c, respectively. Thus, to change the observed τ_θ to τ_0, we have a heat flow correction of 0.005θ and a heat capacity correction of 0.016:

(11.6) $$\tau_0 = \tau_\theta(1 + 0.005\theta)(1 + 0.016)$$

We have applied this method to bare glass sandwiches, using $e_s = 0.026$ in Eq. (11.5), and we obtained $e_g = 0.88$. If the inside surface area of the glass bell jar is, in practice, much larger than that of the glass test pieces, then it may be considered black.

We do not treat here the obvious generalizations of this method. The method can, of course, be adapted to measurements with a painted silver sphere, etc. It is not only a convenient procedure but is often operationally similar to the intended application of its resulting e's.

11.9. Pyrheliometer Development

In addition to the applications indicated above, the investigations that are reported here are a part of the broader program to develop a new pyrheliometer and absolute radiometer. A beginning investigation has already been published (Strong and Lawrence, 1968). The lines along which this pyrheliometer development is planned are outlined in that publication. Reprints of the paper are available on application to the Eppley Foundation for Research, Newport, R. I.

Acknowledgments

The experimental work involved in this report was sponsored by the Office of Naval Research. Various associated personnel were supported by Eppley Student Fellowships.

References

Ångström, K. (1901). Dependence of radiation-absorption of a gas on its density. *Ann. Phys.* **6**, 163–173.

Brasefield, C. J. (1948). Measurement of air temperature in the presence of solar radiation. *J. Meteorol.* **5**, 147–151.

Butler, C. P., Jenkins, R. J., and Parker, W. J. (1964). Absorptance, emittance and thermal efficiencies of surfaces for solar spacecraft power. *Solar Energy* **8**, 1–8.

Dines, W. H., and Dines, L. H. G. (1927). A monthly mean radiation from various parts of the sky at Benson, Oxfordshire. *Mem. Roy. Meteorol. Soc.* No. **11**, 1–8.

Gergen, J. L. (1956). "Blackball": A device for measuring atmospheric infrared radiation. *Rev. Sci. Instrum.* **27**, 453–460 (reference may also be made to Sutton, D. J., and McNall, P. E. (1954). A two-sphere radiometer. *Am. Soc. Htg. Vent. Engr.* **60**, 297–308).

Gergen, J. L. (1957). Atmospheric infrared radiation over Minneapolis to 30 millibars. *J. Meteorol.* **14**, 495–504.

Ney, E. P., Maas, R. W., and Huch, W. F. (1961). The measurement of atmospheric temperature. *J. Meteorol.* **18**, 60–80.

Ramsey, W. Y. (1964). Reflectance of paints from 0.4 to 40.0 microns. *Meteorological Satellite Lab. Rept. No. 31*, U.S. Department of Commerce, Washington, D.C.

Richards, C. H., Stoll, A. M., and Hardy, J. D. (1951). The panradiometer: An absolute measuring instrument for environmental radiation. *Rev. Sci. Instrum.* **22**, 925–934.

Strong, J. (1938). Technique of high vacuum. *In* "Procedures in Experimental Physics," Chapter III. Prentice-Hall, Englewood Cliffs, New Jersey.

Strong. J. (1941). Study of atmospheric absorption and emission in the infrared spectrums *J. Franklin Inst.* **232**, 1–22.

Strong, J. (1957). Balloon observations of earth radiations in the infrared. Contract AF 19 (604) 949, Air Force Cambridge Res. Center, Bedford, Mass.

Strong, J. (1959). Balloon observations of earth radiations in the infrared-III. Contract AF 19 (609) 2238, Air Force Cambridge Res. Center, Bedford, Mass.

Strong, J., and Lawrence, P. W. (1968). Bolometer theory. *Appl. Opt.* **7**, 49–52.

Vernon, H. M. (1932). The measurement of radiant heat in relation to human comfort. *J. Ind. Hyg. Toxicol.* **14**, 95–111.

Winslow, C.-E. A., and Greenburg, L. (1935). Thermo-integrator: New instrument for observation of thermal interchanges. *Trans. Am. Soc. Htg. Vent. Engr.* **41**, 149–151; also in (1935) *Htg. Pip. Air Cond.* **7**, 41–43.

Winslow, C.-E. A., Gagge, A. P., Greenburg, L., Moriyama, I. M., and Rodee, E. J. (1935). The calibration of the thermo-integrator. *Am. J. Hyg.* **22**, 137–156.

Winslow, C.-E. A., Herrington, L. P., and Gagge, A. P. (1936). A new method of partitional calorimetry. *Am. J. Physiol.* **116**, 641–655.

— 12 —
APPLICATIONS OF THERMAL RADIATION MEASUREMENTS IN ATMOSPHERIC SCIENCE

P. M. Kuhn

12.1. ROLE OF RADIATIVE ENERGY EXCHANGE IN WEATHER PROCESSES

The features of longwave (3–100 μm) radiative power transfer in the atmosphere are essentially different from those of shortwave (<3 μm) power transfer. To understand this transfer process as applied to weather, let us briefly consider the main features of this longwave radiation.

Absorption and emission of thermal (longwave) radiation by atmospheric gases are the primary processes. To this must be added the emission of the ground and cloud surfaces. Frequently, the term "terrestrial radiation" is used for radiative transfer in this spectral region.

There is no terrestrial radiation illuminating the outer limit of the earth's atmosphere. Solar radiation at wavelengths beyond 3 μm may be essentially neglected. It can be measured in and at the edges of the atmospheric window (Fig. 12.1) and is important in the study of absorption by atmospheric gases.

Scattering by gases or aerosols (particulates) is not important in infrared transfer processes because of the small size of the aerosols with respect to the wavelength. Even in clouds, the droplets are of the same or smaller magnitude than the wavelength. Considerations have been given in infrared scattering, but its effects are very small compared to those of dominant longwave absorption and emission.

Finally, terrestrial radiation is diffuse, possessing no specular beam.

Considering the thermal effects of radiative divergence and convergence, the atmosphere below 50 km is in thermodynamic equilibrium with the energy of the temperature motion of the gas and aerosol atoms and molecules. The foregoing is a summary of the terrestrial radiation discussion of the optics of the lower atmosphere by Möller (1964).

Figure 12.1 emphasizes the fact that the atmosphere is much more transparent to shortwave than to longwave radiation. Directly affecting weather processes, in the form of a forcing function, is the combination of solar energy,

The principal difference between the two radiometer types is that the conventional radiometer incorporates a pair of independent but similar sensors, while the other has a single (but double parallel surface) sensor. Comparison of infrared cooling observed with the two detectors is indicative of the problems of measuring such atmospheric cooling with the disk radiometer in a purely gaseous atmosphere. Since the latter is similar to the earlier black-ball instrument (Gergen, 1957) both have the same limitations. However, in a particulate atmosphere devoid of water vapor, there is observational evidence that it may be possible to approximate atmospheric longwave cooling by using soundings of air temperature and black-ball or black-disk radiometric observations.

To continue the development of radiometer observations of atmospheric infrared radiation with the routinely operated radiometersonde (Suomi and Kuhn, 1958) and to test the black-disk and black-ball radiometers, ascents were made by University of Wisconsin meteorologists. The results of such ascents, made at night, are based upon a new data-reduction technique in which least-squares approximating polynomials (Johnson, 1965) are used for recovery of maximum information, by filtering the random error component of the radiometersonde system. To illustrate these improvements, one of these ascents is discussed later in this chapter.

Improvements in the Radiometer. Improvements in the design of the Weather Bureau balloon-borne radiometer since its development can be seen by reference to Fig. 12.2. Concurrent ones have been made in the black-disk radiometer (Fig. 12.3). None of the vertical dimensions have been changed because they are critical in the suppression of convection. However, to further minimize any lateral heat transfer, the size of the blackened sensing surface

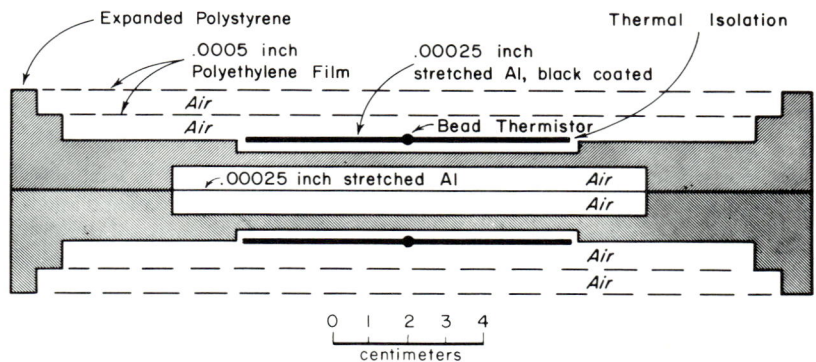

FIG. 12.2. Cross section of the conventional balloon-borne radiometer (double-sensor model).

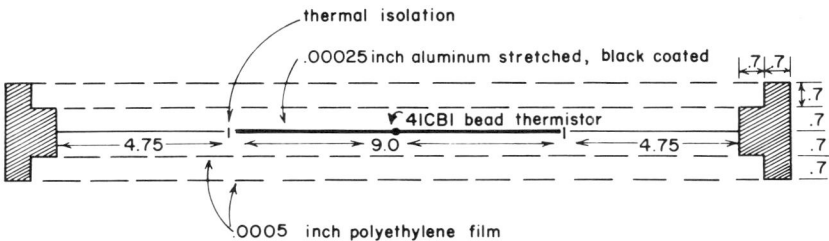

FIG. 12.3. Cross section of the experimental black-disk radiometer (single-sensor model).

has been reduced to a diameter of 9.0 cm. This increases the horizontal distance from the edge of the sensor to the outer edge of the radiometer to 6.1 cm. With maximum temperature gradients, which were observed between the top sensor and the warmer ambient air, at night, the lateral conduction term can be no larger than 2.0 W m^{-2}. The central aluminum shield has been reduced in diameter to further minimize lateral conduction within the assembled device to the outer edge.

In both Figs. 12.2 and 12.3, we see the different configurations of the 0.00025-in., black-coated aluminum sensors. In the case of the single-sensor model as well as the double-sensor model, eight threads suspend the sensing plate (Fig. 12.4) so that it does not touch the expanded polystyrene radiometer trays. By means of this sensor mounting arrangement, the relaxation time (one time constant) has been reduced to 25 ± 2 sec. Formerly, the time constant was 48 ± 5 sec. The depression beneath the thread-suspended sensor, in the two-sensor unit, is aluminized to minimize infrared absorption and transfer.

To prevent heat conduction along the leads to the bead thermistors, which are attached with epoxy-resin to the aluminum sensing disks, No. 40 Advance wire is used. These leads are suspended above the sensors and led out to the edge of the radiometer. Both bead thermistors are epoxied to the upper surface of the respective disk, but each is so small that its shadow from a single radiant beam of nearly grazing incidence to the sensor does not fall off the black-coated disk. Where applicable, all the improvements cited for the conventional radiometer apply to the black-disk radiometer.

The Experiment. Both the integrated total disk radiometer and the standard radiometer, which separately responds to upward- and downward-directed irradiances, can measure the equilibrium irradiance in an assumed plane parallel atmosphere. In the latter instance, this quantity is defined as one-half the sum of the upward- and downward-directed fluxes. For a perfectly "black" disk-type detector responding uniformly to longwave radiation over

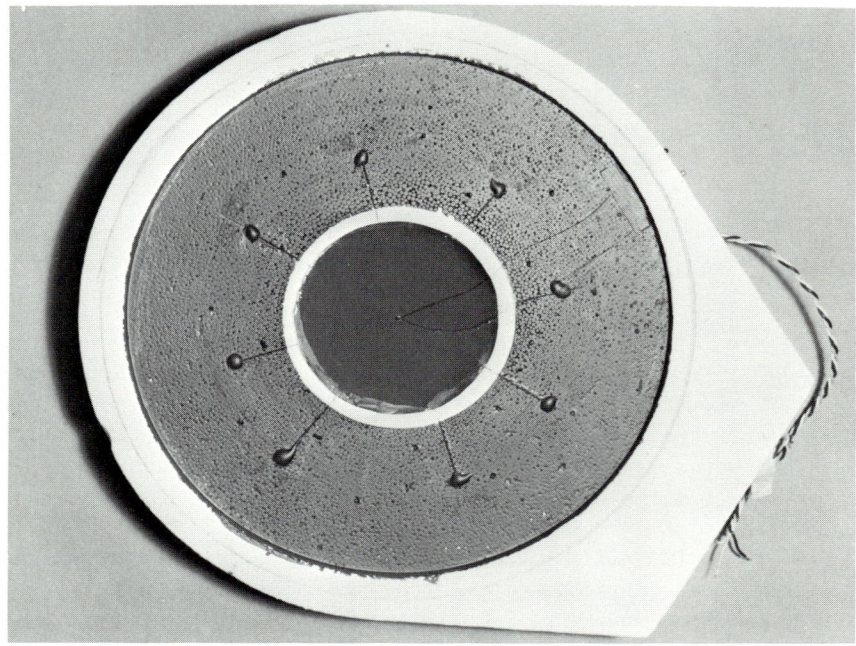

Fig. 12.4. Plane view of the black-disk radiometer.

a wide spectral range, the equilibrium irradiance is given through its equivalent temperature, by (Businger and Kuhn, 1960),

$$(12.1) \qquad W = \sigma T_d^4 = \int_0^{\pi/2} \sigma T_e^4(\theta) \sin \theta \cos \theta \, d\theta$$

$$+ \int_\pi^{\pi/2} \sigma T_e^4(\theta) \sin \theta \cos \theta \, d\theta$$

where W is the irradiance, $\sigma = 5.6686 \times 10^{-8}$ W m^{-2} °K^{-4}, T_d is the equilibrium temperature of a black disk in °K, and T_e is the equivalent radiation temperature defined as follows:

$$(12.2) \qquad b T_e^4(\theta) = \int_0^\infty I_\lambda(\theta) \, d\lambda$$

where $b = 1.804 \times 10^{-8}$ W m^{-2} °K^{-4} (sr)$^{-1}$, θ is the zenith angle to the sensor, and λ the wavelength.

The downward- and upward-equilibrium radiation, measured by the radiometersonde is given by

$$\sigma T_T^4 = 2 \int_0^{\pi/2} \sigma T_e^4(\theta) \sin \theta \cos \theta \, d\theta \tag{12.3a}$$

and

$$\sigma T_B^4 = 2 \int_\pi^{\pi/2} \sigma T_e^4(\theta) \sin \theta \cos \theta \, d\theta \tag{12.3b}$$

where T_T and T_B are the top and bottom equilibrium temperatures of the radiometersonde. The equilibrium irradiance measured by both devices can be equated as follows:

$$\sigma T_d^4 = 0.5\sigma(T_T^4 + T_B^4) \tag{12.4}$$

In other words, one-half the sum of the right-hand terms in Eqs. (12.3a) and (12.3b) is identical to the right-hand terms in Eq. (12.1).

As explained, the conventional and the black-disk radiometersondes observe, respectively, the top and bottom sensor equilibrium temperatures and the total equilibrium temperature [Eqs. (12.2), (12.3a), and (12.3b)] as functions of measured blackbody temperatures, thermal conduction, and lag of the sensing surfaces. These are well known and are reported frequently in the literature. However, by statistical treatment of the data, more information can be extracted from such radiometric ascents than was possible in the past, and confidence limits can be stated for individual estimates of each of the various irradiances and atmospheric infrared cooling.

The downward-directed flux measured by the double-sensor radiometersonde is given in absolute measurement by

$$\sigma T_T^4 = \alpha \sigma T_t^4 + K_i \, \Delta T_i \tag{12.5}$$
$$+ K_t \, \Delta T_t + \gamma \, \Delta T_t / \Delta t$$

where T_t is the upper radiometer sensor temperature, ΔT_i is the temperature difference between the upper and lower sensor surfaces, and K_i is a function of the thermal conductivity of the radiometer insulating material and the reflectance of the polyethylene wind shields of the radiometer, and α is the absorption of the black-coated sensor surface; ΔT_t and K_t are similarly defined for the sensor-to-polyethylene air cell of the radiometer; γ is the lag coefficient of the sensor system, and Δt is an increment of time (Suomi and Kuhn, 1958).

Measurement of the upward-directed flux may be treated in an analogous manner.

To test the assumption that the black-disk radiometric observation of total irradiance, combined with the equilibrium blackbody emittance at air temperature, cannot approximate atmospheric radiative cooling, a dual sensor ascent was undertaken. A secondary purpose in the ascent was to confirm the hypothesis that the single-disk irradiance (i.e., the differential of both surfaces) was equal to one-half the sum of the downward- and upward-standard values. The spherical thermistor detector embedded symmetrically in the sensor plate absorbs and radiates equally. The two types of instrument were balloon-flown during the night of April 1, 1964, at Miami, Florida. The radiometers were deployed 60 m below a 1200-gm neoprene balloon. During the flight, temperature observations T_d of the disk sensor and T_t, T_b, and T_a of the top, bottom, and air sensors of the radiometersonde were recorded. Before presenting the result of these measurements, we should briefly discuss (next section) the aforementioned filtering technique as applied here.

Improvements in the Data-Reduction Technique. The effect of a random instrumental error component in the temperature observation of the radiometersonde, given by an equivalent expression for Eq. (12.5), is

$$\sigma T_T{}^4 = c_1(\eta_t + \epsilon_t)^4 + 3c_2(\eta_t + \epsilon_t)^2$$
$$+ 3c_3(\eta_t + \epsilon_t) + c_4 \frac{d}{dt}(\eta_t + \epsilon_t)$$
$$+ c_2(\eta_b + \epsilon_b)^2 + c_3(\eta_b + \epsilon_b)$$
(12.6) $$+ 2c_2(\eta_a + \epsilon_a)^2 + 2c_3(\eta_a + \epsilon_a)$$

where subscripts t, b, and a refer to top, bottom, and air observed temperatures.

The equations relating the true temperatures to the observed temperatures (η_t, η_b, and η_a) and the random temperature errors (ϵ_t, ϵ_b, and ϵ_a) are

(12.7a) $\qquad\qquad\qquad T_t = \eta_t + \epsilon_t$

(12.7b) $\qquad\qquad\qquad T_b = \eta_b + \epsilon_b$

(12.7c) $\qquad\qquad\qquad T_a = \eta_a + \epsilon_a$

It has been shown by Johnson (1965) that Eq. (12.6) can be derived by employing Eqs. (12.7) in conjunction with Eq. (12.5). Then, it can be seen that the true downward irradiance of Eq. (12.5) is exactly given when the random instrumental error of the three temperature measurements is zero. Since the latter is never zero, approximating polynomials are used to filter the major part of the random error component.

To justify the use of filtering by approximating polynomials, the basic observations must be discrete measurements of a smooth true function on which a random instrumental error is superimposed. The important advantage of using least-squares approximating polynomials for the filtering is that the consequences of an exact fit to each data point are avoided. An exact polynomial fit requires that the estimated function pass through each observation. Since the data series contains a random error component, the estimated function by an exact fit oscillates about the smooth true function. In contrast, an unbiased approximating polynomial can be used to determine a point at each value of the independent variable, which is a better estimate of the smooth true function than the original observation. Therefore, the random error component can be largely filtered from the observation.

In this particular flight, each of the profiles was filtered by fitting a second-degree polynomial to five adjacent observations for each temperature series, thus leaving two degrees of freedom for the random error component to be distributed about the smooth predicted polynomial temperature function. Time is used as the independent variable because it is the most accurate measurement. A filtered estimate is then made for the mid-observation of the five. The next filtered estimate is made by again selecting five observations, adding the new adjacent observation in time, deleting the oldest observation in time from the prior five points, and predicting for the new mid-observation. Thus, filtered estimates are made for the entire series of the original observations of the top, bottom, and air temperatures. As a check, the estimate of variance for the random error component from the residual sums of squares is compared with a laboratory determination of the error variance for each of the individual temperature series. Since the two estimates of variation are nearly equal, we are assured that the functions estimated, by approximating polynomials, provide better estimates of the true irradiance functions than do the original observations. In the probabilistic sense, when the two estimates of variance are equal, the estimated function must converge to the true function (Johnson, 1965). The rms errors of the filtered estimates of (a) the downward and upward irradiances ranged from 1.1 to 1.4 W m^{-2}, (b) the equilibrium irradiance from 0.7 to 1.0 W m^{-2}, and (c) the net irradiance from 1.6 to 2.0 W m^{-2}. The rms error of the filtered estimate for cooling is less than 0.25°C day^{-1}.

The curve σT_A^4, in Fig. 12.5, is a plot of the equivalent blackbody radiation from the filtered estimates of air temperature.

Discussion of the Observations. The profiles of the filtered estimates of σT_T^4 top, σT_B^4 bottom, and σT_d^4 disk (W m^{-2}) are presented in Fig. 12.5. The total irradiance measured by the disk was almost identical to one-half the sum of the downward- and upward-directed irradiances. This must be so if the two instruments are correctly measuring total irradiance. The rms error random differences between the two estimates of the total irradiance was 4.9 W m^{-2}. This represents a reduction in the influence of the random

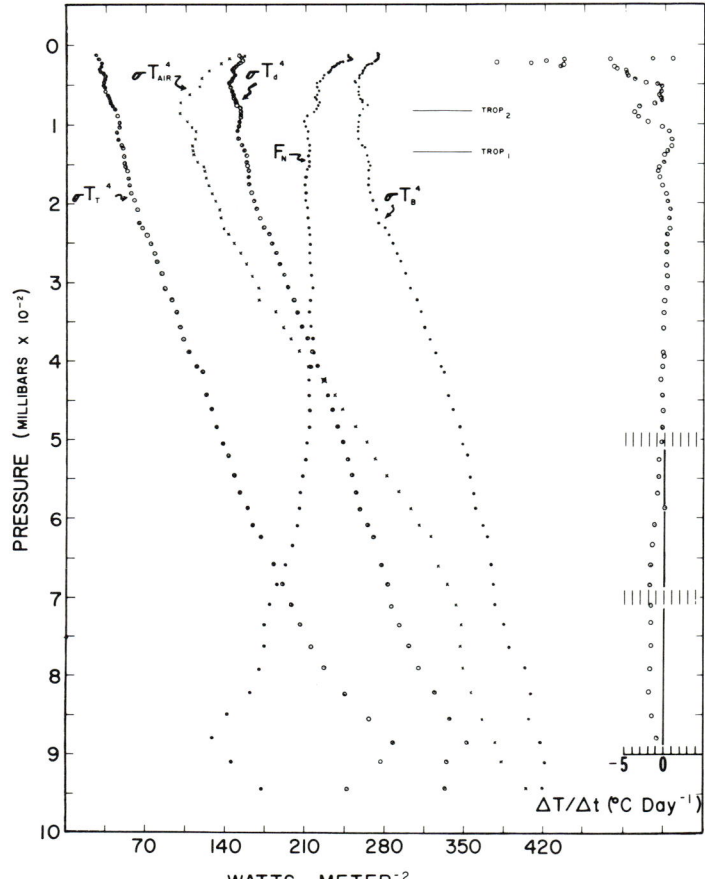

Fig. 12.5. Equivalent blackbody radiant emittance.

error component due to the filtering by the approximating polynomials. Since the estimates were nearly identical, only the profile from the radiometersonde was plotted in the figure.

"Black-disk" cooling may be assumed to be proportional to the difference between the equilibrium blackbody irradiance at air temperature and the total irradiance measured by the black disk at the same level. The latter difference is defined as

$$(12.8) \qquad \Delta T/\Delta t (°C/\text{day}) \propto \sigma T_a^4 - \sigma T_d^4$$

Typically, longwave atmospheric cooling is given by

(12.9) $\Delta T/\Delta t$ (°C/day) $= (g\sigma/c_p \Delta_p)((T_B^4 - T_T^4)_2 - (T_B^4 - T_T^4)_1)$

The symbols g, c_p, and Δ_p are, respectively, the gravitational field strength, the specific heat of air and the pressure increment between levels in the atmosphere. All other terms were previously defined. Subscripts 1 and 2 refer to lower and upper levels separated by the pressure increment Δ_p.

Figure 12.5 shows that the solution $\Delta T/\Delta t$ (°C day^{-1}), of Eq. (12.8), for the disk radiometer, derived from the curve F_n (net irradiance), indicates cooling from the surface to 420 mb. Longwave warming occurs through 20 mb. Solving Eq. (12.9) for the sounding results in cooling to 315 mb, warming to 190 mb, cooling to 140 mb, warming to 105 mb, and strong cooling, to the top of the ascent, from above the tropopause. The magnitude and sign of the radiometersonde-observed cooling generally disagree with that of the disk radiometer.

Figure 12.6 is a plot of the right-hand side of Eq. (12.8) versus the right- or left-hand side of Eq. (12.9), from the filtered irradiance estimates. Open circles indicate observations under cloudless-sky conditions: dotted circles represent cloudy-sky observations. From this and Fig. 12.5, we conclude that

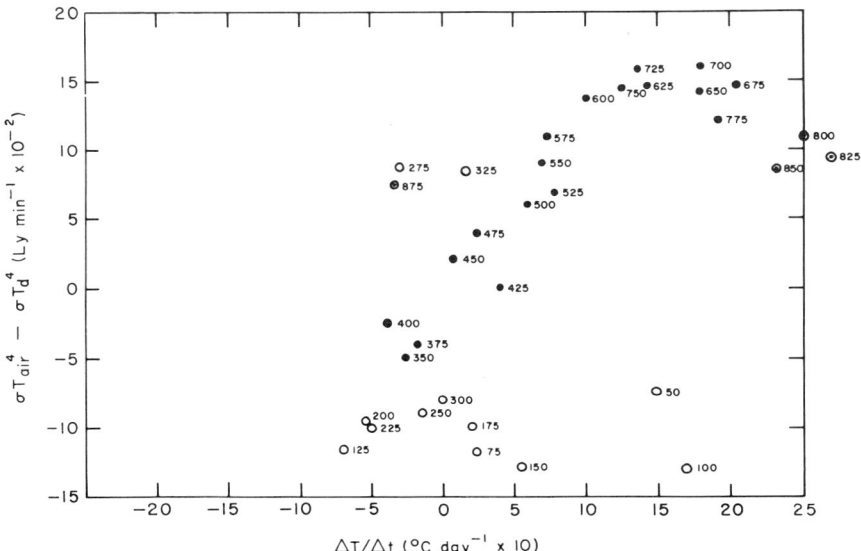

FIG. 12.6. Radiometer cooling versus $(\sigma T_a^4 - \sigma T_d^4)$.

in the atmosphere above Miami on April 1, 1964, the black disk did not correctly record the infrared cooling or warming. There is nothing approaching a linear correlation in evidence in Fig. 12.6 except for the region between 400 and 600 mb.

Figure 12.7 is a profile of infrared radiative cooling, characteristic of cooling found above a moist layer. It is discussed for this reason. The A scale gives the abscissa for the curve denoted by crosses. This is the radiometersonde cooling rate curve. The B scale gives the function $\sigma T_a^4 - \sigma T_d^4$, in relative units. In this ascent made at Madison, Wisconsin, on February 1, 1962, the top of the moist layer was close to 400 mb. It is evident, from the cooling curve obtained from the radiometer observations, that strong divergence of radiation occurred at the top of this moist layer, where the maximum cooling rate was 4.0°C day^{-1}. Since the total irradiance is given by the sum of the upward- and downward-directed irradiances, a curve of cooling from Eq. (12.8) is plotted in relative units. Clearly, there is little correlation between the two curves in the troposphere. In the stratosphere, however, the mean cooling rate of the two curves is approximately the same. The black-ball radiometric device only approximates a black-disk radiometer. The latter and the radiometersonde both measure total irradiance. From these experiments, we conclude that neither the black-disk nor the black-ball radiometers can measure atmospheric infrared cooling with the accuracy of the radiometersonde.

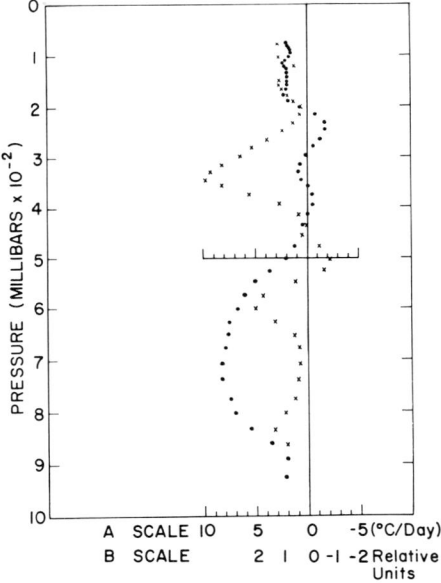

FIG. 12.7. Infrared cooling above a moist layer.

12.2.2. Rocket IR Instrumentation

Instrumentation for the measurement of infrared radiant emittance by Arcas, or similar rockets, can fill the atmospheric gap between the levels of balloon and satellite IR instrumentation. Thirty kilometers is the terminal altitude for most radiosonde balloon ascents, while 450 to 500 km is the average altitude of IR detectors on the Tiros and Nimbus satellite vehicles. Present low-cost, rocket-borne detectors are planned for operation from 90 downwards to 30 km.

The purpose of this experiment will be to determine the presence (or absence) of atmospheric IR radiators between 32 and 80 km (10 and 0.2 mb). If radiant power profiles indicate their presence, an attempt will be made to reduce the noise equivalent power of the radiometers to enable inference of the optical mass of such radiators. The speed of response of effective flat-plate radiometric collectors necessitates the use of the descent-pattern of a parachute-deployed rocket with an upward- and downward-"looking" radiometer.

12.2.3. Satellite IR Instrumentation Capability

Table I, from Butler and Moore (1966), summarizes the data on satellites that have carried and that can carry meteorological sensors.

TABLE I. Satellites capable of carrying meteorological sensors.

Vehicle	Size[a] (in.)	Weight (lb)	Power supply (W)	Payload capacity (lb)	Stabilization	Comments
Tiros	42 D 22.5 H	300	30	100	Spin-stablized + magnetic torquing, also wheel mode	Suitable for experimental use; wheel used in TOS
Nimbus	Sensor ring only 57 OD 40 ID 13 H	726	450	200	3-axis, gas jet, reaction wheel	Intended mainly for experiment test bed
OSO	37 H 92 D	450–550	400	250	3-axis, cold gas, 3 inertia wheels	Measure solar radiation
OGO	68 L 33 W 33 H	1000	500	150–220	3-axis, gas jet + flywheel	Intended primarily for experiment test bed
ATS	66 H 59 D	700	N.A.	200	Spin and gravity-gradient	Test bed for gravity-gradient and other experiments

[a] D—diameter, H—height, OD—outer diameter, ID—inner diameter, L—length, W—width.

Table II, also from Butler and Moore *op cit.*, summarizes the measurements required to "meteorologically" define the state of the atmosphere.

A brief review of the history of satellite weather-observation development is now in order. A quotation from Butler and Moore clearly charts that course:

> The concept of using satellites for weather-observation was discussed as early as 1951 by S. M. Greenfield and W. W. Kellog. In 1954, H. Wexler presented the idea of a satellite weather observatory equipped with infrared detectors, radars, and television cameras, and S. F. Singer specified the requirements for meteorological measurements from a minimum satellite vehicle. However, the first real attempt at image sensing from an earth satellite was made in the cloud-cover experiment on Vanguard 2.

The high-resolution infrared radiometer (HRIR), flown on Nimbus 1, was designed to scan the earth for radiant power in the 3.4–4.2 μm (2950–2400 cm^{-1}) region. It was a highly successful experiment, respresenting a breakthrough, in obtaining night-time earth cloud cover. The HRIR employs a cooled (197°K) lead selenide photoconductive detector and a precise bandpass filter operating in a relatively narrow atmospheric window. The HRIR can measure target temperatures ranging from 210 to 300°K with an accuracy of ± 1°K. The nominal surface resolution is 2–5 nautical miles. The time constant of the detector is 650 μsec. For the electronics, the bandwidth ranges from 0 to 280 Hz at a chopper frequency of 1500 Hz. The addition of an 0.5 μm passband filter in the optics of Nimbus 2 increased its response, during daylight, in the visible part of the spectrum.

We again quote from Butler and Moore in discussing the Nimbus follow-up programs:

> *Nimbus C Medium Resolution Infrared Radiometer (MRIR)*
> The MRIR experiment is similar in many ways to the five-channel radiometer flown on Tiros 2, 4, and 7. It will measure the intensity and distribution of emitted infrared and reflected solar radiation of the earth and atmosphere in five selected spectral intervals.
> The MRIR uses a scanning-mirror optical system, with detectors at the focal point, and mechanical lightbeam choppers. Each detector produces a signal in the spectral band to which it is sensitive. The instantaneous field-of-view is 2.85°, corresponding to about 30 mi. on the surface of the earth from the nominal orbit altitude of 600 nautical miles. It scans at 8 Hz. For global coverage, the data will be recorded continuously night and day for playback at the CDA stations.
> The five spectral intervals and their specfic functions are:

(1) Water-vapor band—6.5 to 7.0 μm—provides information on atmospheric structure and water-vapor distribution.

(2) Atmospheric window—10 to 11 μm—measures the temperature of the earth in a band where the atmosphere is essentially transparent. In addition, MRIR maps showing isolines of radiant emittance can be interpreted like cloud cover maps (such as those prepared from the 8–12 μm Tiros data) to provide low-resolution backup to the AVCS and HRIR observations.

(3) Stratospheric temperatures—14 to 16 μm—measures the emission from the CO_2 absorption band.

(4) Terrestrial radiation —7 to 30 μm—measures the earth's total longwave infrared emission.

(5) Albedo—0.2 to 4 μm—measures the energy reflected from the earth in the near-infrared region.

By using the sun-synchronous orbit and stabilized sensors, we should obtain much more information on the heat budget of the earth from MRIR channels four and five than we obtained with the Tiros radiometers.

12.3. Effects on Radiative Atmospheric Heat Budget during Artificially Induced Cirrus Metamorphoses

Walter Orr Roberts, in speaking of climate control to a November, 1965, International Symposium on Electromagnetic Sensing of the Earth from Satellites, stated:

> For about a week at least, this month, I am convinced that there was, over a substantial part of the United States from the Rocky Mountains east, a persistent high cirrus cloud cover. And I am also convinced from watching the sky over Boulder, and from observations during a couple of trips I had to take to the East during this time, that a major fraction, perhaps all, of this cirrus cloud found its immediate origin in airplane jet vapor trails. I have a couple of quite striking photographs showing these as they appear in Boulder. What degree of radiative influence did these cirrus clouds have on the thermal balance of the atmosphere? Were the temperature changes sufficient to have subsequent influences of major proportions on the weather? If the airplanes had not been flying through the region, would the cirrus clouds have formed anyway? These are questions to which we do not know the answers. But the cirrus blanket was probably a significant modifier of the heat sources and sinks of this period, and thus of the weather.

Figures 12.8 and 12.9 illustrate the before and after effects, respectively, visually, of artificially induced cirrus. Quantitatively, as our concluding

TABLE II. Satellite radiometry.

Quantity	Measurement required	Measurement technique	Specific sensors	Status
Cloud cover	Night and day coverage	Television, IR, UV	AVCS television	Flown on Nimbus; will fly on TOS
			APT television	Flown on Nimbus; will fly on TOS
			APT with tape recorder[a]	Under development
			Photomultiplier cloud camera[b]	Under development
			Dielectric-tape camera	Under study
			Image-dissector camera	Under study
			HRIR	Has flown on Nimbus
			Cameras on manned spacecraft	Have flown
Temperature	Complete vertical profile, surface to 100,000 ft	IR, microwave	Infrared spectrometer[c]	Under development; will fly on Nimbus
			Infrared interferometer[e]	Under development; will fly on Nimbus
Pressure density	Vertical profile	IR, microwave lasers	None	Not developed; Ground laser experiments in progress
Humidity	Vertical profile, integrated total in column	IR, microwave	None	Not developed

12. THERMAL RADIATION MEASUREMENTS

		In situ devices		
Winds	Speed, direction several layers, 750–10 mb		IRLS GHOST EOLE	Under development
Precipitation	Distribution rate	TV, passive and active microwave	Radar	(See cloud cover) Active, developed for ground use
Heat balance	Solar input, earth radiative output	IR, visible spectrum	Tiros radiometer TOS radiometer	Flown on Tiros (Developed) will fly on TOS
Ozone	Vertical profile integrated total in column	UV	UV spectrometer	Not developed
Snow and ice cover, sea state	Horizontal distribution, magnitude	TV, IR microwave	TV (See cloud cover)[d]	IR, microwave not developed
Sferics	Location, quantity, magnitude	rf	Sferics detector	Under development

[a] Will replace AVCS and APT.
[b] Will fly on ATS (also Nimbus satellite flight—see Chapter 13, this volume).
[c] Balloon flight has been made.
[d] Correlation with sun glitter has not been quantitatively established.

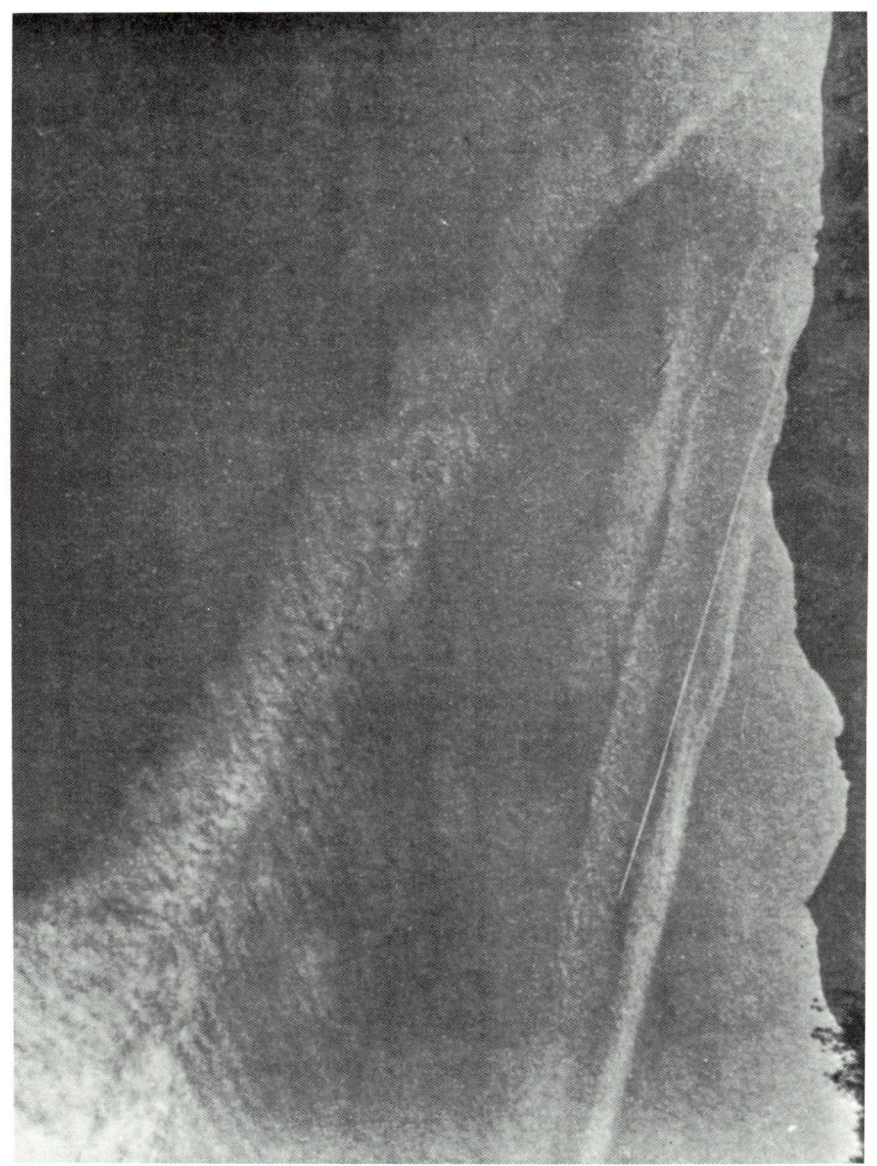

Fig 12.8. Cirrus cloud viewed before artificial induction.

12. THERMAL RADIATION MEASUREMENTS 349

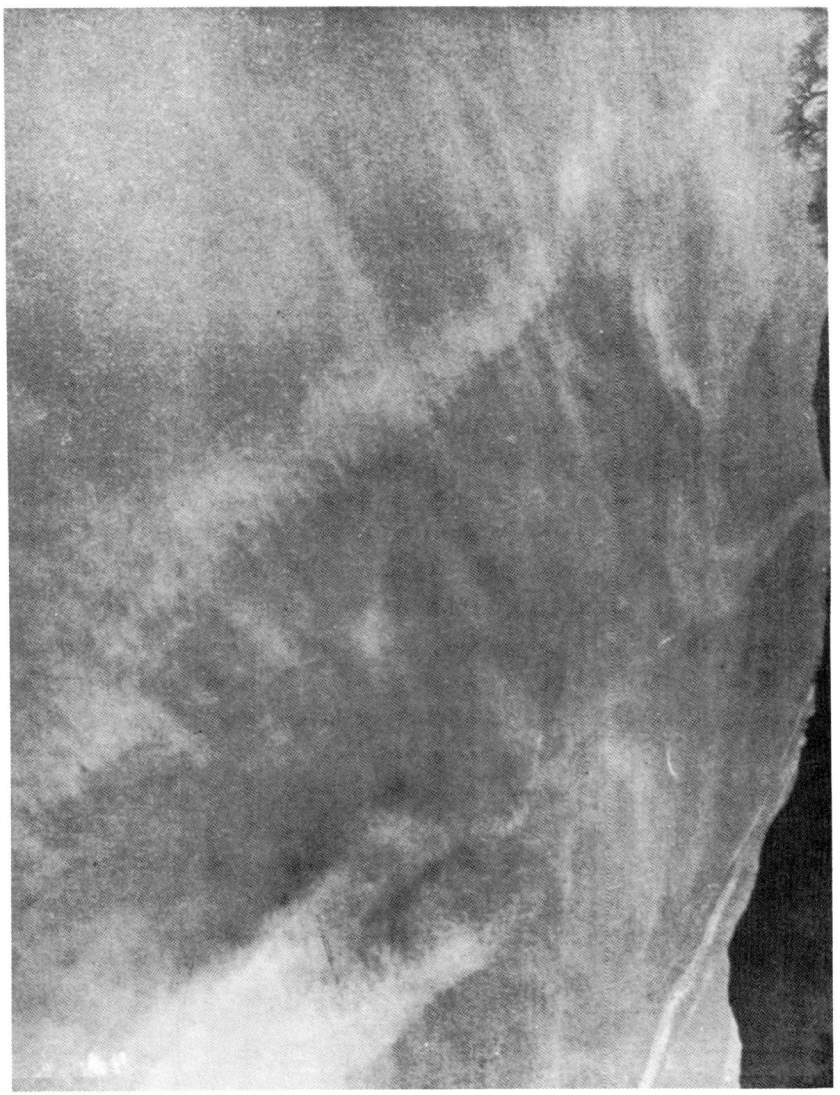

FIG. 12.9. Same as Fig. 12.8., after artificial induction.

illustrations, let us consider Fig. 12.10 and 12.11. The former presents the profile in the vertical of air temperature, and the upward, downward, and net irradiances before the onset of a cirrus cloud layer. It represents cloudless-sky conditions and resulting terrestrial irradiance.

Figure 12.11 illustrates the same parameters after interpositioning of a cirrus overcast between 20 and 23,000 ft. A 33% reduction in the upward irradiance at the top of the sounding is immediately evident. A reduction in the average longwave cooling rate for the column of 0.7°C day^{-1} is the direct result of the cirrus formation. The consequences on the mean temperature at

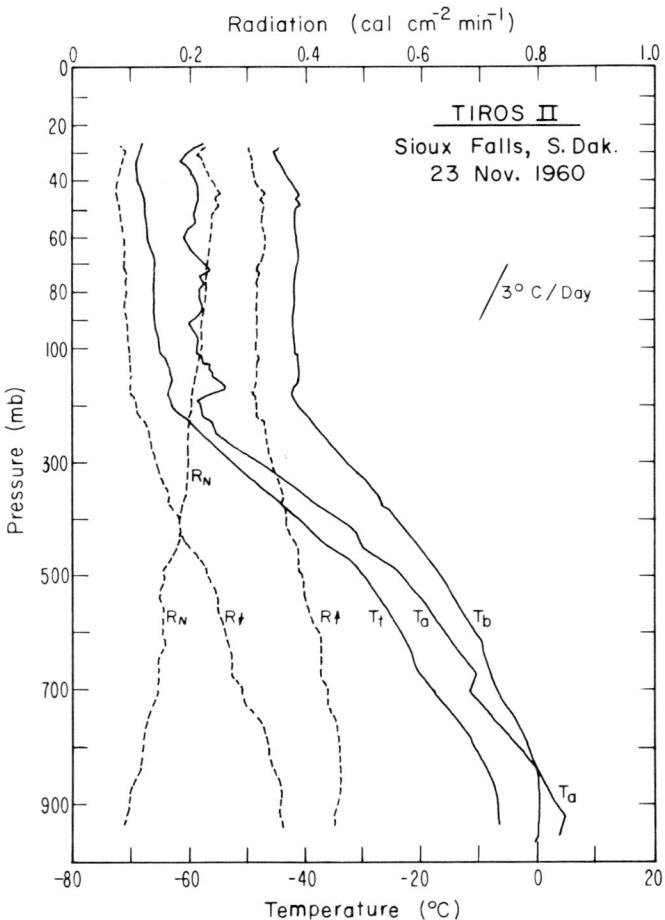

FIG. 12.10. Vertical profiles of air temperature and upward, downward, and net irradiance before the onset of a cirrus cloud layer.

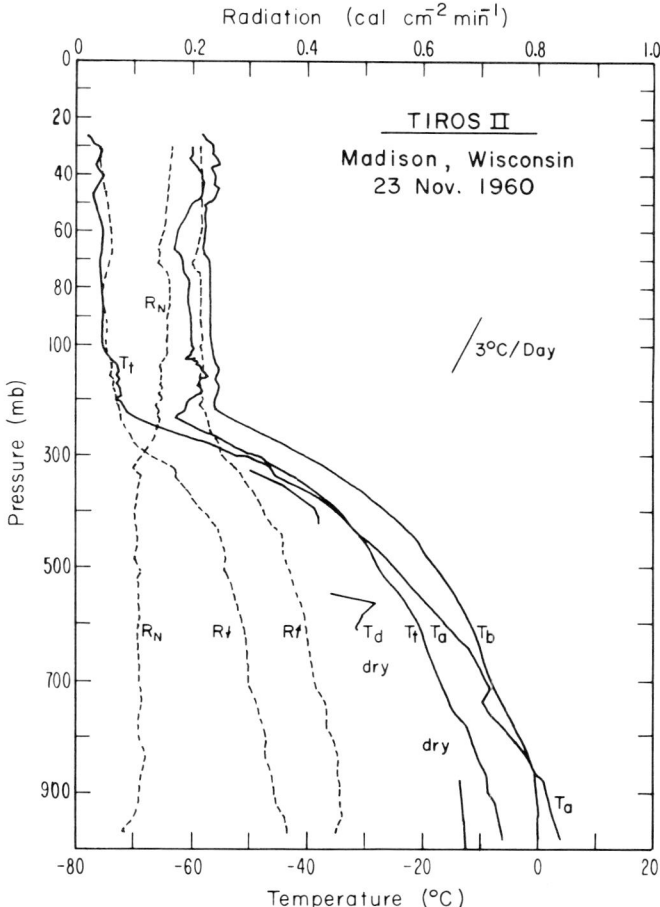

FIG. 12.11. Same as Fig. 12.10, after cirrus development.

the earth's surface, for an extensive geographical region such as one-half the U.S., is obvious. The implications of local effects of this nature, spread over such an area, on weather modification are also obvious. The requirement for low-resolution, modestly accurate IR survey measurements in the several applications we have discussed also seems evident.

We have discussed, in moderate detail, certain longwave experiments and their applications to atmospheric radiant power transfer processes and their ultimate relationship to weather. The instruments employed, because of mass observational and budget requirements, have been simple, for the most part. They include low-resolution, broadband detectors and low-cost, bandpass chopper bolometer radiometers. However, when such measurements are combined with transfer approximations, the results are encouraging and pilot the way to more precise but limited observations. Satellite radiometry, an enormous topic in itself, has been purposely excluded because of the voluminous literature on the subject in recent years.

12.4. Observed and Calculated Irradiances for the Real Atmosphere

Observed radiometersonde upward radiant emittances, in cloudless atmospheres, are shown to be consistently lower in magnitude above 400 mb than calculated radiant emittances for the same soundings. Conversely, downward-observed radiant emittances are larger in magnitude than calculated profiles for the same soundings above 400 mb. This difference is explained in terms of a contribution of aerosol to the emission and transmission in these assumed water-vapor and carbon-dioxide atmospheres.

For a few aircraft-verified cloudy atmospheres in the troposphere, by equating calculated irradiances to observed ones, assuming a correct hygrometer determined mass of H_2O and assuming CO_2 mass, a functional relationship between cloud effective emissivity, cloud type and sky condition was obtained. This enabled a full solution to be derived of the power transfer approximation equation, in terms of spectrally integrated flux emissivities and transmissivities from the "top" of the atmosphere to the surface through a cloud layer. The upward radiant flux H is expressed mathematically as

(12.10)
$$H = \int_0^{\epsilon_f(p, H_2O)} B(T(p)) \, d\epsilon_f(p, T, H_2O) + \int_0^{\epsilon_f(p, CO_2)} B(T(p)) \, d\epsilon_f(p, T, CO_2)$$
$$+ (1 - \epsilon_f(\text{cloud}))(1 - \epsilon_f(p_0, H_2O) - \epsilon_f(p_0, CO_2)) B(T(p_0))$$

where ϵ_f is an emissivity integrated over all wave numbers and all azimuthal and zenith angles. The blackbody emittance at temperature T and pressure p is given by $B(T(p))$; H_2O and CO_2 symbolize water vapor and carbon dioxide gas; p_0 is the surface pressure, and ϵ_f (cloud) is a cloud "blackness" (absorptance) parameter. The values of ϵ_f for water vapor were obtained *in situ* (Kuhn, 1963) and have been used by Zdunkowski *et al.* (1965). The values of ϵ_f for carbon dioxide are those of Howard *et al.* (1955).

Due to inaccuracies in determining a gross cloud depth, and until more verified samples of cloud depth are obtained, cloud emissivity ϵ_f^*(cloud) was determined as a function of cloud type and cover and is equivalent to 1.0 $- (\tau + \rho)$ where τ and ρ are, respectively, cloud transmittance and reflectance. This relationship is given below:

TABLE III. Cloud type and cover.

		ϵ_f^* (cloud)
High	overcast	0.45
Mid		0.65
Low		0.90
High	1/2 sky cover	0.20
Mid		0.30
Low		0.50

These results (from (Eq. (12.1)) appear more reliable than earlier determinations of cloud "blackness" (Kuhn, 1963). This is because the power transfer approximation equation is now solved for the entire atmosphere, through a cloud layer, and equated to the observed irradiance by determining the satisfying cloud emissivity in an iterative predictor corrector computer procedure. Needless to say, high-speed computers are essential to the solutions we have discussed.

12.5. Infrared Remote Sensing of the Atmosphere

12.5.1. Background

The radiative power transfer equation can be solved, iteratively, to infer atmospheric water vapor from remote radiant power measurements. The object here is to propose and describe this technique for deducing water-vapor quantities employing observations of radiant power as input to the radiative power transfer equation. With more sensitive bolometer radiometers, having selective bandpass filters, greater accuracy than is reported can be obtained. Readily available equipment was used in the pilot study.

The observations were made in two spectral regions, viz., 4.39–20.83 μm and 7.35–13.16 μm. Results of a similar method, utilizing balloons, were given by Kuhn (1966) and Kuhn and Cox (1967). The approach differs from those of satellite radiometric inferences of atmospheric water vapor and temperature (Houghton, 1961; Wark, 1961; Möller, 1962; and King, 1958). The latter observations are necessarily limited to observations from a fixed satellite orbit. They employ highly sensitive sensors receptive to radiant power in one or more different spectral intervals.

This new procedure requires measurements of temperature, height, and spectral irradiance at a number of aircraft holding levels. Two radiometers sensitive over two different spectral intervals consistute the sensor capability. One instrument monitors the air–surface interface temperature in the relatively transparent position of the atmospheric spectrum, while the other measures the spectral component of upward-directed terrestrial irradiance as a function of height in the broadband, earth-atmosphere, self-emission region. Ideally, the second radiometer would be sensitive over the strongly absorbing atmospheric water-vapor band centered at 6.7 μm. However, funding and ready availability have, to date, prevented use of such a special purpose bolometer radiometer.

12.5.2. Radiant Power Computations

Kuhn and McFadden (1967) reviewed and applied Kings' (1958) techniques for a simple inverting of the filtered radiative transfer integral. In the following, we summarize these procedures.

For a plane parallel atmosphere, containing no scatterers, in local thermodynamic equilibrium and consisting of gaseous water vapor and carbon dioxide, the upward radiant flux H through any reference level above the surface may be obtained by the following finite difference form of the radiative transfer equation,

$$(12.11) \quad H = -\sum_{j=1}^{m} \Delta \nu_j \sum_{i=1}^{n} B_i(\nu, T(p))\phi(\nu) \, \Delta\tau_i(\nu, U^*(p, H_2O), T)$$

$$- \sum_{j=1}^{m} \Delta \nu_j \sum_{i=1}^{n} B_i(\nu, T(p))\phi(\nu) \, \Delta\tau_i(\nu, U^*(p, CO_2), T)$$

$$+ \sum_{i=1}^{m} B_i(\nu, T(p_0))(\tau_i(\nu, U^*(p, H_2O), T)\tau_i(\nu, U^*(p, CO_2), T)$$

The effects of other radiatively active gases such as ozone and nitrous oxide in the wavelength regions considered are insignificant, amounting to less than 1 % of the upward irradiance, and are therefore neglected.

Equation (12.11) is employed in deducing the quantity of atmospheric water vapor from measurements of upward irradiance for a particular spectral interval. An iteration procedure, resulting in a direct solution of the water-vapor transmissivity $\tau_w(\nu, p, T)$, is used. Since the water-vapor transmissivity is a function of the quantity of water vapor w beneath a given reference level, and since the amount of water vapor is a function of the mixing ratio, the solution yields the mixing ratio.

The iteration procedure requires calculations of irradiance with a stepped series of trial values of water vapor [Eq. (12.11)] until the difference between calculated and observed upward irradiance is minimized. Carbon dioxide emission and transmission are calculated assuming a constant mixing ratio. A first approximation for the water-vapor profile is described by a power-law expression of the form

(12.12) $$W = W_0(p/p_0)^\lambda$$

after Smith (1966). λ, the exponent of a power-law expression, changes the watervapor profile as required in the iteration. For tropical soundings, W_0 is assumed to be 5.0 g/kg, certainly a lower limit. For midlatitude summer soundings, W_0 is assumed to 0.6 g/kg. For midlatitude winter soundings, W_0 is assumed to be 0.06 g/kg. In other words, low values are chosen to start the computations.

The iteration procedure, involving repeated solutions of Eq. (12.11), requires minimizing the quantity

$$\sum_i (H_o - H_c)_i^2$$

where H_o and H_c are the observed and calculated fluxex, respectively. Stepwise changes in λ, [Eqs. (12.11, 12.13)] for subsequent successive increases in H, provide the repeated input of the "trial" water-vapor quantities required for Eq. (12.10). The computer evaluates H from an initial "minimum" profile of W_0 shaped by an initial value of 0.5. The λ values are stepped upward to a maximum value of 3.0, and then the process repeats with a new stepped increase in W_0. Negative values of λ will allow the mixing ratio profile to increase with altitude above the surface. The changing of the entire water-vapor profile by the power function approximation of Eq. (12.11) imposes a stabilizing constraint on the solution of Eq. (12.10). This same procedure has been used by Kuhn and Cox (1967) to infer stratospheric water-vapor profiles. It should be noted that the constraint of Eq. (12.11) does not preclude convergence at every level, within reason, since the stepwise variation of λ allows, literally, almost any characteristic shape of the W-profile to obtain. Average computer solution time is 0.3 min (CDC-1604) for ten levels, over the spectral range 4.39–20.83 μm (2280–480 cm^{-1}).

A further application of the infrared technique for remote sensing inference of total atmospheric water-vapor content, from the surface to a prescribed level, without the requirements of actual aircraft sounding, was suggested by Suomi after King (1958). Briefly, this approach involves the sensing of the upward component of irradiance at a fixed level but at several different nadir angles. The primary measurements would be made in the water-vapor vibration-rotation band centered at 6.3 μm (1587 cm^{-1}) and in the window region, 10–12 μm. The atmosphere beneath flight level is assumed horizontally homogeneous. Thus, a choice of several nadir viewing angles would result in an equal number of "air masses," giving optical thickness of U^* and multiples of U^*. The irradiances sensed at the different angles are functions of an assumed lapse of temperature. The irradiances are also a function of these unknown optical air masses. Since one measures a difference in the irradiances directly, and since (although the absolute value is unknown) one does know the relationship between the magnitudes of the optical depths of the various atmospheric columns observed at the different nadir angles, a set of equations exists which enables a solution to be obtained of the absolute magnitude of the total optical depth. Such a technique, an adjunct to this study, offers interesting possibilities when only the total mass of water vapor in the atmospheric column is desired, as for example, in climatic surveys over inaccessible areas.

References

Businger, J. A., and Kuhn, P. M. (1960). On the observation of total and net atmospheric radiation. *J. Meteorol.* **17**, 400–405.

Butler, H. T., and Moore, H. S. (1966). The meteorological instrumentation of satellites. GSFC X-480-65-368, NASA-Goddard Space Flight Center, Greenbelt, Maryland.

Fleagle, R. G., and Businger, J. A. (1963). "An Introduction to Atmospheric Physics," 346 pp. Academic Press, New York.

Gergen, J. (1957). Atmospheric infrared radiation over Minneapolis to 30 millibars. *J. Meteorol.* **14**, 495–504.

Houghton, J. T. (1961). The meteorological significance of remote measurements of infrared emission from atmospheric carbon dioxide. *Quart. J. Roy. Meteorol. Soc.* **86**, 102–104.

Howard, J. N. (1960). "Handbook of Geophysics," U.S.A.F., Rev. Ed., 16–3. MacMillan, New York.

Howard, J. N., Burch, D. L., and Williams, D. (1955). Near-infrared transmission through synthetic atmospheres. *Geophys. Res. Papers (U.S.)*, 40.

Johnson, D. R. (1965). The role of terrestrial radiation in the generation of available potential energy. Ph.D. Thesis, University of Wisconsin, Madison, Wisconsin.

King, J. I. F. (1958). The radiative heat transfer of planet Earth. *In* "Scientific Uses of Earth Satellites," ed. J. van Allen. pp. 133–136, University of Michigan Press, Ann Arbor, Michigan.

Kuhn, P. M. (1963). Soundings of observed and computed infrared flux. *J. Geophys. Res.* **68**, 1415–1420.

Kuhn, P. M. (1966). Use of radiometers on balloons for moisture determination. *J. Spacecraft and Rockets* **3**, 754–756.

Kuhn, P. M., and Cox, S. K. (1967). Radiometric inference of stratospheric water vapor. *J. Appl. Meteorol.* **6**, 142–149.

Kuhn, P. M., and McFadden, J. D. (1967). Atmospheric water vapor profiles derived from remote sensing radiometer measurements. *Monthly Weather Rev.* **95**, 565–569.

Möller, F. (1962). Einige vorläufige Auswertungen der Strahlungsmessungen von Tiros II. *Arch. Meteorol. Geophys. Biokl.* **B12**, 78–94.

Möller, F. (1964). Optics of the lower atmosphere. *Appl. Opt.* **3**, 157–166.

Smith, W. L. (1966). Note on the relationship between total precipitable water and surface dew point. *J. Appl. Meteorol.* **5**, 726–729.

Suomi, V. E., and Kuhn, P. M. (1958). An economical net radiometer. *Tellus* **10**, 160–163.

Wark, D. Q. (1961). On indirect temperature soundings of the stratosphere from satellites. *J. Geophys. Res.* **66**, 77–82.

Zdunkowski, W., Henderson, D., and Hales, J. V. (1965). The influence of haze on infrared radiation measurements detected by space vehicles. Final Rept. U.S. Weather Bur. Contract CWB-10648, Washington, D.C.

— 13 —
RECENT ADVANCES IN SATELLITE RADIATION MEASUREMENTS

R. A. Hanel

13.1. Brief Summary of Satellite Radiometry

Many radiometric experiments have been performed from satellites and space probes. Mapping of the earth by television cameras and by infrared radiometers has allowed recognition of cloud patterns, identification of frontal systems, and tracking of major storms. Quantitative results from radiometers were used in heat balance studies, in stratospheric temperature and upper troposphere water-vapor analyses, and in many other applications.

The television pictures taken by Tiros and Nimbus are well known. An example of a Nimbus picture (Fig. 13.1) shows the northeast coast of the United States. Long Island, Cape Cod, and clouds toward the south may be recognized. Individual picture frames have often been assembled into montages such as the one shown in Fig. 13.2. The Mediterranean Sea contrasts well with clouds and the bright desert areas in northern Africa. Similar montages of the whole globe have become valuable aids to weather forecasters. An Automatic Picture Transmission system was specifically designed for the benefit of local users. The APT system broadcasts instantaneous cloud pictures directly to users within the antenna range of the satellite, and so minimizes the time between the actual picture taking and availability at the forecasters' work table. These APT pictures may be received with relatively inexpensive equipment (Stampfl and Stroud, 1964).

Recently, computer programs were developed (Bristor et al., 1966) to remove the distortion noticeable at the boundaries between individual frames of Fig. 13.2. Mercator and polar sterographic cloud maps may be generated. An example of these projections (Fig. 13.3), using pictures from the Tiros Operational Satellite system (ESSA), shows the periodic wave nature of low-pressure systems, centered around the poles.

During December 1966, a scanning radiometer designed by Suomi and Parent from the University of Wisconsin (e.g., McQuain, 1967) was launched

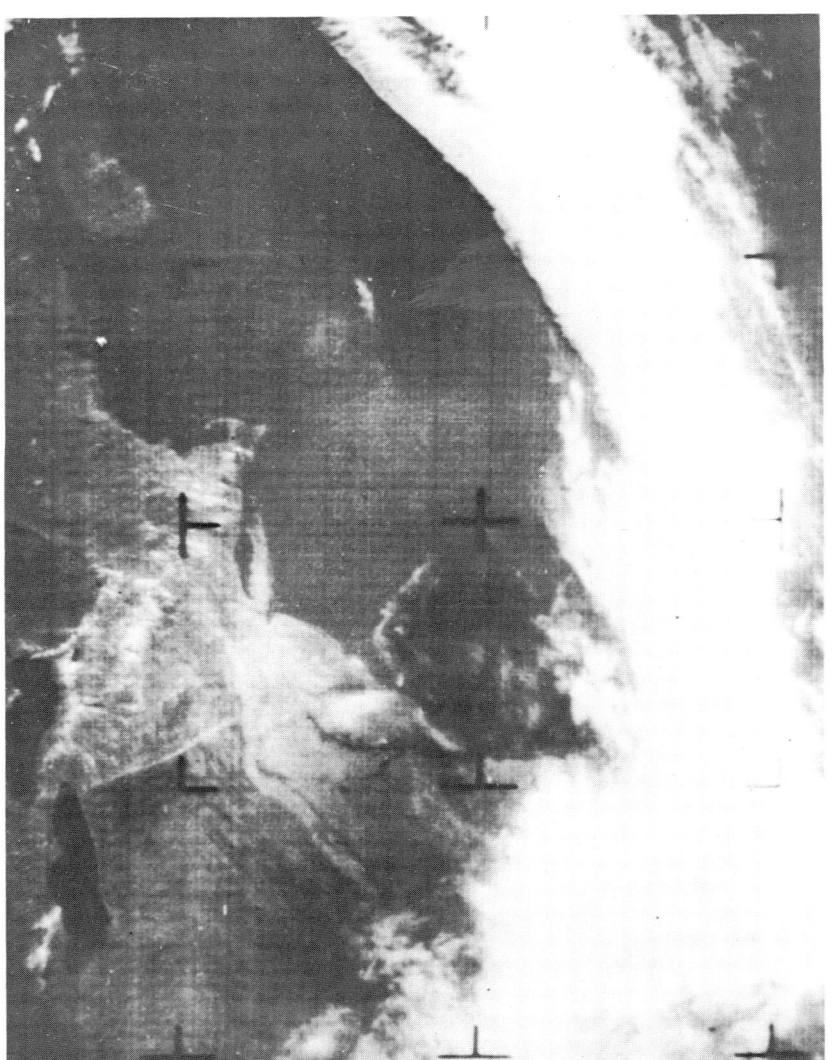

FIG. 13.1. Television picture from Nimbus II. Visible are the northeast coast of the U.S. including Nova Scotia and the Great Lakes. More to the south is Long Island and the coastal line to North Carolina. Heavy clouds shield Georgia and South Carolina.

FIG. 13.2. Mosaic of Nimbus II pictures of the Mediterranean Sea. Spain, Italy, Greece, and the North African coast may be observed.

on the Advanced Technology Satellite. This radiometer produced pictures from earth-synchronous altitudes. At this altitude (35,871 km) the orbital period of the satellite and the rotation rate of the earth are equal, and the spacecraft appears approximately stationary over a particular location on the equator. Specularly reflected sunlight may be noticed in Fig. 13.4, as well as cloud structure associated with intertropical convergence zone. Fujita, from the University of Chicago, arranged a sequence of these ATS pictures into a film strip which demonstrated, impressively, cloud motion in a time-lapse fashion. Within limitations, wind velocities may be inferred from the apparent motion of the clouds. The extensive interchange of air masses between the northern and southern hemisphere, which was recognized by this experiment, came as a surprise to many meteorologists. Most recently, a similar scanning radiometer on ATS 3 equipped with yellow, red, and blue filters obtained color maps of the earth. A U.S. Department of Defense satellite (Dodge) also obtained a color picture of the earth by three-color photographs (*Nat. Geogr. Mag.*, 1967).

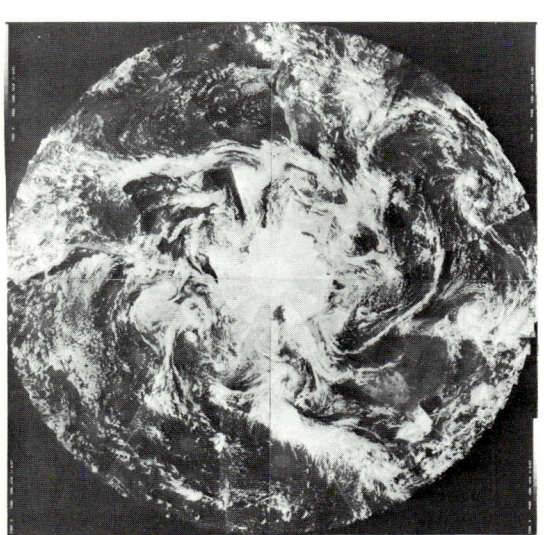
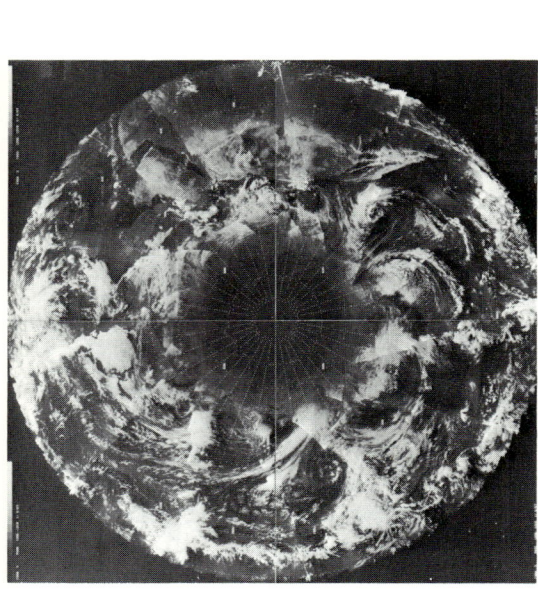
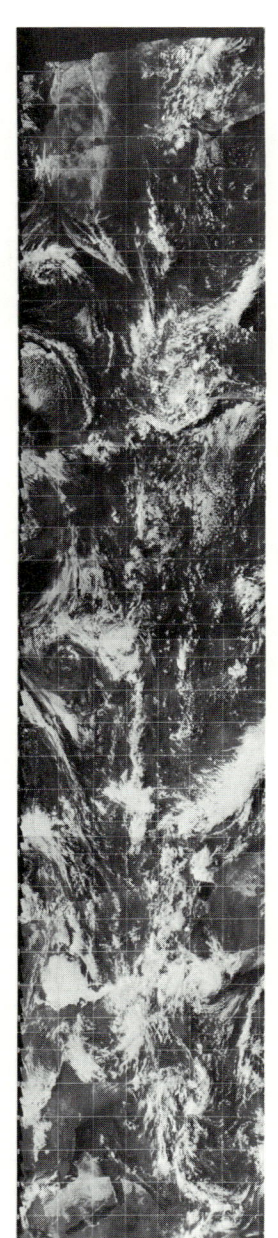

Fig. 13.3. Polar stereographic and Mercator maps derived from the Tiros Operational Satellite system (TOS) by computer program (Bristor et al., 1967) for January 6, 1967.

13. ADVANCES IN SATELLITE RADIATION MEASUREMENTS 363

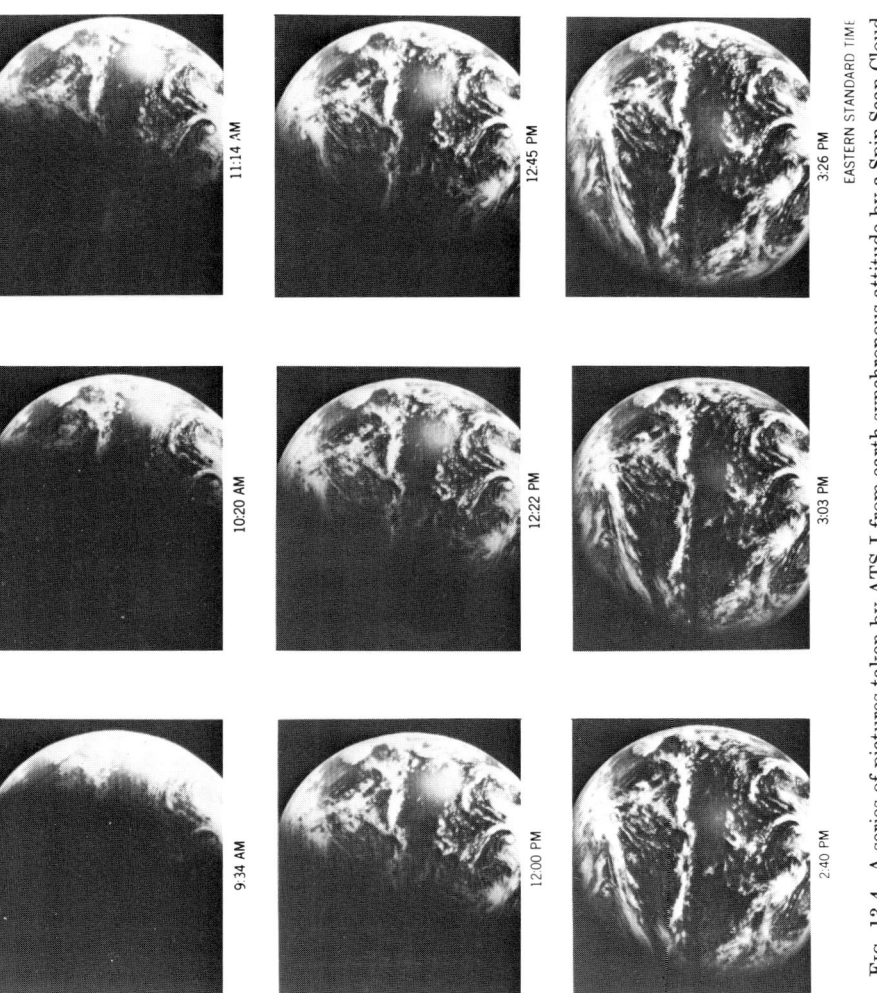

FIG. 13.4. A series of pictures taken by ATS-I from earth synchronous attitude by a Spin Scan Cloud Camera. The glint of reflected solar radiation may be observed.

Several five-channel, medium-resolution infrared radiometers (MRIR) were flown on Tiros and Nimbus, which measured reflected solar radiation and terrestrial infrared emission (Bandeen et al., 1961; Nordberg et al., 1966). Figure 13.5 shows pole-to-pole maps recorded in five wavelength regions by Nimbus II. The first channel (6.4–6.9 μm) is indicative of the amount of water vapor in the atmosphere. Bright areas represent clouds or large amounts of water at high altitudes. In the darker areas, radiation from the surface is attenuated to a lesser degree by water vapor and the dark areas represent, therefore, generally less humid regions or areas of high surface temperatures. The channel sensitive to 10–11 μm responds to radiation from

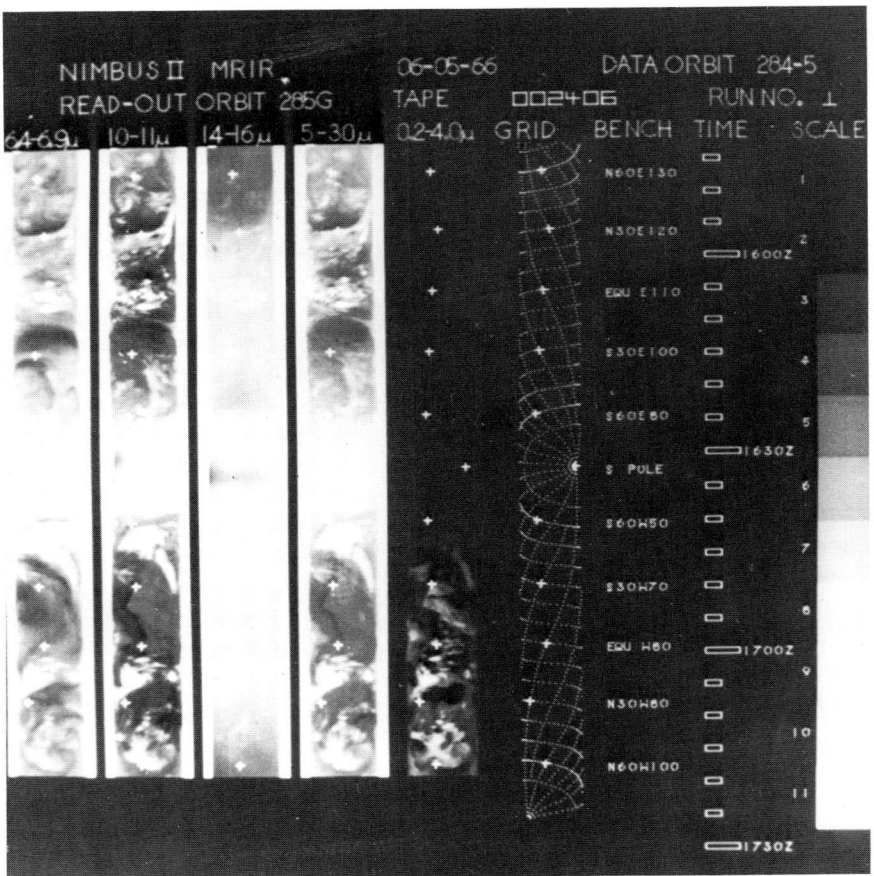

FIG. 13.5. Multichannel display from the Nimbus medium-resolution infrared radiometer. A grid map, the time code, and a gray scale are reproduced on the right side of the picture. (See text for further details.)

the surface or cloud tops. At this wavelength, the clear atmosphere is relatively transparent. In the third map (14–16 μm CO_2 band), the radiation originates mostly in the lower stratosphere. Only high altitude clouds may be recognized. The 5–30 μm channel responds to almost all of the thermally emitted radiation. Data from this and the last channel, which measures reflected solar radiation from 0.2 to 4 μm, may be used in heat balance studies. The difference between the absorbed solar energy and the thermal radiation emitted toward space is the net gain or loss of energy in a particular area. Equalization processes between the tropical zone, where the atmosphere and oceans gain energy, and the high latitudes where the losses dominate, provide the driving forces of the global circulation.

The pictorial presentations are primarily for illustrative purposes; for numerical analyses, the data are normally processed by computers. Analyses of the heat balance, moisture content of the upper troposphere, and other studies have been carried out by a number of investigators, recently, for example, by Raschke and Pasternak (1967).

Another experiment flown on Nimbus is the high-resolution infrared radiometer (HRIR) which is sensitive in the 3.5–4.2 μm atmospheric window. The experiment provided cloud information from the dark side of the globe similar to the 8–12 μm window channel of the MRIR, but with a higher spatial resolution of about 8 km. On the day side, reflected sunlight interferes with the measurement of thermal emission in the 4 μm window. The eastern U.S. is shown in Fig. 13.6, with hurricane Inez in the lower part of the picture. The equivalent brightness temperatures recorded by the HRIR instrument over hurricane Gladys are plotted in Fig. 13.7 (Raschke and Pasternak, 1967). The relatively warm eye of the hurricane and the spiral structure of the cloud pattern may be recognized.

But these infrared data contain more than just the locations of clouds. Since thick clouds have a high emissivity in the infrared, the measured brightness temperature is close to the true atmospheric temperature near the upper cloud boundary. A comparison of the equivalent blackbody temperatures, as measured in the atmospheric windows (8–12 μm or 3.5–4.1 μm), with climatological temperature profiles provides, therefore, a coarse estimate of the height of the upper limits of clouds. These cloud height estimates, the charts of stratospheric temperatures from the 14–16 μm channel (MRIR) (Bandeen et al., 1963; Nordberg et al., 1965) and derivations of the mean humidities in the upper troposphere (Raschke and Bandeen, 1967), from the water-vapor channel (MRIR), were the first attempts at deriving atmospheric parameters as a function of altitude. In general, however, radiometric mapping with relatively high-spatial, but low-spectral, resolution yields primarily two-dimensional patterns with rather limited information on the vertical distribution of temperature and water vapor. A full understanding of

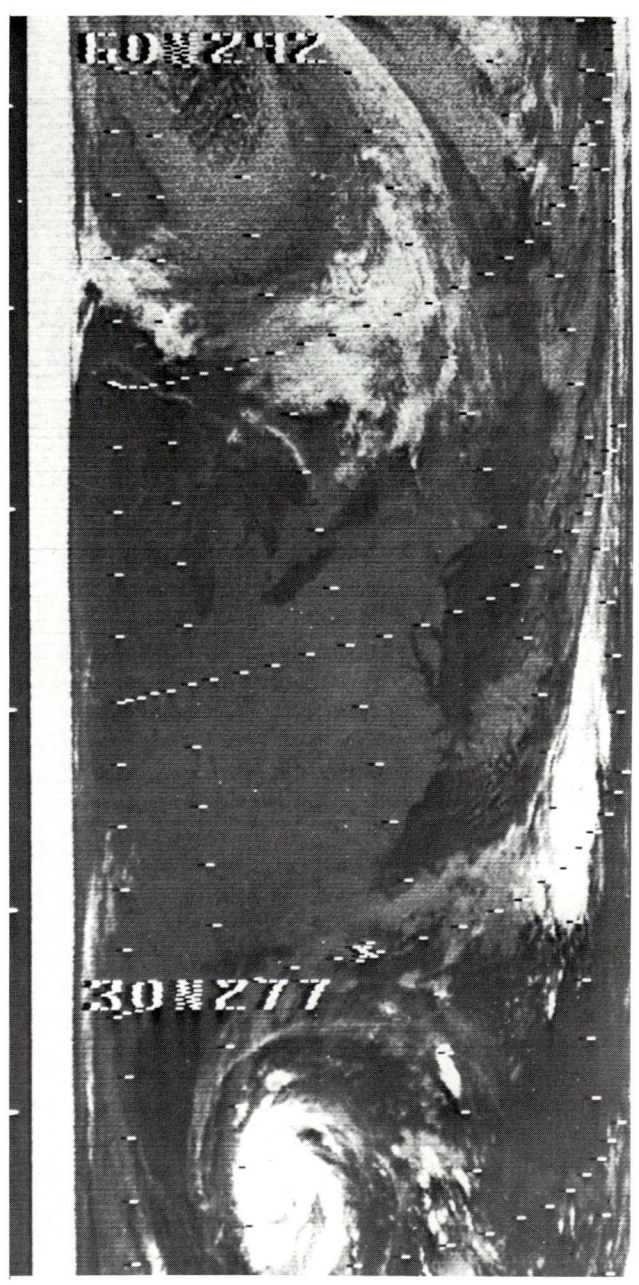

FIG. 13.6. Map generated by the high-resolution infrared radiometer (3.5–4.2 μm). The east coast of the U.S., Canada and the Gulf of Mexico are shown. Long Island, the Great Lakes, and the Chesapeake Bay area may be identified. Hurricane Inez is visible in the south. The latitude and longitude grid points are generated by computer.

13. ADVANCES IN SATELLITE RADIATION MEASUREMENTS 367

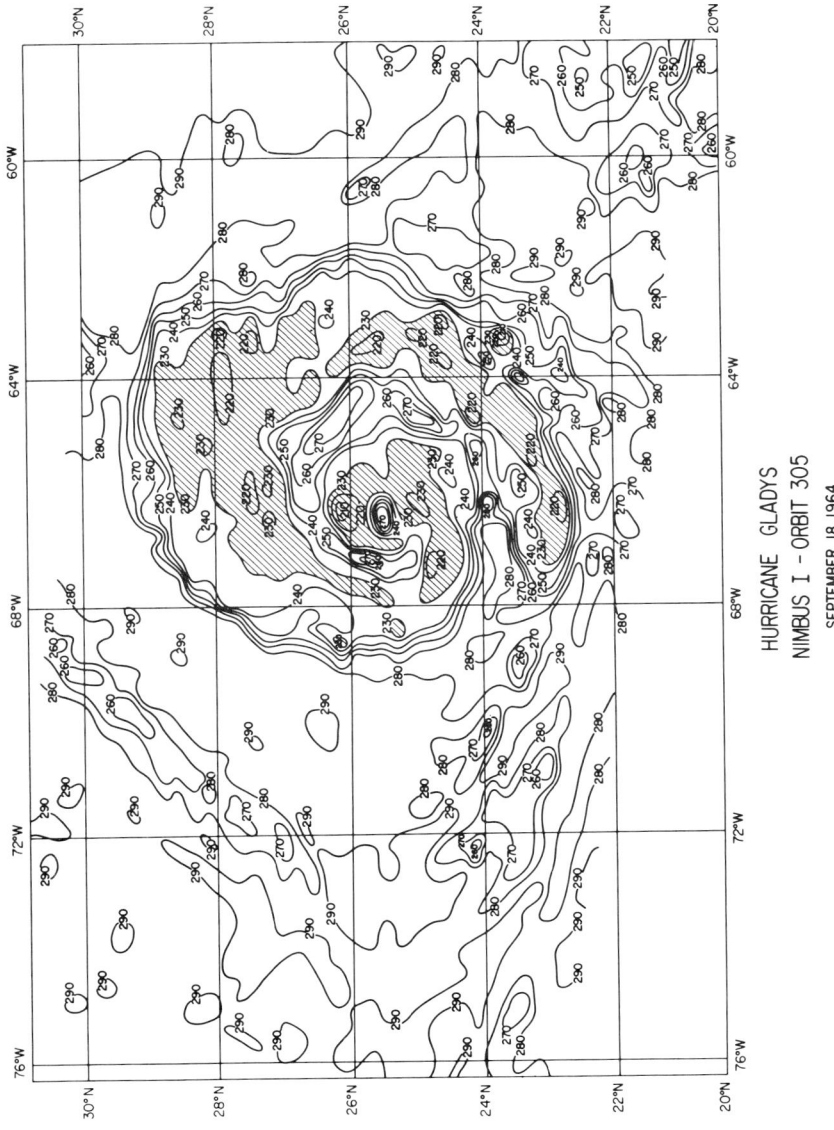

FIG. 13.7. Computer map of hurricane Gladys. Contours of equal brightness temperatures have been drawn. The eye of the hurricane and the spiral structure can be identified (Raschke and Pasternak, 1967).

meteorologically interesting phenomena can be obtained only by three-dimensional models. The most promising tool of modern meteorology is the study of the atmospheric behavior by solving, by high-speed computers, the equations which describe the conservation of mass, momentum, and energy. The numerical models used to predict the general circulations, also known as numerical forecasting (see, e.g., Thompson, 1961), require a knowledge of atmospheric conditions on a world-wide scale, in order that the equations which govern atmospheric processes may be integrated and the numerical results be compared to reality. The initial conditions include the horizontal and vertical distribution of temperature, winds, humidity, clouds, and, if possible, ozone and surface pressure also. To date, however, only the surface temperature, cloudiness, and, to a lesser degree, the mean stratospheric temperatures and upper tropospheric humidities have been sensed by remote techniques from satellites. The acquisition of the remaining quantities represents a real challenge to the development of future experiments.

13.2. Physical Principles of Vertical Sounding

The need for vertical soundings of the atmosphere has been recognized by many investigators. Before a discussion of the future satellite experiments which are concerned with this problem is presented, it is instructive to consider the principles of remote sensing and some relevant instrumental concepts first.

13.2.1. Temperature Profile

One major objective of the vertical sounding experiments is to measure atmospheric as well as surface temperatures. It was realized early that the thermal emission spectra, as well as the limb functions of stellar and planetary atmospheres, contain information on the temperature profile. King (1956) suggested the measurement of the limb function to determine atmospheric temperatures on earth, from a satellite, and Kaplan (1959) suggested a spectral scan in the 15 μm CO_2 band for the same purpose. King's method suffers somewhat from the lack of horizontal stratification in the earth's atmosphere. Kaplan's method has been explored extensively (Wark, 1961; Yamamoto, 1961; King, 1963; Wark and Fleming, 1966).

At wavelengths longer than about 4 μm, where reflected solar radiation becomes negligible compared to thermal emission, the outgoing specific intensity $I_{\nu\mu}$, which may be measured by means of satellites, consists primarily of three components: radiation emitted by the surface and attenuated by the atmosphere; the downward flux F_ν^- diffusely reflected by the surface; and radiation from the atmospheric gases themselves:

$$(13.1) \quad I_{\nu\mu} = \varepsilon_\nu B_\nu(T_s)\tau_s + (1-\varepsilon_\nu)\pi F_\nu^- \tau_s + \int_0^{\tau_s} B_\nu(T_{(p)}) \, d\tau_{\nu\mu}$$

The temperature is T and the subscript s refers to the surface. The wavenumber ν (cm^{-1}) is the reciprocal value of the wavelength λ (cm), μ is the cosine of the zenith angle, and τ stands for the transmissivity. The surface emissivity is ε_ν and B_ν is the Planck function (blackbody radiance). Equation (13.1) is a solution of the radiative transfer equation for an atmosphere in local thermodynamic equilibrium bounded by a lower surface. If the surface emissivity deviates considerably from unity, then the term with the downward flux reflected by the surface must be considered also. Fortunately, on earth, this is rarely necessary, except over deserts where the "reststrahlen" effect in certain minerals may lower the effective emissivity. In this case, a more elaborate analysis has to be adopted.

For an atmospheric constituent of known and uniform mixing ratio, such as CO_2 and O_2, the transmissivity is a known function of the pressure p. In general, τ is a function of wavenumber, zenith angle, and the pressure- and temperature-corrected optical path. In the region of the atmospheric windows, the total transmissivity to the surface is high and the intensity is primarily governed by the surface temperature and emissivity; hence, the first term in Eq. (13.1) dominates. For more opaque intervals, τ is small and the first two terms in the above equation contribute little. In all cases, if the black surface is considered an isothermal extension of the atmosphere, Eq. (13.1) simplifies to

$$(13.2) \qquad I_{\nu\mu} = \int_0^\infty B_\nu(T(\log p))(\partial \tau / \partial \log p)\, d \log p$$

where the integration may now be carried out from zero to infinity. Equations (13.1) and (13.2) apply strictly to monochromatic radiation. In applications, a finite interval $\Delta\nu$ must be considered since instruments are limited in their spectral resolution. Fortunately, the form of Eq. (13.2) is still valid, but the weighted average transmissivity $\bar{\tau}_\nu$, within the element of spectral resolution $\Delta\nu$, must be taken instead of the monochromatic value of the transmissivity. Sometimes, it is convenient to introduce the logarithm of the pressure as the independent variable.

In a physical sense, Eq. (13.2) may be understood with the aid of the weighting function $\partial \tau / \partial \log p$ shown for several five-wavenumber intervals in Fig. 13.8 (Kunde, 1966). In each spectral interval, different atmospheric layers contribute different amounts to the outgoing radiation. The maximum contribution comes from the region where the optical depth is unity. Higher layers are too transparent to emit effectively and layers much below are shielded so that their radiation does not reach the upper atmospheric boundary.

The problem of finding the temperature profile requires the inversion of Eq. (13.2). From the known weighting function and the measured intensity

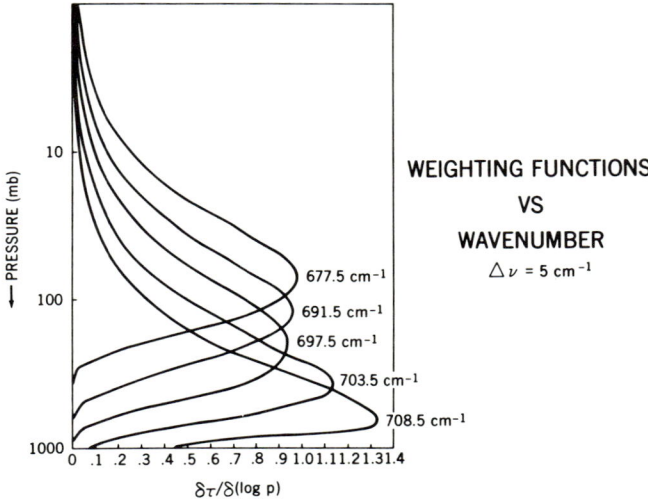

Fig. 13.8. Weighting functions used in the computation of the vertical temperature profile. Weighting functions were calculated for five wavenumber intervals [after Kunde (1966)].

I_ν, $B(T(\log p))$ has to be found. It is evident from the form of Eq. (13.2) that a scan in frequency, where ν is the variable and $\mu = 1$ (Kaplan, 1959), and a scan in viewing angle, where ν is constant and μ the variable (King, 1956), are formally identical problems.

Several inversion methods have been discussed in the literature (Kaplan, 1961; Yamamoto, 1961; Wark, 1961; King, 1964; Twomey, 1966; Wark and Fleming, 1966). In the inversion method of Wark and his co-workers, the Planck function is first linearized in ν and then expanded into an orthogonal series in the independent variable, generally the logarithm of pressure. The choice of the orthogonal functions is important. The series must converge rapidly since only a few coefficients can be obtained, as explained below. For each spectral interval for which an intensity measurement is secured, one obtains one equation. This set of linear equations is solved for the unknown coefficients of the expansion of $B(T(\log p))$. Since B is a unique function of T, T is obtained as a function of $\log p$. Clearly, the number of coefficients cannot exceed the number of independent equations, which depends on the shape of the weighting functions as well as on the precision of the measurement. The weighting functions shown in Fig. 13.8 apply to spectral intervals five wavenumbers wide. Slightly narrower weighting functions may be obtained by an increase in spectral resolution. Unfortunately, even with infinite spectral resolution, the width is still of the order of one pressure scale height (8 km).

Small errors in measured intensities may result in oscillatory solutions of large amplitude; therefore, in a practical application of this technique, other criteria must often be evoked to obtain physically meaningful results. Sufficient smoothing must be applied to avoid instability, but an excessive amount would destroy the details inherent in the measurements. To secure the best temperature profiles, *a priori* information on the temperatures, for example, from measurements on the previous days or from climatological expectation, should be used. The latter approach was followed by Alishouse *et al.* (1967—see also Obukhov, 1960) by developing vertical temperature soundings obtained from conventional radiosondes into a set of numerical orthogonal functions. Satellite data are to be used to determine the coefficients in this rapidly convergent orthogonal series.

The 15 μm CO_2 band is not the only spectral region which may be examined for the purpose of temperature sounding. The strong CO_2 band near 4.3 μm (McClatchey, 1965) and the oxygen lines grouped around 0.5 cm (Meeks and Lilly, 1963) offer alternate choices. As a consequence of the properties of the Planck function, the specific intensity expressed in relative terms varies more rapidly with temperature at 4 μm than at 15 μm, but this requires a proportionally larger dynamic range in the instrument. Also the total energy at 4 μm is only a small fraction of the energy available at 15 μm. Furthermore, reflected solar radiation interferes on the day side with the temperature deduction in the lower troposphere, and the deviation from blackness of some natural surfaces is more pronounced at 4 μm than it is at longer wavelengths (Hovis, 1966). Detector sensitivity and cooling requirements also enter into discussions on the preference of bands.

The microwave approach promises the possibility of extending the temperature measurements below clouds, a region inaccessible to infrared soundings, although the effect of large droplets and ice crystals on the outgoing intensity at 0.5 cm is not negligible. Difficulties also arise from the low emissivity of water surfaces, which depends also on the view angle and roughness of the sea surface.

13.2.2. *Water Vapor and Ozone Distribution*

Suppose the analysis of the specific intensities in an absorption band of CO_2 and in an atmospheric window has been performed, and the atmospheric temperatures are known down to the surface. Equation (13.2) may be integrated by parts and yields

$$(13.3) \qquad I_\nu = B_\nu(\tau = 1) + \int_0^{X_s} \frac{\partial B_\nu(T_{(x)})}{\partial x} \tau_\nu(u_{(x)}) \, dx$$

where $u_{(x)}$ is the mass of the optical gas between level x and the top of the atmosphere. Aside from $B(\tau=1)$, Eq. (13.3) is formally similar to Eq. (13.2).

Numerical methods similar to those used in the temperature problem can now be applied to a spectral region where water vapor or ozone are the major atmospheric absorbers, and their vertical distribution may then be obtained (Yamamoto, 1966; Conrath, 1967).

The ozone analysis will have to be limited to the region below about 30 km. To investigate the O_3 distribution above this level, by this technique, a much higher spectral resolution than 200 (5 cm^{-1} at 1000 cm^{-1}) would be required.

The distribution of water vapor may be derived by studying the 6.3 μm band or, even better, the rotational bands which show sufficient strength beyond 20 μm. Weaker lines in the atmospheric window are of interest too. Water vapor may also be studied in the microwave region at 1.35 cm. The ozone distribution may be inferred from measurements near 9.6 μm or in the ultraviolet part of the spectrum (Dave and Heath, 1965). Absorption coefficients for water vapor and ozone are available [for a summary see, e.g., Goody (1964)], but the accuracy and reliability of tabulated values are rather limited and better experimental data as well as theoretical work are needed.

13.2.3. Minor Atmospheric Constituents

Minor constituents such as methane (CH_4) or nitrous oxide (N_2O) also show absorption bands in the thermal emission spectrum. Weak absorbers may be identified by statistical techniques, which use the cross-correlation functions generated for the spectrum measured from space and the spectrum calculated from numerical models of the atmosphere which take the particular constituent and other gases as well into account. The detection and tracing of some air pollutants may be feasible by this technique. Needless to say, the detection of minor constituents and the temperature soundings may be applied to other planetary atmospheres as well.

The theory of the temperature sounding from satellites has been derived for absorbing atmospheres only. Scattering processes are usually avoided in calculations for reasons of simplicity. Atmospheric particles, such as aerosols or cirrus clouds, may cause difficulties not currently fully understood in the derivations of temperature and humidity profiles.

In the next section, several future experiments to be flown on Nimbus will be discussed. The main attention will be given to those experiments which address themselves to vertical sounding, although other experiments of interest will be mentioned.

13.3. Vertical Sounding Experiments

The physical principles of instruments will be examined first, then the signal-to-noise ratio and, finally, the calibration for a particular instrument will be analyzed.

A grating spectrometer, SIRS (Wark and Fleming, 1966; Hilleary et al., 1966) and an interferometer, IRIS (Hanel and Chaney, 1965) are undergoing final preparations for Nimbus B. Improved versions of these experiments and a radiometer, designed by Houghton and Smith from England, are being prepared for Nimbus D.

The SIRS experiment is an Ebert–Fastie grating instrument with a common entrance slit, a fixed grating, and several immersed bolometers (one for each wavelength measured) which serve as exit slits. Seven channels measure the spectral intensity within strategically located parts of the 15 μm CO_2 band, with a spectral resolution equivalent to 5 cm^{-1}, while another channel is sensitive in the atmospheric window near 11 μm. The latter is used to infer surface temperatures and to assess cloudiness. The incoming radiation is mechanically chopped, so that the detectors see an alternating signal proportional to the difference in radiance between earth and stellar background, which is considered as a near-zero reference. Occasionally, the instrument is exposed to a calibration blackbody. The alternating signal at the detectors is amplified, synchronously rectified, and presented to the spacecraft telemetry system (Dreyfus and Hilleary, 1962; Hilleary et al., 1966). SIRS has had several successful balloon flights which showed the feasibility of this approach (Hilleary et al., 1965).

The second instrument to be flown on Nimbus B for the purpose of vertical sounding is a Michelson interferometer, IRIS. This instrument measures the emission spectrum between 500 and 2000 wavenumbers (20–5 μm) with a spectral resolution of 5 cm^{-1}. For Nimbus D, a spectral range from 200 to 1600 cm^{-1} (50–6.3 μm) and a somewhat higher resolution are contemplated.

Since the use of a Michelson interferometer as a spectrometer is not widespread, a brief explanation of the underlying principles may be helpful [see, e.g., Michelson (1927); Connes (1961); Mertz (1965)].

The essential part of the interferometer is the beamsplitter which divides the incoming radiation into two approximately equal components (Fig. 13.9). After reflection at the fixed and moving mirrors, respectively, the two beams interfere with each other with a phase proportional to the optical path difference between both beams. The recombined components are then focused onto the detector where the intensity is recorded as a function of the path difference, δ. Whenever the recombining components are in phase, the interferometer is essentially transparent for the particular wavelength. Whenever the path difference is such that the recombining beams are out of phase, the interferometer is opaque for that wavelength and, moreover, if the beamsplitter is made of a dielectric, nonabsorbing material, the interferometer acts as a reflector. The interferometer may be considered to be a chopper or, better, a modulator in which the intensity at each wavenumber is modulated at a different frequency. The amplitudes registered at the

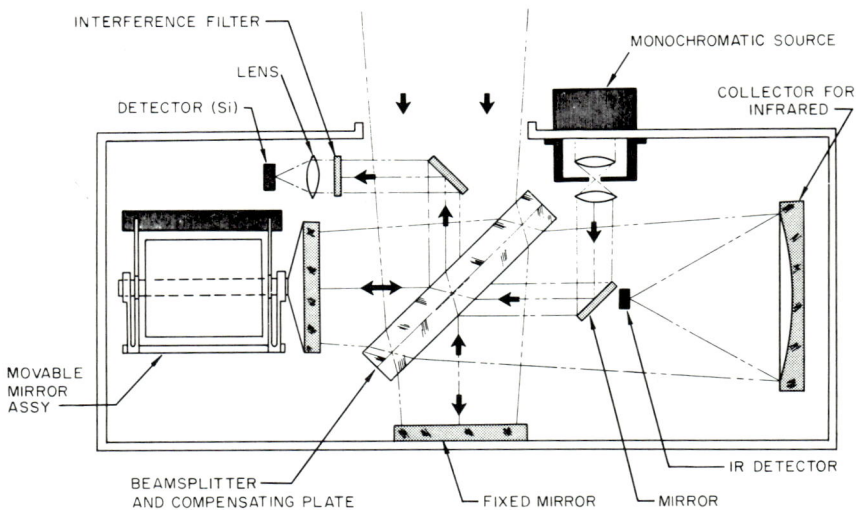

Fig. 13.9. Schematic of Michelson interferometer. The monochromatic source is a neon-discharge tube filtered to emit 5852 Å energy.

detector are proportional to the intensity and the frequencies to the wavenumber-mirror velocity product. For monochromatic radiation, a sinusoidal signal appears at the detector, while for a continuous spectrum, the superposition of many amplitudes of various frequencies takes place. This combined signal is called the interferogram. Neglecting constant terms, it may be expressed

$$(13.4) \qquad i(\delta) = \int_{\nu_1}^{\nu_2} I_\nu \cos(2\pi\nu\delta)\, d\nu$$

After amplification and quantization by an analog to digital converter, the signal is transmitted to the ground where the spectrum is reconstructed by applying the inverse transformation

$$(13.5) \qquad I_\nu' = \int_{-\delta_{max}}^{+\delta_{max}} i(\delta) \cos(2\pi\nu\delta)\, d\delta$$

The original intensity I_ν can be recovered only within certain limits of accuracy and resolution. One limitation is the small range over which $i(\delta)$ can be recorded. Others are caused by the finite solid angle of acceptance and the noise properties of the detector. Proper sampling intervals of the interferogram are derived by a fringe control system from the quasi-monochromatic 5852 Å line of a neon-discharge bulb. A breadboard model was jointly designed by personnel of the University of Michigan and the Goddard Space Flight

Center. This instrument has been flown on a balloon by the University of Michigan. Results were reported by Chaney et al. (1967) and Hanel and Chaney (1966).

A typical emission spectrum of a clear atmosphere, taken by the breadboard of the IRIS instrument, is shown in Fig. 13.10. The specific intensity was measured from an altitude of approximately 33.5 km. The strong bands of CO_2, O_3, and H_2O are clearly visible, besides many weaker features of the same gases and also of CH_4. The nearly constant brightness temperature observed in the window corresponds to the surface temperature. This spectrum was used by several investigators to derive atmospheric temperatures (Chaney et al., 1967; Conrath, 1967), and humidities (Conrath, 1967). Examples are given in Figs. 13.11 and 13.12, respectively. Radiosonde data from approximately the same area and time permit a comparison between actual and inferred conditions.

The third instrument, being scheduled for Nimbus D, which is concerned with vertical temperatures sounding, is a multichannel radiometer developed by Houghton and Smith at Oxford and Reading Universities, respectively. The instrument uses narrow-band interference filters to isolate certain regions of the 15 μm CO_2 spectrum. The energy is then passed through an absorption cell containing CO_2 gas which removes the radiation in the line centers. This considerably reduces the range of absorption coefficient associated with the

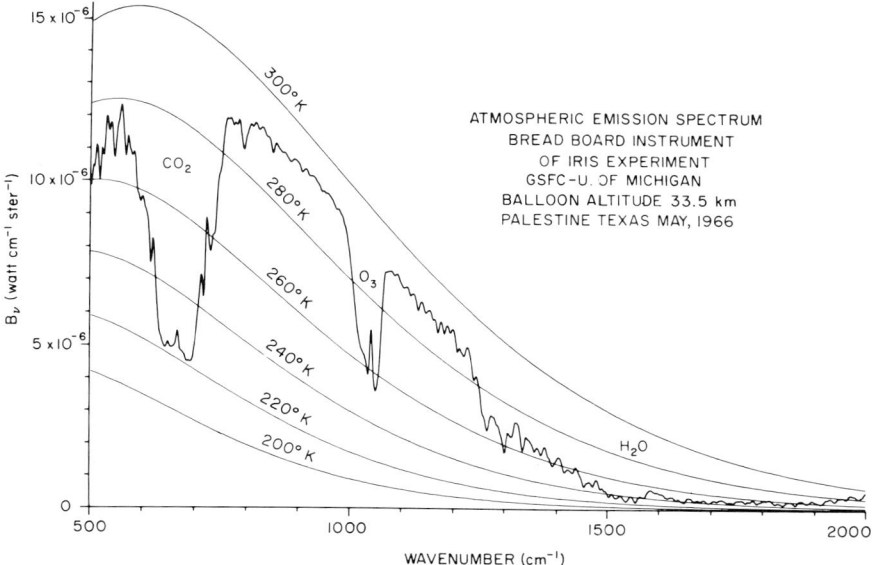

FIG. 13.10. Atmospheric emission spectrum obtained from a balloon-borne IRIS breadboard instrument.

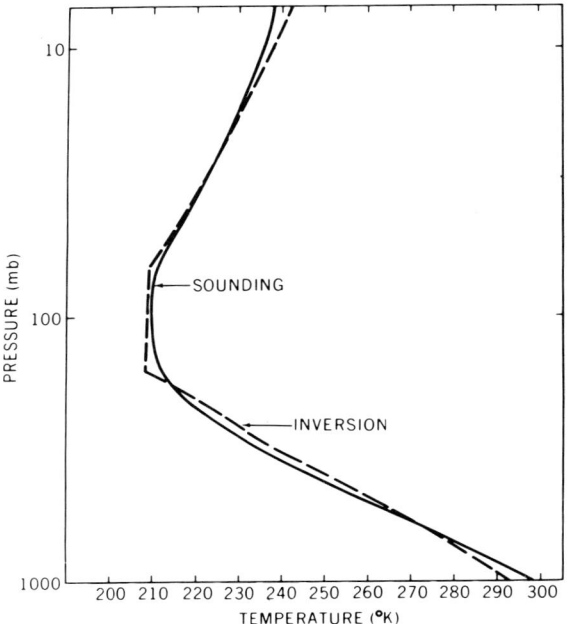

Fig. 13.11. Comparison between vertical sounding computed from the spectrum shown in Fig. 13.10 and radiosonde data [after Conrath (1967)].

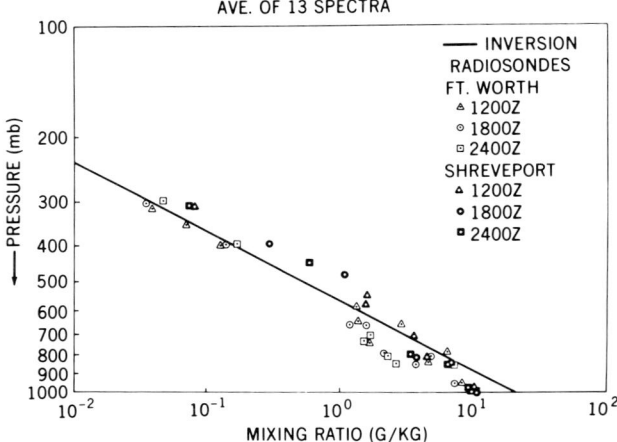

Fig. 13.12. Comparison of computed water-vapor mixing ratio and radiosonde data taken in the vicinity of the launch site [after Conrath (1967)].

atmospheric emission being measured in a particular interval, and results in a weighting function of shape and width comparable with the monochromatic case. In two channels concerned with the stratospheric emission, a double CO_2 cell is used, and an optical differencing process is performed by a vibrating reed chopper. The resulting signal at the detector now corresponds to the strong line centers themselves and is associated with a weighting function at a greater height than can be obtained directly without very high spectral resolution. By suitable choice of CO_2 amounts and spectral intervals, a set of optimum weighting functions has been obtained to cover the range from ground level to almost 60 km; thus, the instrument combines the advantage of energy grasp associated with a bandwidth of several wavenumbers with the effective resolution of the order of 0.1 cm^{-1}.

13.4. Intercomparison of the Instruments

As will be apparent from the above discussion, several devices have been considered and a greater number of proposals have been brought forward to measure the specific intensity at several wavelength intervals near 15 μm, and in the atmospheric window. For instructive purposes, the suitability of such instruments to carry out these measurements will be analyzed by studying the obtainable signal-to-noise ratios. These hypothetical instruments do not correspond to the discussed experiments but should be taken only as representatives of their classes. The discussion on the signal-to-noise ratio also provides an opportunity to demonstrate some of the physical principles of radiometry in the infrared.

In general, all radiometric devices, may they be radiometers, spectrometers, or interferometers, have a number of common elements. They all have an entrance aperture A, which receives radiation within a certain solid angle Ω, and optical components which then channel the radiant energy onto a detector. A comparison of different approaches shall be based on the signal-to-noise ratio, within a narrow wavenumber interval $\Delta \nu$, which can be achieved by each method under otherwise identical conditions.

The radiant signal energy falling on the detector within the spectral interval $\Delta \nu$ is

(13.6) $$P_{\Delta \nu} = I_{\Delta \nu} A_d \Omega_d$$

$I_{\Delta \nu}$ is the intensity at the detector, A_d is the effective detector area, and Ω_d the solid angle subtended by the source at the detector. On the other hand, the equivalent noise power at the detector is (e.g., Smith et al., 1968)

(13.7) $$P_N = (A_d \, \Delta f)^{1/2} / D^*$$

The frequency interval Δf is the effective bandwidth of the system and D^* (cm Hz$^{1/2}$ W^{-1}) the detectivity as defined for certain infrared detectors. For unimmersed thermistor bolometers, D^* is about 2×10^8; for mercury or copper-doped germanium detectors, D^* is about a hundred times greater, or 2×10^{10}. The above analysis tacitly assumes that the instrument is detector noise limited which is normally true in the infrared. In practice, a small allowance must also be made for noise generated in the preamplifier and, possibly, in the detector bias circuit. The signal-to-noise ratio is the ratio of signal power to noise power

(13.8) $$S/N = \eta_1(P_{\Delta \nu}/P_N)$$

The factor η_1 is an efficiency factor. For example, η_1 is about $1/2\sqrt{2}$ for a conventional chopper system, where it accounts for the conversion from peak-to-peak to rms values. If the radiation reflected from the chopper blade or reflected during the period of destructive interference (in the case of the Michelson interferometer) is channeled to a second detector, then the signal is doubled. The noise power, however, is increased by a factor of $\sqrt{2}$ only, assuming no statistical correlation between the noise generated in both detectors. The factor η_1 is then $\frac{1}{2}$. The conversion from peak to rms values is necessary for all devices to be discussed, since they use either a chopper or some other means to effect a sinusoidal modulation of the incoming radiation while the noise power is given by the rms value.

Before inserting Eq. (13.6) and (13.7) into (13.8), several other substitutions will be made. First, instead of $I_{\Delta \nu}$ at the detector, one may use $I_\nu \Delta \nu \eta_2$, the specific intensity available within the field-of-view times the resolved wavenumber interval (considered to be small) times the optical efficiency of the particular instrument. The optical efficiency includes reflection and transmission losses of the optical elements. Losses due to nonblackness of the detector are included in the D^* value.

Secondly, in all optical systems, the product $A_d \Omega_d = A \Omega$ is invariant. Without the subscript, the area and solid angle are referred to the entrance aperture and with the subscript d, to the area and solid angle at the detector. It is assumed that all instruments are properly designed, so that the effective $A\Omega$ is optimized and A is large enough not to violate diffraction laws for the field-of-view and wavelength used.

The third substitution

(13.9) $$\Delta f \simeq 1/\tau$$

replaces the effective noise bandwidth with the integration or dwell time τ of the system.

The signal-to-noise ratio becomes finally,

(13.10) $$S/N = \eta_1 \eta_2 I_\nu \Delta \nu \, D^*(A\Omega\Omega_d \tau)^{1/2}$$

Equation (13.10) applies when the source fills the field-of-view completely. For meteorological missions, this is nearly always the case. For astronomical applications, this is rarely the case. If the field-of-view of the instrument is larger than the target (star or planet), then I_ν must be interpreted as the average intensity within the field of view.

As mentioned before, η_1 is approximately 0.35 for a chopped system and an interferometer which does not recover the reflected beam. The optical efficiency η_2 varies from instrument to instrument depending on the number of reflecting surfaces and the transmission characteristics of filters. It goes without saying that the optical efficiency η_2 is to be maximized by selecting a proper layout and good quality optical elements. In mapping instruments, such as the High Resolution Infrared Radiometer (HRIR) on Nimbus for example, Ω, the instantaneous field-of-view is generally small and since a high spatial resolution implies a fast scan rate, τ is then small also. The bandwidth $\Delta\nu$ is made as large as permissible for the particular application, and the best detector (D^*) is chosen and illuminated by the fastest condensing optics practical. The collecting area then determines the signal-to-noise ratio.

For the grating instruments and interferometers, $\Delta\nu$ is small. High spectral resolution is generally the desired goal if these instruments are used. Again, a good detector (high D^*) and a fast condensing optic (large Ω_d) are as important as high optical efficiency. It is in A, Ω, and τ, where the conventional spectrometer and the Michelson interferometer differ most, as will be shown later.

As a numerical example, the theoretical signal-to-noise ratios obtainable from a grating instrument, from a Michelson interferometer, a filter-wedge spectrometer, and an interference-filter radiometer will be compared. The filter-wedge spectrometer uses a continuously tunable interference filter as the dispersive element (e.g., Hovis et al., 1967). The specific intensity I_ν of the target, the spectral resolution $\Delta\nu$, the D^* value of the detector, and the solid angle Ω_d at the detector are assumed to be equal for all instruments to be compared, although the grating instrument with individual detectors may run into difficulties if the spectral intervals are grouped too closely together. In all cases, the best detector may be selected to fit the spectral range and operating temperature requirement, and the detector shall be considered to be illuminated by the fastest condensing optic practical (e.g., $\Omega_d = 0.66$ for an F number of 1). The instruments mentioned deviate somewhat in optical efficiency, but more so in the collecting area, the solid angle of acceptance, and the time constant. In a comparison of the different devices, the term $\eta_2(A\Omega\tau)^{1/2}$ becomes the important factor.

Table I summarizes these parameters for typical instruments of the above nature. The first four columns, showing estimated weight, field-of-view, spectral range covered, and total number of spectral intervals to be resolved,

TABLE I. Comparison of instruments.

	1	2	3	4	5	6	7	8	9	10	11
	Typical weight (lbs)	Field of view (°)	Spectral range (μm)	Number of intervals resolved	A (cm^2)	Ω (sr)	τ (sec)	$(A\Omega\tau)^{1/2}$	η_2	$\eta_2(A\Omega\tau)^{1/2}$	Bit rate (bit/sec)
Michelson interferometer	20	8	5–20	300	10	0.016	10	1.26	0.3	0.38	4200
Grating spectrometer	75	12	6 channels near 15	6	0.5	0.035	6	0.32	0.6	0.19	12
Filter-wedge spectrometer	8	26	13–17	36	0.12	0.15	12/36 =0.33	0.077	0.4	0.031	70
Hypothetical 6-filter radiometer	8	26	6 channels near 15	6	1.0	0.15	10	1.22	0.4	0.49	12

are self-explanatory. In all cases, a width of the resolved spectral intervals of 5 cm^{-1} is assumed. The larger solid angle of the filter-wedge and the six-channel radiometers would require a small telescope for near-earth satellite altitudes. The weight of such a telescope, or telescopes, is included in the figures of the table.

As may be seen in column five, the large collecting area of the interferometer is one major advantage. In the spectrometer, the slit width is inversely proportional to the spectral resolution and cannot be enlarged without a simultaneous increase in overall size. The small area (0.2 × 0.6 cm) of the filter-wedge instrument was chosen as a convenient value consistent with the overall weight quoted. The same comment applies to the aperture of the multichannel radiometer, where 1 cm^2 was chosen as a convenient value rather than a necessary limit.

The solid angle of the interferometer is limited by the path difference between on-axis and off-axis rays. This may be expressed by the well-known formula

$$(13.11) \qquad \Omega_{\max} \leqslant 2\pi \, \Delta\nu/\nu_{\max}$$

The value quoted in column 6 in Table I applies to a $\nu_{\max}$ of 2000 cm^{-1} and $\Delta\nu = 5$ cm^{-1}. The solid angle of the spectrometer is given by the square of the F number times the cosine of the angle θ under which the grating is used. For narrow ranges, $\cos\theta$ can be made close to one, but the F number of grating instruments is limited, for practical purposes, by aberrations to values slower than $F/4$ ($\Omega = 0.05$).

The large angle of acceptance of the filter wedge and interference filter is an advantage. The same physical reasons limit the solid angle of the interferometer and filter wedge, but since the angles made by the off-axis rays with the optical axis are smaller in optically denser materials, the limiting solid angle measured outside the interference filter is given by

$$(13.12) \qquad \Omega_{\max} \leqslant n 2\pi \, \Delta\nu/\nu$$

where n is the effective refractive index of the layers used in the construction of the filter.

In column 7, the integration or dwell time of the interferometer is shown as 10 sec. This dwell time is limited by the orbital speed of the spacecraft and the effectiveness of the applied image motion compensation. A conventional grating spectrometer with a single detector could allow only 1/300 of the dwell time of the interferometer, for each spectral interval, if one wanted to cover the same spectral range. The large dwell time obtained in spite of the use of only one detector is the other major advantage of the interferometer and is sometimes called the multiplex or Felgett advantage.

The grating instrument must use several detectors to achieve comparable integration times. For a limited number of channels, this is quite feasible, although the calibration of several detectors and preamplifiers is somewhat more difficult. In all multidetector devices, great care must be taken to match the field-of-view of all channels.

The filter-wedge spectrometer using a single detector has the same disadvantage as the grating instrument; the dwell time is restricted to less than one over the number of spectral intervals. The multichannel radiometer has a long integration time similar to the multidetector spectrometer which, in effect, is a multichannel radiometer using a grating as a dispersive filter.

The optical efficiency is probably the most difficult parameter to assess. However, if properly designed, the individual instruments should not differ by great amounts. The efficiency of properly designed beamsplitters, interference filters, and gratings is of the same order; furthermore, all instruments have two or three reflecting surfaces in the light path. Instruments with a narrow spectral range can be optimized slightly better than those instruments encompassing a wider spectral range. The relative values of η_2 shown in column 9 reflect this fact.

The term $\eta_2(A\Omega\tau)^{1/2}$, shown in column 10, may be considered a factor-of-merit. For the specific task of temperature sounding, an interference filter radiometer seems to be the most efficient approach. The interferometer is then followed by the grating spectrometer and, finally, the filter wedge concludes the list. In the selected design, the filter-wedge instrument suffers from the single-detector mode.

Of course, the interferometer obtains 300 spectral intervals compared to 36 for the filter wedge and six for the other two devices. The vast increase in information must necessarily be paid for in the higher bit rate shown in column 11. A comparison of the instruments would also be unfair unless it is mentioned that the data reduction for the interferometer requires a Fourier transformation on a digital computer. Although straightforward today (see, e.g., Forman, 1966), in the past it has hampered the practical application of Fourier spectroscopy.

The instrument is only one link in a long data acquisition chain. Another important consideration, especially pertinent to this lecture series, is the problem of absolute calibration. In the unfriendly environment of space, absolute calibration is a difficult task indeed.

13.5. Calibration of the Interferometer

This section does not deal with the different laboratory methods used to check and measure the performance and stability characteristics of infrared devices. It shall be assumed that this is done before the device is placed in orbit. Here, the technique of on-board calibration shall be explored. The nature and complexity of the necessary calibration technique depend very much on the device as well as on the use of the data. On-board calibration is mandatory if radiometric accuracy is of consideration. The harsh launch environment may cause sudden changes in the responsivity of instruments, especially if critical optical alignment must be maintained. A slow, but continuous drift may be caused by a gradual change in the reflectivity and emissivity of optical surfaces. Outgasing and redeposition of volatile elements or, in the case of manned spacecrafts, the occasional dump of waste material, can easily have deteriorating effects on optical elements. For these reasons, on-board calibration is an absolute necessity for measurements where accuracy approaching the $\pm 1\%$ mark is desired. This figure is about the minimum required for a successful temperature sounding.

As an example, the on-board calibration procedure planned for the Nimbus B IRIS experiment will be discussed. The calibration of a Michelson interferometer appears, at first sight, somewhat more complex than that of a spectrometer since the spectrum is not directly accessible. Only after a Fourier transformation may the spectrum be observed. For this reason, the ground-based computer must be considered and treated as part of the instrument. The detector, amplifiers, analog-to-digital converter, digital tape recorder, transmission link, receiver, and ground station, and, finally, the computer form a long and complicated chain. In orbit, the calibration, as well as the data collection, must be carried out over this chain.

For the purpose of calibration, the instrument is exposed at regular intervals to a built-in calibration blackbody and to the cold interstellar background, by rotating the very first optical element, a flat mirror. Care is taken to maintain approximately the same angle of incidence on the mirror for all modes of operation. The calibration interferograms are processed in the same manner as the interferograms from the atmosphere which is the target of interest.

The output amplitude in the transformed interferograms is a number which corresponds to a voltage level at an analog-to-digital converter in the instrument. However, one does not wish to depend on the long-time stability of this converter. For that reason, the computer output for each wavenumber shall be taken as a relative value only. Let it be called $C_{(\nu)}$. The response of the instrument, as for all thermal devices, is proportional to the difference in radiance between the device and the target

(13.13) $$C_{(\nu)} = r_\nu (B_{\text{target}} - B_{\text{instrument}})$$

The concept of responsivity r_ν, well known for detectors, is applied here in a more general nature. Since $C_{(\nu)}$ is taken to be a dimensionless number, r_ν has the dimension of W^{-1} cm (sr). One obtains a set of three equations; one for the target (index 1), one for the cold blackbody (index 2), and one for the warm blackbody (index 3). Under the assumption that the responsivity r_ν is independent of the target brightness and that the detection and amplification is a linear process, the three equations may be solved to yield B_1 as well as r and B_{instr}. If one uses the interstellar background as the cold reference ($\sim 4°K$), then B_2, for all practical purposes, is zero and the equations simplify to

$$B_1 = B_3(C_2 - C_1)/(C_2 - C_3)$$
(13.14) $$r = (C_2 - C_3)/-B_3$$
$$B_{\text{instr}} = B_3 \, C_2/(C_2 - C_3)$$

All quantities are functions ν. The equation for B_1 is used to reduce the atmospheric spectra. Neither the responsivity nor the instrument temperature are contained explicitly in this equation. The calibration spectra C_2 and C_3 are the average of many individual spectra, so that the random effects in these spectra are greatly reduced. The temperature of the warm blackbody is carefully monitored by two independent elements.

The responsivity values are used for two purposes. First, the sample standard deviation (e.g., Wilson, 1952) of the responsivity is determined for each orbit, where $\overline{r_\nu}$ is the mean responsivity per orbit and k the number of calibrations ($k \approx 25$ per orbit). Thus

(13.15) $$s_\nu = \left(\sum_{i=1}^{k} (r_i - \bar{r})^2/k - 1 \right)^{1/2}$$

The r_i's are the responsivities computed from each calibration pair (hot and cold blackbodies). The average responsivity per orbit is called $\bar{r}$ and k is the number of calibration pairs per orbit. Instead of the standard deviation of the responsivity, the noise equivalent radiance (NER) is calculated from the calibration spectra. The NER is calculated from

$$\text{NER} = sB_3/\bar{r}$$

The same level of the NER may be expected to exist in the individual atmospheric spectra.

The NER gives the short-time repeatability of the instrument; a comparison of the average orbital responsivity for each spectral interval from orbit-to-orbit, and from day-to-day, yields the long-term drift.

The instrument temperature T_i, calculated from B_i, and the one measured by the thermistors imbedded in the housing should be in close agreement. A deviation from this agreement is used as a caution flag which calls for a special investigation if it should occur.

13.6. Future Experiments

Neither the expected results from the interferometer, the grating spectrometer, nor the multichannel radiometer (discussed in Section 13.3) will provide all the parameters required to specify the initial conditions in general circulation models. Temperatures, humidity and ozone are within the realm of vertical sounding experiments, but probably the most important parameter, the wind field, and somewhat less important, the surface pressure are still missing. Several measurement systems address themselves to this problem. These experiments are EOLE (Centre National d'Etudes Spatial, France), IRLS (Interrogation, Recording, and Locating System, NASA, GSFC, and NCAR), and OPLE (Omega Position Location Experiment, NASA, GSFC). A summary of these global horizontal sounding techniques is given in a National Academy of Science Report (1966). In all of these experiments, "in situ" detectors on balloons or ocean platforms collect atmospheric and oceanographic data. The platforms are to be interrogated regularly by the satellite passing overhead. Temperatures and (by tracking the balloons) wind fields may also be obtained. Buoys may, in addition to the air temperature, sense surface pressure and water temperatures. Here, the balloon and buoy experiments would be complementary to the remote sounding experiments, since they provide those ingredients to the calculations of the general circulation which are difficult to measure by remote soundings.

Other radiometric experiments will be concerned with obtaining very high spatial resolution from earth synchronous altitudes. The design of an infrared scanner (10–12 μm) to map cloudiness, both day and night, from earth synchronous altitudes, with a spatial resolution equivalent to that obtained by HRIR or better, is going to be a difficult task. In order to demonstrate what might be expected from an infrared map taken from synchronous altitudes, HRIR data for one day (July 11, 1966) were arranged on a globe (Warnecke, 1967). The photograph (Fig. 13.13) was then taken from a distance simulating the satellite altitude. After this experiment is performed in reality in the 10–12 μm window, cloud motions may be followed in a fashion similar to the sequence of frames shown for the ATS scan camera, but day and night side maps would show equally well without the disturbing appearance of the terminator in the field-of-view. Many other radiometric experiments are in an early design phase or ready to be placed on satellites. A summary of these experiments is given in a paper by Tepper (1967).

Whatever the specific goal of individual experiments, the technique of remote sensing of the earth's atmosphere and, hopefully, the applications of these methods to other planetary atmospheres must be based on the theory of radiative transfer, on the one hand, and sound and accurate radiometric practices, on the other. Then, remote sensing from satellites will become an indispensible tool of meteorology and atmospheric science.

Fig. 13.13. Mosaic of HRIR pictures from Nimbus II mounted on a globe and photographed from a distance corresponding to a synchronous altitude [after Warnecke (1967)].

Appendix

On April 14, 1969, the weather satellite Nimbus III placed into orbit two of the vertical sounding experiments which were discussed in Section 13.3. The satellite infrared spectrometer (SIRS) and the infrared interferometer spectrometer (IRIS) have both achieved their scientific objectives.

Orbital averages of spectra of the calibration sources used in the IRIS (Fig. 13.14a) and a precise temperature measurement of the warm calibration blackbody permit the calculation of the responsivity of the instrument shown in (b) of the same figure. A single interferogram produced the atmospheric emission spectrum plotted in (c), together with the noise equivalent radiance which was computed from the repeatability of individual calibration spectra.

Temperature profiles derived from both experiments agree well with conventional radiosondes launched below the satellite for comparison purposes—see Fig. 13.15 (Wark and Hilleary, 1969; Hanel and Conrath, 1969). IRIS also provides water vapor and ozone information. The success of the atmospheric sounding experiments on Nimbus III is a major step forward towards realizing a "Global Atmospheric Research Program."

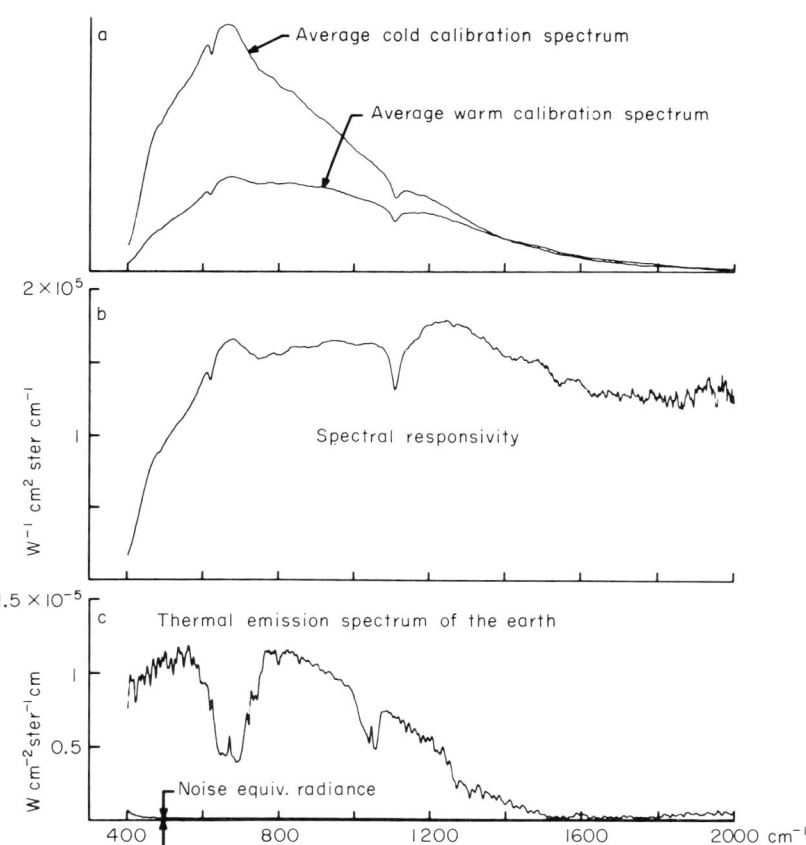

FIG. 13.14. In-flight calibration of the IRIS (Nimbus III) experiment, the derived instrument spectral responsivity, and the thermal emission spectrum of the earth computed from a single interferogram.

McQuain, R. H. (1967). ATS-I camera experiment successful. *Bull. Am. Meteorol. Soc.* **48**, 74.
Meeks, M. L., and Lilly, A. E. (1963). The microwave spectrum of oxygen in the earth's atmosphere. *J. Geophys. Res.* **68**, 1683–1703.
Mertz, L. (1965). "Transformations in Optics." Wiley, New York.
Michelson, A. A. (1927). "Studies in Optics." Univ. of Chicago Press (Phoenix Books, 1962), Chicago, Illinois.
National Academy of Science (1966). National Research Council Publication 1290. The feasibility of a global observation and analysis experiment. Rept. Panel on Intern. Meteorol. Cooperation to the Comm. on Atmospheric Sci.
National Geographic Magazine (1967). November, 726–731.
Nordberg, W., Bandeen, W. R., Warnecke, G., and Kunde, V. (1965). Space Research V (P. Muller, ed.) North-Holland Publ. Amsterdam.
Nordberg, W., McCulloch, A., Foshee, L. L., and Bandeen, W. R. (1966). Preliminary results from Nimbus II. *Bull. Am. Meteorol. Soc.* **47**, 857–872.
Obukhov, A. M. (1960). Statistical orthogonal expansions of empirical functions. *Isv. Geophysica.* Engl. Transl. by Am. Geophys. Union, Nov. 1960, 288–291.
Raschke, E., and Bandeen, W. R. (1966). Space Research, VII. North-Holland Publ. (R. L. Smith-Rose, ed.). Amsterdam.
Raschke, E., and Bandeen, W. R. (1967). A quasi-global analysis of tropospheric water vapor content from Tiros IV radiation data. *J. Appl. Meteorol.* **6**, 468–481.
Raschke, E., and Pasternak, M. (1967). The global radiation balance of the earth-atmospheric system obtained from radiation data of the meteorological satellite Nimbus II. *Proc. Internat. Space Sci. Symp., VIII*, London, July 1967.
Smith, R. A., Jones, F. E., and Chasmar, R. P. (1968). "The Detection and Measurement of infrared Radiation," 2nd ed. Oxford Univ. Press (Clarendon), London.
Stampfl, R. A., and Stroud, W. G. (1964). Automatic picture transmission TV camera system for Meteorological satellites. *J. Soc. Motion Picture Television Engrs.* **73**, 130–134.
Tepper, M. (1967). Space technology developments for the World Weather Watch. *Bull. Am. Meteorol. Soc.* **48**, 94.
Thompson, P. D. (1961). "Numerical Weather Analysis and Prediction," Macmillan, New York.
Twomey, S. (1966). Indirect measurements of atmospheric temperature profiles. II Mathematical aspects of the inversion problem. *Monthly Weather Rev.* **94**, 363–366.
Wark, D. Q. (1961). On indirect soundings of the stratosphere from satellites. *J. Geophys. Res.* **66**, 77.
Wark, D. Q., and Fleming, H. C. (1966). Indirect measurements of atmospheric temperature profiles. I Introduction. *Monthly Weather Rev.* **94**, 351–362.
Wark, D. Q., and Hilleary, D. T. (1969). Atmospheric temperature: Successful test of remote probing. *Science* **165**, 1256–1258.
Warnecke, G. (1967). Private communication.
Wilson, E. B. (1952). "An introduction to Scientific Research," Chapter 9. McGraw-Hill, New York.
Yamamoto, G. (1961). Numerical methods for estimating the stratospheric temperature distribution from satellite measurements in the CO_2 band. *J. Meteorol.* **18**, 581.
Yamamoto, G. (1966). Private communication.

— 14 —

THE DESIGN AND CONSTRUCTION OF EVAPORATED MULTILAYER FILTERS FOR USE IN SOLAR RADIATION TECHNOLOGY

E. E. Barr

14.1. Introduction

Interference filters are not new. Ancient peoples undoubtedly observed colors resulting from the presence of oil on water; Hooke, for instance, in the seventeenth century, suggested that a change in mode or phase was necessary to explain these color changes. He did not, however, connect this proposal with the concept of waves. Newton's ideas on the corpuscular nature of light propagation dominated thought during this period, and it was not until the late years of the eighteenth century that opposing views gained ground. Young advanced the theory of light as a wave motion in 1802. This was strengthened and molded, by Fresnel and Maxwell, into what we now know as the classical theory of light.

Analogous to current activity in maser optics, many workers found numerous interference bands caused by different geometric arrangements. The discovery of fringes, due to the interference of the reflected beams from the opposing sides of a parallel plate, was made by Haidinger (1849); and it is these that are important in the development of interference filters. By taking into account the contribution of multiple reflected beams between the parallel faces of the plate, Airy (1833) produced the mathematics necessary for the determination of the relative intensity distribution within the fringes. Such knowledge led to the invention of the Fabry–Perot interferometer (1897) towards the end of the nineteenth century. This instrument exhibits the highest resolution of any at present.

The reflecting layers for Fabry–Perot plates were initially of silver deposited from solution or by sputtering. Significant improvement in the efficiency of such films had to await the development of high-speed vacuum pumps. The early development of these, as well as the basic techniques for the vacuum deposition of films, can be attributed notably to Strong, Cartwright, and Turner (Cartwright and Strong, 1931; Cartwright and Turner, 1938).

These scientists, together with their contemporaries in Europe (viz., Hass, in Germany, and Holland, in England), improved the procedures so much that reflection reducing as well as highly reflective films were vacuum-deposited for the optics industry in general. Military applications in World War II made a significant impact on the growth of thin film activities. During this period, multilayer stacks consisting of alternate layers of high- and low-index material were studied. These proved to be highly efficient reflectors but were very limited in their free filter range.

The solid Fabry–Perot filter was invented, in Germany, by Geffken (1939). This consisted of a transparent layer sandwiched between two semitransparent silver layers, all three layers being deposited on one side of a glass substrate. Multiplex filters of this type are very much in use today. Polster (1949) substituted reflecting dielectric stacks for the silver layers of Geffken's filter. The bandwidth and peak transmission are very much narrower and larger, respectively, than those of the silver filter. This, of course, is due to the very small absorption within the dielectric stacks.

Intensive work, both mathematical and experimental, was undertaken by many groups throughout the U.S.A. and in Europe. The most powerful tool was the application of matrix methods to the problems inherent in thin film stacks, namely the proper phase and amplitude relationship to avoid irregularities for different wavelengths. Computers are widely used to provide parametric values or, given this information, to derive the integrated results. The current direction in development is towards the automated deposition unit, thus minimizing human error.

Whatever the degree of sophistication, we must not lose sight of the fact that we are dealing with interference similar in nature to that existing between the plates of a Fabry–Perot interferometer and the same advantages and disadvantages prevail. The theory of this interferometer is given in the next section in elementary detail.

14.2. Theory of the Fabry–Perot Interferometer

Consider two quartz plates with very flat surfaces mounted in parallel alignment, but separated by a distance d. d must be nearly the same everywhere between the plates and is usually achieved using three quartz dowels whose lengths are made as nearly alike as possible. Once these spacer dowels are inserted, a further improvement in the parallelism of the two inner surfaces can be made by varying the force on the dowels, so that compression changes their lengths—a tricky business. To make an instrument which exhibits high resolution, the inner surfaces are coated equally with highly reflecting, semi-transparent films. The plates are shown schematically in Fig. 14.1.

A simple, single-period interference filter is the same, except that the spacer is solid and in intimate contact with the reflecting stacks. Vacuum deposition techniques afford a convenient means of assuring uniform thickness of the spacer and, therefore, the substrate backing for the filter need not be polished to anything like the flatness of good F–P interferometer plates.

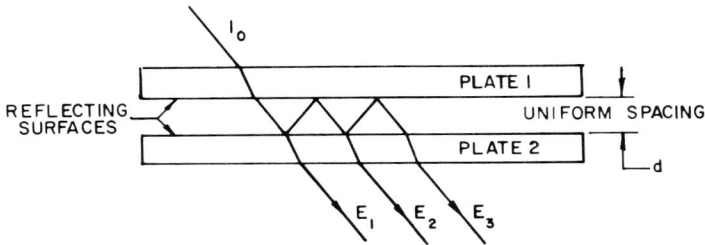

FIG. 14.1. Schematic of Fabry–Perot interferometer plates.

A glance at the direction of the light rays in Fig. 14.1 will indicate the idea of multiple reflections within the spacer. The summation of E_1, E_2, E_3, etc., equals the transmittance and this summation is given, by the Airy formula, as

$$(14.1) \qquad T = \frac{T_0^2}{(1-R_0)^2 + 4R_0 \sin^2(\delta/2)}$$

where

T_0 is the transmission of plate 1 or 2,
R_0 is the reflection of 1 or 2, and
δ is the phase retardation between successive beams such as E_1, E_2.

By dividing the numerator and denominator of the right-hand member of Eq. (14.1) by $(1-R_0)^2$, we have a more convenient form, viz.,

$$(14.2) \qquad T = \frac{T_0^2}{(1-R_0)^2} \cdot \left[1 + \frac{4R_0 \sin^2(\delta/2)}{(1-R_0)^2}\right]^{-1}$$

If $\delta/2 = 0$ or any multiple of π, the *maximum* transmittance is given by

$$(14.3) \qquad T_{\max} = \frac{T_0^2}{(1-R_0)^2}$$

If, in making a filter or coating F–P plates, dielectric materials with no absorption are used, $R_0 + T_0 = 1$. Solving for T_0, we have $T_0 = 1 - R_0$ and $T_0^2 = (1-R_0)^2$. Substituting in Eq. (14.3), we have $T_{\max} = 1$ or 100%.

For values of $\delta/2 = \pi/2$ or any odd multiple of $\pi/2$, $\sin^2(\delta/2) = 1$ and the transmission *minimum* is

$$(14.4) \qquad T_{\min} = \frac{T_0^2}{(1 + R_0)^2}$$

As R_0 approaches 1, T_0 approaches 0 since $T_0 = 1 - R_0$. In principle, $T_{\min}$ can approach zero as closely as one wishes but can never be zero.

A further simplification of Eq. (14.2) is normally made prior to calculation. It will be seen in Eq. (14.2) that the right-hand member contains the quantity $4R_0/(1-R_0)^2$. This fraction is usually designated as F and is called the coefficient of finesse or simply the reflection finesse.

For future reference, it might be well to make a table of calculated values of F and $F^{1/2}$ as a function of arbitrary values of R_0. Note that as R_0 approaches 1, F becomes very large, as does $F^{1/2}$. We shall see the importance of this presently. Substituting F from Table I and $T_{\max}$ from Eq. (14.3) into Eq. (14.2), we have

$$(14.5) \qquad T = \frac{T_{\max}}{1 + F \sin^2(\delta/2)}$$

and

$$(14.6) \qquad T_{\max}/T = 1 + F \sin^2(\delta/2)$$

TABLE I. Calculated values of the coefficient of finesse F and $F^{1/2}$ as functions of arbitrary values of the reflection coefficient R^0.

R^0	F	$F^{1/2}$
0.3	2.45	1.565
0.4	4.4	2.098
0.5	8	2.83
0.6	15	3.873
0.7	31	5.568
0.8	80	8.945
0.85	151	12.29
0 9	360	18.97
0.92	575	23.98
0.94	1044	32.63
0.96	2400	49.00
0.98	9800	99.00
0.99	39,600	199.00
0.995	159,000	398.8
0.999	3,996,000	2000.
0.9995	15,992,000	4000.
0.9999	399,960,000	20,000.

At the 1/2 power points $T_{max}/T = 2$, at the 1/10 power points $T_{max}/T = 10$, and so on. Considering the 1/2 power points, or what we call the bandwidth, we have

$$1 + F \sin^2(\delta/2) = 2$$

or

(14.7) $$\sin(\delta/2) = \pm(1/F^{1/2})$$

As $F^{1/2}$ becomes large, the sine of the phase angle becomes small and the angle itself becomes small; the smaller the phase angle, the narrower the bandwidth.

Let $P-1$ and $P-2$ be the reflecting surfaces on either side of a transparent plate (Fig. 14.2). A ray I strikes $P-1$ at A. Part is reflected and travels along AF and the remainder is refracted into the spacer layer. Part of the latter is reflected at P, travels along PC and after another reflection loss at C, emerges and travels along CG, parallel to EF. Obviously, the path difference of these two rays is: $AP + PC - AE$, where line EC is constructed perpendicular to both emerging rays. After the refracted ray CG emerges, there is no additional path change between the two rays. We need only calculate $(AP + PC - AE)$ using quantities we know. These known quantities include the index of refraction n, the thickness of the spacer d, the angle of incidence i, the angle of refraction r, and the wavelengths involved.

From Snell's law ($\sin i = n \sin r$, where one medium is air)

$$AP = d/\cos r = PC$$

$$AP + PC = 2d/\cos r = \text{pathlength between } P-1 \text{ and } P-2$$

$$AE = AC \cos(\angle CAE) = AC \cos(90 - i) = AC \sin i,$$

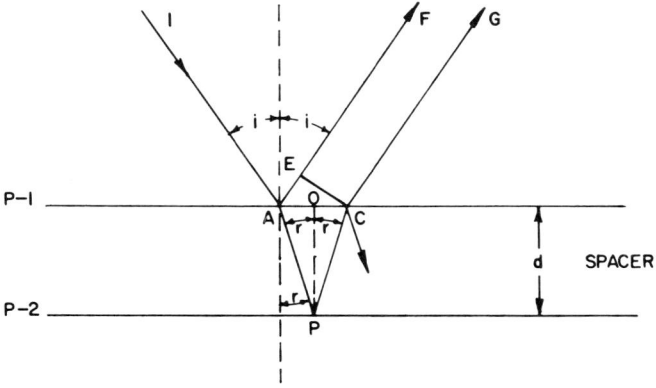

FIG. 14.2. Diagram of multiple reflections between F–P plates.

and
$$\tan r \text{ (in } \triangle APO) = AO/d, \quad AO = AC/2$$

Therefore,
$$AC = 2d \tan r$$

and
$$AE = 2d \tan r \cos(90 - i)$$
$$= 2d \tan r \sin i$$

Hence,
$$\text{path difference} = AP + PC - AE,$$

and

(14.8) $$\text{phase difference} = 2\pi \left[\frac{AP + PC}{\lambda'} - \frac{AE}{\lambda} \right]$$

where λ is the wavelength in the surrounding medium (air) and λ' the wavelength within the spacer of index n. The ratio $\lambda/\lambda' = n$.

Substituting the above identities in Eq. (14.8), we have

$$\text{reflection phase difference} = 2\pi \left[\frac{2d}{\lambda' \cos r} - \frac{2d \tan r \sin i}{\lambda} \right]$$
$$= 4\pi d \left[\frac{1}{\lambda' \cos r} - \frac{\sin r \sin i}{\lambda \cos r} \right]$$

Since $\sin i = n \sin r$ and $\lambda = n\lambda'$, we have

$$\text{reflection phase difference} = 4\pi d \left[\frac{1 - \sin^2 r}{\lambda' \cos r} \right]$$

and since $1 - \sin^2 r = \cos^2 r$

(14.9) $$\text{reflection phase difference} = (4\pi d/\lambda') \cos r = \delta_r.$$

Equation (14.9) gives the phase difference between the reflected beams AF and CG. According to this equation, if the plate thickness d approaches zero, the phase difference also approaches zero and we should get strong reflection. This cannot be true because a plate of zero thickness is no plate at all and the reflection is zero. This means that there must be an additional phase change due to something other than the plate thickness. Since, for $d = 0$, there can be no reflected light, this phase change must be π or 180°. Therefore, in place of Eq. (14.9) we have

(14.10) $$\text{reflection phase difference } \delta_r = 2\pi(2d/\lambda')\cos r \pm \pi$$

If $(2d/\lambda')\cos r = 0, 1, 2, 3$, etc., the reflection is minimum and the phase difference is $\pm\pi, 2\pi \pm \pi, 4\pi \pm \pi, 6\pi \pm \pi$, etc. As to whether the phase difference is $+\pi, -\pi, +3\pi, -3\pi$, etc., it does not make any difference because the addition or subtraction of 2π to the phase does not change the relationship of interfering waves. Since for minimum reflection the transmission must be maximum, the phase difference between successive transmitted beams is therefore

(14.11) $\qquad$ transmission phase difference $\delta_t = 2\pi(2d/\lambda')\cos r$

For all the successive transmitted beams to be in phase and, therefore, transmit a maximum, $(2d/\lambda')\cos r$ must be 0, 1, 2, 3, etc.

So far, we have considered light within the plate. Measurements made outside the plate must take into account the index of refraction n of the plate. Since the wavelength outside the plate is λ and $\lambda/\lambda' = n$, we can multiply and divide the right-hand member of Eq. (14.11) by n, and have

(14.12)
$$\delta_t = 2\pi(2nd/n\lambda')\cos r$$
$$= 2\pi(2nd/\lambda)\cos r$$

nd is called the optical thickness of the plate and it is this quantity which is of importance in interference. Suppose we choose $nd = N\lambda_0$, where λ_0 is a particular fixed wavelength and N is any integer. Then Eq. (14.12) becomes

(14.13) $\qquad \delta_t = 4\pi(N\lambda_0/\lambda)\cos r$

where λ can be any wavelength.

Looking back to Eq. (14.2), we see that for maximum transmission $\delta/2$ must equal 0 or any multiple of π. From Eq. (14.13),

(14.14) $\qquad \delta_t/2 = 2\pi(N\lambda_0/\lambda)\cos r$

Since for $\delta_t/2 = 0$, we have no plate at all, we can forget this condition. To satisfy the remaining conditions for maximum transmission, $(\lambda_0/\lambda)\cos r$ must equal 1, 2, 3, etc. For normal incidence light, $\cos r = 1$ and transmission bands will occur for $\lambda = 2N\lambda_0, N\lambda_0, N\lambda_0/2, N\lambda_0/2$, etc. For other than normal incidence, we must have

(14.15) $\qquad \lambda/\lambda_0 = \cos r$

and

(14.16) $\qquad \cos r = (1/n)(n^2 - \sin^2 i)^{1/2}$

Substituting the value of $\cos r$ of Eq. (14.16) in Eq. (14.15), we have

(14.17) $\qquad \dfrac{\lambda_i}{\lambda_{0\perp}} = \dfrac{1}{n}(n^2 - \sin^2 i)^{1/2}$

where $\lambda_{0\perp}$ is the wavelength position of a particular passband at normal incidence and λ_i is the wavelength of the same passband when viewed at an incidence angle of i. If one knows the wavelength position at normal incidence, the wavelength position at any incidence angle i is

$$(14.18) \qquad \lambda_i = \lambda_{0\perp} \frac{(n^2 - \sin^2 i)^{1/2}}{n}$$

where n is the index of refraction of the plate or spacer layer.

Equation (14.18) gives the wavelength position of the band as a function both of the refractive index of the spacer layer n and of the angle of incidence i. The ratio $\lambda_i/\lambda_{0\perp}$ is given in Table II as a function of i for different values of n.

TABLE II. Calculated values of the ratio of the wavelength position of the band head of a F–P filter for light incident at angle λ_i and its wavelength position for light incident at normal incidence λ_0, for different values of the refractive index n of the spacer layer.

i (in deg)	n				
	1.35	1.5	1.75	1.95	2.2
0.5	0.999979	0.999983	0.999987	0.999990	0.9999921
1.0	0.999916	0.999934	0.999950	0.999959	0.999969
2.0	0.99965	0.99972	0.99980	0.999838	0.99987
3.0	0.99924	0.99939	0.99955	0.99963	0.99972
4.0	0.99866	0.99893	0.99920	0.99935	0.99950
5.0	0.99791	0.99831	0.99875	0.99899	0.99921
10.0	0.99169	0.99327	0.99506	0.99602	0.99685
15.0	0.98144	0.98500	0.98899	0.99114	0.99306
20.0	0.97636	0.97365	0.98069	0.98448	0.98784
25.0	0.94973	0.95948	0.97037	0.97621	0.98137
30.0	0.92888	0.94281	0.95828	0.96654	0.97383
45.0	0.85184	0.88190	0.91465	0.93188	0.94693

The relative transmittance within a particular Fabry–Perot band can be calculated using Eqs. (14.7) and (14.14). This calculation has been made for a first-order ($\frac{1}{2}$-wave spacer) band and the results are shown in Fig. 14.3.

Several different values of the reflectance R_0 were chosen to illustrate its effect on the $\frac{1}{2}$-bandwidth ω. The ratio of the bandwidths for values of $T = 0.1$ and 0.5 may be thought of as the shape factor and its value is approximately 3. The minimum off-peak transmission decreases as R_0 increases. Since we have neglected absorption, its value is expressed as the ratio $(1 - R_0)^2/(1 + R_0)^2$. The contrast ratio[1] can only be large if th band-

[1] The contrast ratio is the ratio of the transmittance within the band to the transmittance outside the band.

14. DESIGN/CONSTRUCTION OF EVAPORATED MULTILAYER FILTERS   399

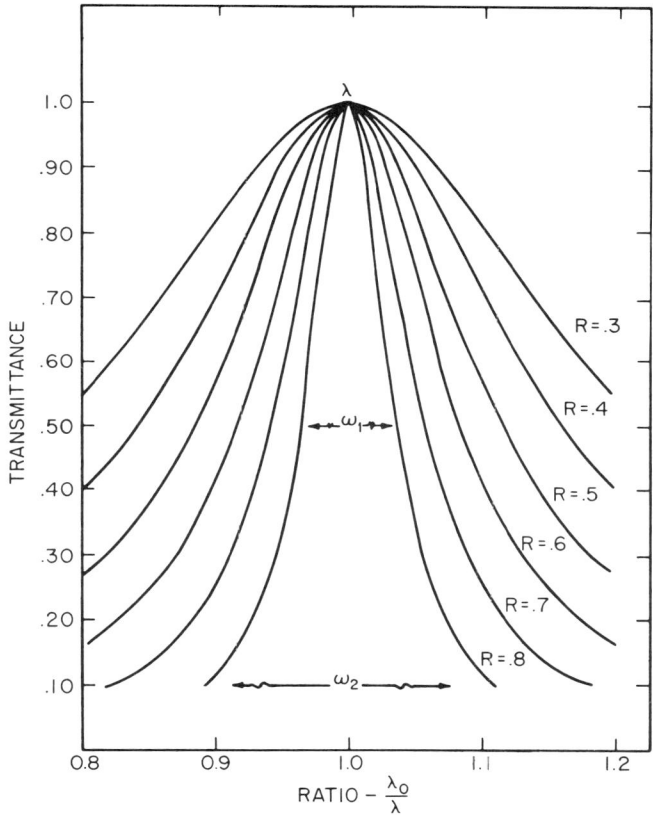

FIG. 14.3. Relative transmittance within a particular F–P band (first-order, $\frac{1}{2}$-wave spacer).

width is quite small. The objective in Fabry–Perot interferometry is not so much a matter of high contrast but rather the achievement of high resolution. To use a F–P band as a filter, high contrast is of the greatest importance. To satisfy this condition, a single period (R_0–spacer–R_0) F–P filter must exhibit a bandwidth of a few angstroms. The minimum transmission and the bandwidth (expressed as a decimal fraction of the center band position) are listed in Table III for different values of R_0.

The data in Table III can be used as a guide in designing F–P filters and especially as a warning to those who define filters in terms of half-width and wavelength position of the center band.

TABLE III. The bandwidth of a F–P filter, expressed as a decimal fraction of the wavelength position of the band and the coefficient of minimum transmittance, calculated for different values of the reflection coefficient R_0.

Reflection coef. R_0	Bandwidth $\Delta\lambda/\lambda_0$	Coefficient of minimum transmission $(1 - R_0)^2/(1 + R_0)^2$
0.3	44×10^{-2}	29×10^{-2}
0.4	32	18
0.5	23	11
0.6	17	63×10^{-3}
0.7	11	32
0.8	71×10^{-3}	12
0.9	34	28×10^{-4}
0.96	20	42×10^{-5}
0.99	13×10^{-4}	25×10^{-6}
0.999	32×10^{-5}	25×10^{-7}

The magnitude of the reflection coefficient of a filter is uniquely found from the value of its bandwidth and wavelength position. A spectrogram of a single-period F–P filter is shown in Fig. 14.4. The band is centered on the wavelength 6563.1 Å and exhibits a $\frac{1}{2}$ width of 1.2 Å. The ratio of bandwidth to wavelength is 18×10^{-5}, corresponding to a reflectance coefficient (Table III) greater than 0.999.

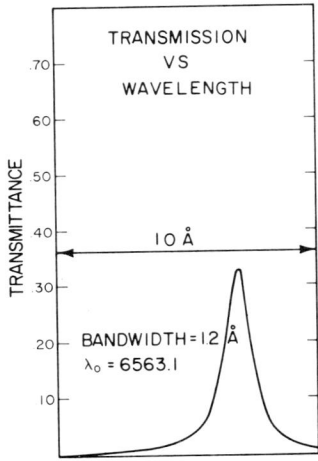

FIG. 14.4. Spectrogram of a single-period F–P filter.

14.3. Filter Contrast and Bandwidth

It would appear at this point of our discussion that a F–P filter is only appropriate for applications requiring very narrow band filters. However, if one considers two or more low contrast filters or etalons used in series, the resulting transmittance will be the product of the transmittance of the individual elements. If the elements are mounted with a spacing between, the whole idea is defeated unless the filters can be tipped to avoid multiple reflections. Since the bands shift with angle, this procedure is difficult, to say the least. Etalons have been and will again be employed in this fashion, but multiple F–P filters can be deposited in intimate contact on one side of a suitable substrate, thus avoiding reflections between elements (Smith, 1958). A single- and a double-period F–P filter are shown in Fig. 14.5. The wider band is the single period. The inner band results from the deposition of a duplicate design directly on top of the first. Actually, these are different filters. A single period was made and followed by a double period at the same wavelength. The double-period curve is practically a multiple of the single, but the most striking feature is the shape and rejection of the double filter shown (in 10 % full scale) in Fig. 14.6. A comparison of the bandwidths for different decimal values of peak transmission is presented in Table IV.

The above technique is obviously useful and can be expanded to include a great many designs and periods, but before we explore these possibilities let us examine the dependence, on the angle of the incident light, of the band position of filters similar to those discussed in Fig. 14.5.

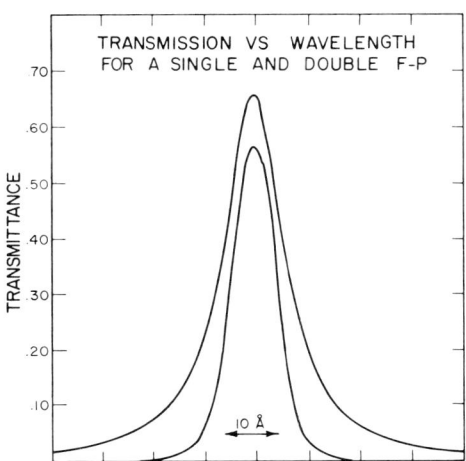

Fig. 14.5. Single- and double-period F–P filters.

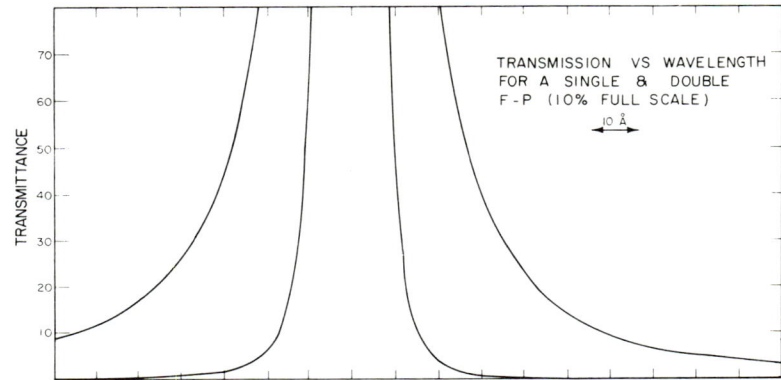

Fig. 14.6. Comparison of the profiles and rejection values of the single- and double-period F–P filters.

Table IV. The full-width of a single- and a double-period F–P filter, measured for different decimal values of the peak transmittance.

Decimal fraction of peak T	Bandwidth Single-period F–P filter	Double-period F–P filter
0.5	14 Å	10 Å
0.1	42	20
0.01	~140	35
0.001		65
0.0001		~100

Equation (14.18) states that the band position is a function both of the angle of incidence and the effective index of refraction of the spacers between the reflection stacks. This effective index is not necessarily the same as that found for a single layer of the spacer material deposited on glass, but it is readily determined by measurement of the band position for a few different angles of incidence. A good way of doing this is to locate the filter on a precision turntable in front of a monochromator. With the fore optics adjusted to produce parallel light, the filter may be tilted to any desired angle. Figure 14.7 shows a double-period, 10 Å bandwidth filter with cryolite spacers. The center-band position is indicated for incidence angles of 0, 5, 10, and 15°. Consulting Table II, we find that the wavelength change with angle closely matches that given for $n = 1.45$. This filter will not accept a very large field angle, 3 to 4° at best, but it can be effectively used to scan across an emission line to obtain a good ratio of line to background energy.

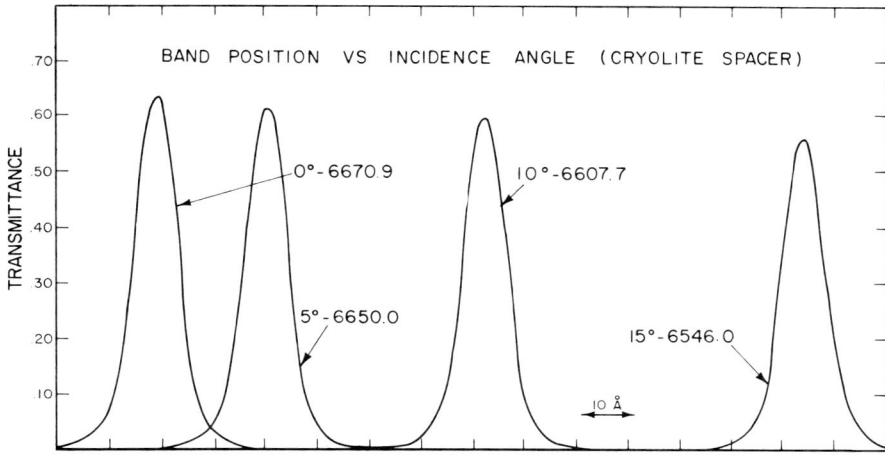

FIG. 14.7. Double-period 10 Å bandwidth filter with cryolite spacers.

Figure 14.8 presents the same information for a filter with zinc sulfide spacers, and here the change in peak wavelength with angle is much less rapid. The wavelength change closely matches that predicted in Table II for $n = 1.95$. This filter can be effectively used in a somewhat larger field than that employing cryolite spacers. Since the peak of these filters has a significant width, the band can always be positioned slightly on the longwave side of the spectrum line and thus optimize conditions for the largest acceptance angle.

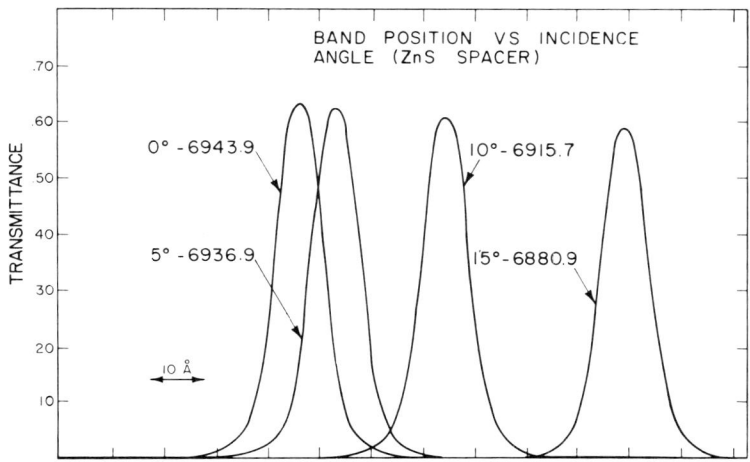

FIG. 14.8. Similar to Fig. 14.7 but with zinc sulfide spacers.

If an interference filter is measured in a conventional spectrophotometer, a field angle is imposed. There are field limits but no information on the uniformity of energy within the field. In general, however, most of the energy is concentrated within a field somewhat smaller than that dictated by the limits. Measurements made with the filter normal to the axis of the beam are reasonably accurate unless the filter is quite narrow. But, if the filter is rotated about the axis, great departures from true values may become evident. An example is shown in Fig. 14.9: the field limit was 15°.

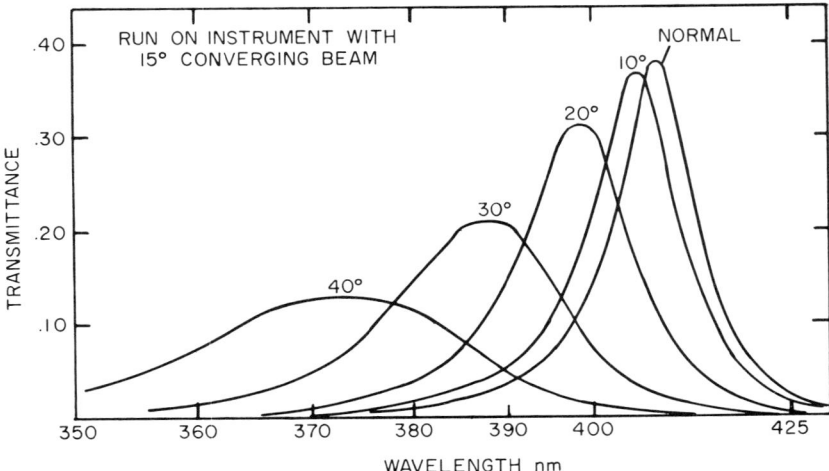

FIG. 14.9. Effect of angle of incidence of light beam on filter transmission.

Lissberger's formula (Lissberger, 1959; Lissberger and Wilcox, 1959) provides a means of calculating the λ-position and transmission for a filter in a particular field, provided data similar to those given in Figs. 14.7 and 14.8 are obtained by measurement. Here

$$(14.19) \quad T_{\lambda,\alpha} = \frac{\int_0^\alpha t\lambda, \theta \sin\theta \, d\theta}{\int_0^\alpha \sin\theta \, d\theta}$$

where $t\lambda, \theta$ is the transmission for light of wavelength λ incident at angle θ, and α is the half-field angle.

14.4. Temperature Dependence of Band Position

The effect of temperature on the spectral position of the pass band of a filter is significant in applications requiring very narrow bands. Figure 14.10(a) is a plot of the center-band position of a 10 Å-wide filter as a function of temperature. The fact that the plotted points do not fall on a straight line is probably due to a hysteresis effect. The thermocouple was placed in contact with one of the outer surfaces of a cemented pair of flats with the film stack between. Assuming a linear change in wavelength with temperature, the coefficient of this filter made with cryolite spacers is 0.37 Å/°C over the range of temperature −60 to +70°C. This change in band position with temperature cannot be explained on the basis of thermal expansion. By taking reasonable values for the thermal expansion coefficients of the substrate and spacer material, the shift in wavelength of the filter band, with temperature, could not be in excess of 0.1 Å/°C. We therefore conclude that a temperature change causes a change in the index of refraction. Standard crystallographic tables substantiate this view.

Filter bands at different wavelengths exhibit different values of the wavelength shift with temperature. Using cryolite spacers, the coefficient varies from a low value of about 0.1 Å/°C at 400 nm to a high of approximately 0.5 Å/°C at 1000 nm. The use of zinc sulfide spacers reduces these numbers by approximately 50%, except that at short wavelengths zinc sulfide and cryolite yield approximately the same result.

Figure 14.10(b) shows the effect of a temperature change of 50°C on two relatively wide-band filters, one (Filter B) positioned in the blue portion of the visible spectrum and the other (Filter A) positioned in the near ultraviolet. There is no significant change in transmission. The change in wavelength position is very slight, relative to the bandwidth.

14.5. Wide Bandpass Filters

The wide-band filters shown in Fig. 14.10(b) are multiplex Fabry–Perot filters; each period of the composite film stack is a wide F–P filter which, alone, has low-contrast features. The spectrograms shown in Fig. 14.11 are representative of wide filters consisting of stacked F–P elements. The wide, low-contrast filter on the single-period element is simply a three-layer sandwich of cryolite, $\frac{1}{2}$-wave thick, with a zinc sulfide layer $\frac{1}{4}$-wave thick on each side. To couple five of these together, a $\frac{1}{4}$-wave thickness of cryolite is deposited between each period. Letting the symbols H and L represent a $\frac{1}{4}$-wave optical thickness of zinc sulfide and cryolite, respectively, we can write the design as $(HLLHL)^5$. The phase changes resulting from the change in optical thickness, as one moves away from the center-band position λ_0, account for the satisfactory shape of the passband.

The free range of this filter (FFR) is limited, extending only from a value of approximately 0.8 to 1.3 λ_0. Alone, this element has very limited usefulness.

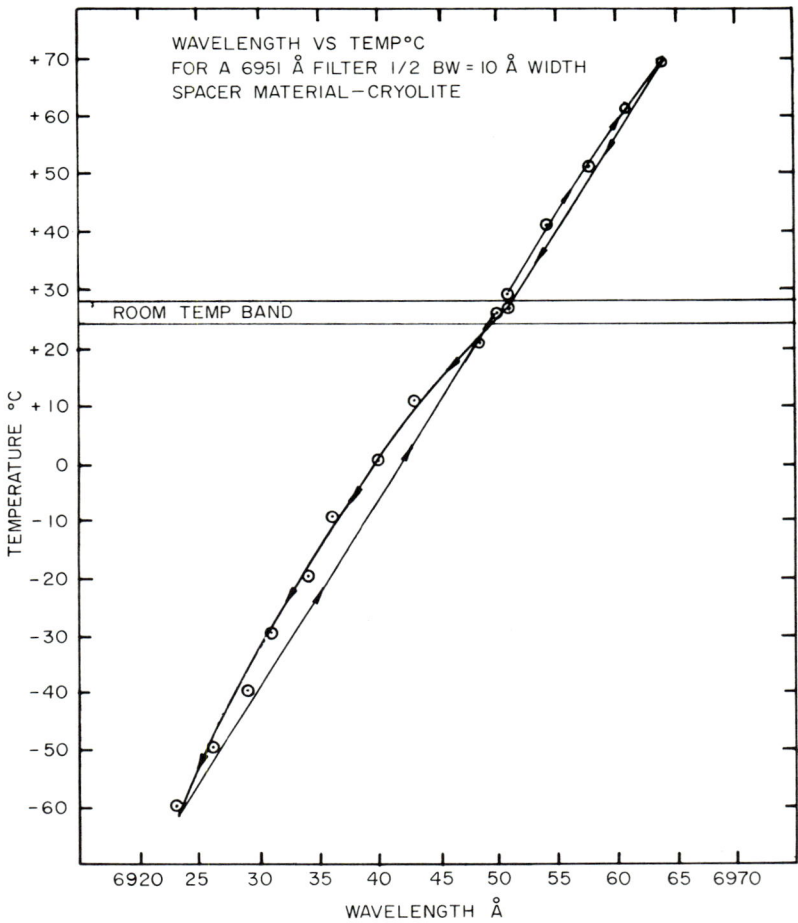

Fig. 14.10. (a) Effect of temperature on the spectral position of the pass band of a narrow filter.

To extend the FFR, a second element constituted of metal as well as dielectric layers is made and if the spacers are of first order, the FFR extends indefinitely towards longer wavelengths (Berning and Turner, 1957). Figure 14.12 is a spectrogram of one such filter. It consists of three silver layers separated by a $\frac{1}{2}$-wave of cryolite. Two layers of dielectric (HLL) are deposited first and last, to provide a somewhat rectangular band head and improved transmission. Most materials of high-refraction index absorb in the ultraviolet. Silver absorbs as well in this region. The best materials available at present for the

14. DESIGN/CONSTRUCTION OF EVAPORATED MULTILAYER FILTERS   407

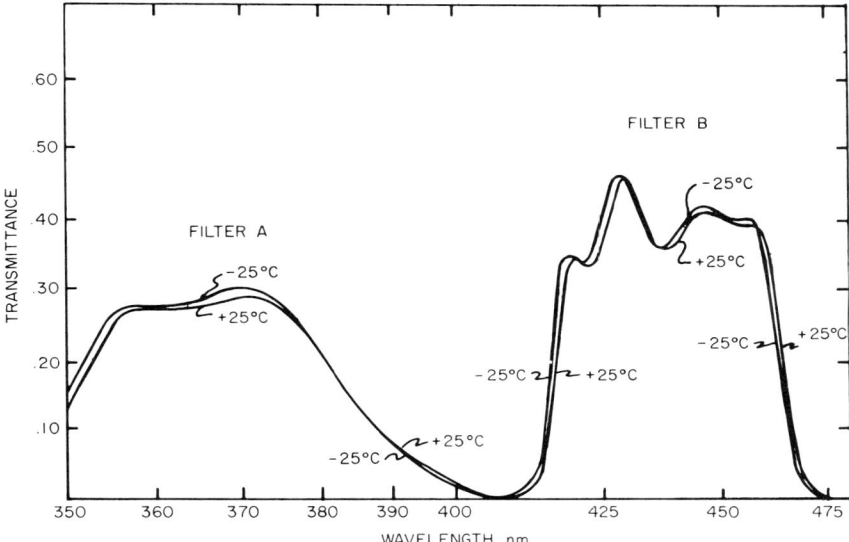

FIG. 14.10. (b) Effect of a temperature change of 50°C on two relatively wide-band filters.

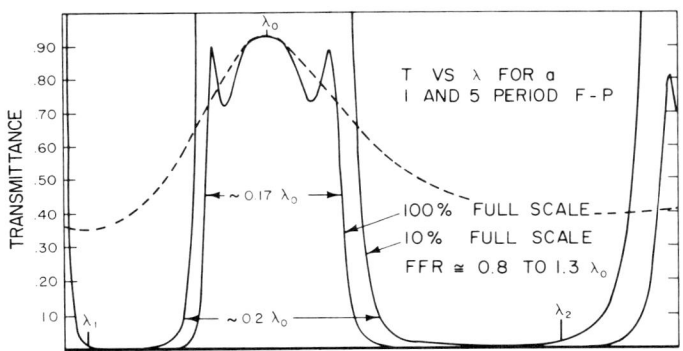

FIG. 14.11. Typical spectrograms of wide filters consisting of stacked F–P elements.

manufacture of broadband UV filters are aluminum in combination with various fluorides.

Figure 14.13 shows a spectrogram of a completed filter. The free filter range extends from X-ray to X-band. The secondary transmission outside of the band does not exceed $T_{\max}/10^4$. The second- and higher-order bands of the silver-dielectric element are removed by incorporating a low-pass color glass

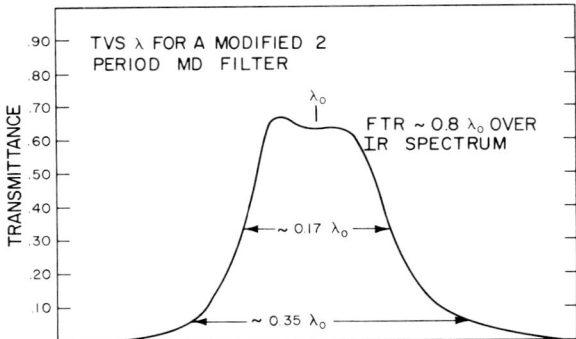

FIG. 14.12. Spectrogram of a filter with extended free range (incorporating metal in addition to dielectric layers).

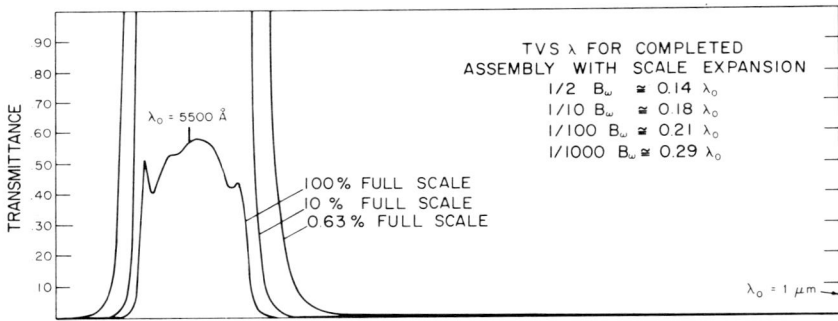

FIG. 14.13. Assembly of such a filter as depicted in Fig. 14.12.

within the assembly. The bandwidth, shown as a fraction of the center-band position, is the widest possible using zinc sulfide and cryolite.

Filters with still wider bands are made by coupling high- and low-pass filters. A film stack consisting of high- and low-index layers, each $\frac{1}{4}$-wave thick, will reflect a region of the spectrum and transmit on either side of this band. Such stacks are used to make high- or low-pass filters. An example of this technique is shown in Fig. 14.14. The band profile is not as sharp as that in Fig. 14.13, but for very wide filters this feature is not so important. In order to extend the FFR of such a filter, two or more overlapping stacks may be necessary to cover the wavelength region for which the detector is sensitive.

The goal of the filter manufacturer is to provide filters which satisfy the requirements in the fields of photometry and radiometry. The Eppley Laboratory, for instance, provides broadband thermopile detectors with radiometric equipment. The secondary transmission outside of the passband

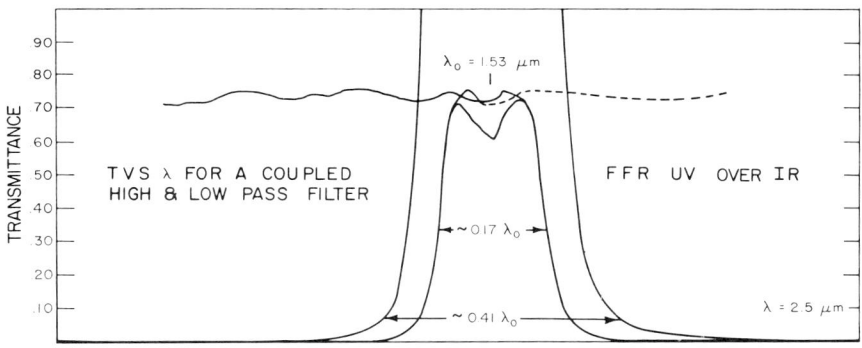

FIG. 14.14. Wide bandpass filter produced by coupling high- and low-pass filters.

of a filter must be very low to function properly with this type of detector. A great deal of development in collaboration with Eppley is continuing. This involves the problem of providing spectral purity and high transmission, on the one hand, and high contrast on the other. The Jet Propulsion Laboratory (Pasadena, California) has funded part of this work, particularly that done to improve UV filters.

As ever increasing variety of light sources, detectors, and applications forces the manufacturer to improve his technique. For a particular application, a few or all of the following general list of requirements might be significant.

1. Free choice of spectral position of band and bandwidth.
2. Rectangular shape of band.
3. High degree of surface uniformity of band characteristics.
4. High transmission efficiency.
5. Free filter range beyond the wavelength limits of the detector.
6. Negligible secondary transmission outside of the band.
7. Wide field of view.
8. Low sensitivity to ambient or source temperature.
9. Negligible degradation from nuclear or UV radiation.
10. Mechanical ruggedness.
11. Stability in the presence of the vapor or the liquid of common solvents, particularly water.
12. Good image quality.

Tests and measurements by the manufacturer and user are gradually forcing improved quality. A particular test procedure has been formulated by Eppley and this is outlined in the next section.

14.6. Physical Properties of Narrow Bandpass Filters

The tests described below were developed, by the Eppley Laboratory in collaboration with Thin Films Products, for the evaluation of the principal physical properties (mechanical, thermal, optical) of narrow bandpass filters intended for meteorological use. This covers application under outdoor conditions, at the ground and aloft, and indoors in solar simulation systems. It is, therefore, essential that the filters be weatherproof, be resistant to atmospheric moisture and other liquids, be able to withstand the effects of ultraviolet radiation, and be sufficiently robust to survive all envisaged exposure conditions.

These tests are as follows:

1. *Spectrophotometric examination of filter transmittance:*

(a) determination of principal bandpass transmittance to within an accuracy of $\tau = 0.005$ at any point over the relevant wavelength interval, with sufficiently high spectrophotometer resolution commensurate with the bandpass of the particular filter under test (the bandpass is normally here defined by the $\tau = 0.01$ limits of transmittance); and

(b) examination of secondary transmission to optical density levels of 3–4 (i.e., corresponding to filter transmittance values of $\tau = 0.001$–0.0001), employing expanded instrument scale techniques, over the wavelength range 0.2–3 μm.

2. *Mechanical shock test.* The filter is vibrated with a displacement of 0.125 in. at a frequency of 40 Hz, corresponding to an acceleration of approximately 10 g, for a period of 10 min, continuously.

3. *Thermal shock test.* The filter is first immersed in dry ice (approximately $-60°$C) for a period of 10 min and then is immediately transferred to a bath of hot water (approximately $+60°$C) where it is kept for a further 30 min.

4. *Solarization test.* The filter is exposed to a strong ultraviolet source (a mercury arc with a quartz envelope) for about 30 min, the UV irradiance being roughly 5–10 times that of the extraterrestrial sun (Fig. 14.15 shows a typical solarization effect on a UV filter containing lead chloride).

5. *Repeat of transmission test 1(a).* On comparison of these two filter transmission curves, if there is divergence greater than 2 % in filter transmission the filter is rejected.

6. *Filter factors.* After adoption of the basic filter transmission data, including fixation of the filter bandpass limits, the filter factor is computed (by machine methods) for the particular source to which it has to be exposed.

A further test should be mentioned. This is concerned with the all important question of secondary transmission in an ultraviolet filter (particularly

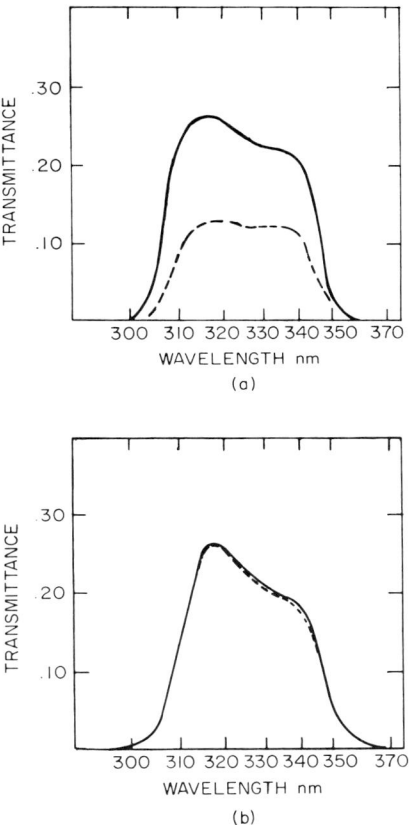

FIG. 14.15. (a) Solarization of a UV filter containing lead chloride; (b) almost complete immunity from such effects on a differently constructed filter.

where the bandpass is very narrow) that is to be used for solar measurements. In such instances, this energy leakage is mainly in the red region of the spectrum where the sun's radiation is many times greater in intensity than in the ultraviolet wavelengths. It is difficult, if not impossible, to detect a filter transmittance of $\tau < 0.0001$ (optical density 4), employing normally commercially available spectrometers. The practice introduced makes use of a barrier-layer photocell, sensitive in the ultraviolet and with peak response in the upper visible region, and a series of color glass, broad bandpass filters capable of complete rejection of energy below specified wavelengths. For example, the photocell is first exposed to the solar beam with only the ultraviolet filter over it. Then, a selected broad filter is superimposed over the cell-UV filter combination. Should the second signal be significant, as determined by the ultimate measurement requirements, the UV filter is rebuilt.

REFERENCES

Airy, G. B. (1833). On the phaenomena of Newton's rings when formed between two transparent substances of different refractive power. *Phil. Mag.* **2**, 20–30.

Berning, P. H., and Turner, A. F. (1957). Induced transmission in absorbing films applied to bandpass filter design. *J. Opt. Soc. Am.* **47**, 230–239.

Cartwright, C. H., and Strong, J. (1931). An apparatus for the evaporation of various materials in high vacua. *Rev. Sci. Instrum.* **2**, 189–193.

Cartwright, C. H., and Turner, A. F. (1938). Reducing the reflection from glass by evaporated films. *Bull. Am. Phys. Soc.* **13**, 10(A).

Fabry, C., and Perot, A. (1897). Sur les franges des lames minces argentées et leur application à la mesure de petites épaisseurs d'air. *Ann. Chim. Phys.* **12**, 459–501.

Geffken, W. (1939). Deutsches Reichs Pat. No. 716153.

Haidinger, W. (1849). Schwarzen ungelbe Parallelinien am Glimmer. *Poggendorf Ann. Phys.* **77**, 217.

Lissberger, P. H. (1959). Properties of all-dielectric interference filters. I. A new method of calculation. *J. Opt. Soc. Am.* **49**, 121–125.

Lissberger, P. H., and Wilcox, W. L. (1959). Properties of all-dielectric interference filters. II. Filters in parallel beams of light incident obliquely and in convergent beams. *J. Opt. Soc. Am.* **49**, 126–130.

Polster, H. D. (1949). A dielectric interferometer filter. *J. Opt. Soc. Am.* **39**, 1054 (A. 15).

Smith, S. D. (1958). Design of multilayer filters by considering two effective interfaces. *J. Opt. Soc. Am.* **48**, 43–50.

Appendix: Radiation Terminology, Symbols, Units, and Conversion Factors

TERMINOLOGY AND SYMBOLS

The first requirement of a terminology in thermal radiation, as in other physical disciplines, is definition of the inherent fundamental concepts in a manner consistent with the established usage in associated aspects in other fields of science. That is to say, the basic quantities employed in precision radiometry should be the same for all scientific and technological branches. However, the names of these quantities and the symbols used to denote them vary. For example, the physicist, meteorologist, engineer, agricultural scientist, and metrologist, etc. usually follow the nomenclature and the related symbols established, generally by long tradition, for their own particular fields of interest; and there is hence a high degree of familiarity in these respects between the author and the intended reader. But, radiometry, increasingly so, is becoming an important tool in the physical sciences. It is therefore to be expected that there will be an unusually diverse terminology and symbology.

In this volume, the authors of the several chapters have employed the special names and mathematical symbols with which they have been most accustomed. This is a perfectly acceptable practice in most journals of learned societies, provided that, in the absence of a more coherent standardization, there are sound reasons for the particular usage.

It is to be hoped that, in future, radiometric terminology will be universally standardized. A likely set of definitions and symbols for general adoption are those contained in U.S. Standard Z7.1-1967, "Nomenclature and Definitions

TABLE I

Quantity	Symbol	Definition	Units (commonly used)
Radiant energy	Q		erg, joule, watt-hour
Radiant density	W	$\dfrac{dQ}{dV}$	joule per cubic meter, erg per cubic centimeter
Radiant flux	Φ	$\dfrac{dQ}{dt}$	erg per second, watt
Radiant flux (at a surface)	M		
Radiant emittance	M	$\dfrac{d\Phi}{dA}$	watt per square centimeter
Irradiance	E		
Radiant intensity	I	$\dfrac{d\Phi}{d\omega}$	watt per steradian
Radiance	L	$\dfrac{d^2\Phi}{d\omega\, dA} \cos\theta$	watt per steradian and per square centimeter

for illuminating Engineering" RP-16. These recommendations, for the most part, have been adopted by the International Commission on Illumination (CIE). This standard terminology for radiation quantities so used in this volume is given in Table I. The term radiant emittance has been employed extensively for radiant flux at a surface, especially when reference is made to blackbodies. Problems so arising are discussed in Chapter 6.

Not only do the different authors here introduce different names and symbols for the same physical quantity, as explained, but, additionally, the same symbol is sometimes used for different quantities in the same chapter where there is justification. For example, σ is indicative both for the Stefan–Boltzmann constant of radiation and for electrical conductivity. Thus, the reader should examine the author's defining equations, to ascertain the sense in which interpretation should be made. In most cases, the intended usage is obvious. In some instances, the same major symbol with a number of subscripts or superscripts may be encountered. The quantity is that denoted by the symbol and ramifications of the appendages must be determined from the text.

The defining equations and the fundamental quantities of radiometry, including the values of the most important constants, are treated at length in Chapters 3, 4 and 6 and are listed below.

c	speed of light in vacuum	k	Boltzmann constant
c_1	first radiation constant	h	Planck's constant
c_2	second radiation constant	e	base of natural logarithms
σ	Stefan–Boltzmann constant		

Some symbols are used by all authors to denote the same quantity these are:

ϵ	emissivity (also e)	α	absorptance
λ	wavelength	ρ	reflectance
ν	frequency	ω, Ω	solid angle
τ	transmittance	T	absolute temperature

Boldface type denotes a vector quantity. The terminology $\exp(x)$ denotes the base of natural logarithms to the power inside the parentheses.

Units

The preferred units for radiation intensity, i.e., radiant flux per unit area or irradiance, are milliwatts per square centimeter (mW cm^{-2}) and such derivatives as Wm^{-2}, and for radiation quantity either milliwatt-hours per square centimeter (mW hr cm^{-2}) or joules per square centimeter (J cm^{-2}). In meteorological practice, the calorie per square centimeter (e.g., cal cm^{-2} min^{-1}, cal cm^{-2} hr^{-1}, cal cm^{-2} day^{-1}, kcal cm^{-2} yr) has long been in use, with its designation, in some countries, as the langley, viz., ly min^{-1} = cal cm^{-2} min^{-1}, etc. For most purposes, the difference between the various standard calories can be ignored, but it should be pointed out that the small or gram calorie is strictly the unit concerned here. The conversion tables which follow are extracts from those published in the "Guide to Meteorological Instrument and Observing Practices," 3rd Ed., No. 8, TP3. World Meteorological Organization, Geneva, Switzerland, 1969.

Conversion Factors

1. *Radiant Flux per Unit Area*

Unit	mW cm^{-2}	cal cm^{-2} min^{-1}
erg cm^{-2} sec^{-1}	10^{-4}	1.433×10^{-6}
W m^{-2}	0.1	1.433×10^{-3}
kW m^{-2}	100	1.433
mW m^{-2}	1	0.01433
cal cm^{-2} min^{-1}	69.8	1
mcal cm^{-2} sec^{-1}	4.19	0.0600

2. *Quantity of Radiation per Unit Area*

Unit	mW hr cm^{-2}	J cm^{-2}	cal cm^{-2}
erg cm^{-2}	2.78×10^{-8}	10^{-7}	2.39×10^{-8}
kW hr m^{-2}	100	360	86.1
mW hr cm^{-2}	1	3.6	0.861
J m^{-2}	2.78×10^{-5}	10^{-4}	2.39×10^{-5}
J cm^{-2}	0.278	1	0.239
cal cm^{-2}	1.163	4.19	1

Note: The internationally recommended units are joule (J), watt (W) and meter (m), where 1 joule = 10^7 erg = 1 watt-second = 2.778×10^{-7} kilowatt-hours = 2.389 kilocalories.

QC
806
A3
v.14
1970

APR 7 1970